Advances in Structural Mechanics Modeled with FEM

Advances in Structural Mechanics Modeled with FEM

Editors

Angelo Marcello Tarantino
Carmelo Majorana
Raimondo Luciano
Michele Bacciocchi

MDPI • Basel • Beijing • Wuhan • Barcelona • Belgrade • Manchester • Tokyo • Cluj • Tianjin

Editors

Angelo Marcello Tarantino
University of Modena and
Reggio Emilia
Italy

Carmelo Majorana
Department of Civil,
Environmental and
Architectural Engineering
(DICEA), University of Padova
Italy

Raimondo Luciano
Engineering Department,
University of Napoli Parthenope
Italy

Michele Bacciocchi
Dipartimento di Economia,
Scienze e Diritto (DESD),
University of San Marino
San Marino

Editorial Office
MDPI
St. Alban-Anlage 66
4052 Basel, Switzerland

This is a reprint of articles from the Special Issue published online in the open access journal *Materials* (ISSN 1996-1944) (available at: https://www.mdpi.com/journal/materials/special_issues/Adv_Struct_Mech_Model_FEM).

For citation purposes, cite each article independently as indicated on the article page online and as indicated below:

LastName, A.A.; LastName, B.B.; LastName, C.C. Article Title. *Journal Name* **Year**, *Volume Number*, Page Range.

ISBN 978-3-0365-0990-7 (Hbk)
ISBN 978-3-0365-0991-4 (PDF)

Contents

About the Editors vii

Angelo Marcello Tarantino, Carmelo Majorana, Raimondo Luciano and Michele Bacciocchi
Special Issue: "Advances in Structural Mechanics Modeled with FEM"
Reprinted from: *Materials* **2021**, *14*, 780, doi:10.3390/ma14040780 1

Hoang Nam Nguyen, Tran Thi Hong, Pham Van Vinh and Do Van Thom
An Efficient Beam Element Based on Quasi-3D Theory for Static Bending Analysis of Functionally Graded Beams
Reprinted from: *Materials* **2019**, *12*, 2198, doi:10.3390/ma12132198 5

Bo Zhang, Yong Li, Nicholas Fantuzzi, Yuan Zhao, Yan-Bao Liu, Bo Peng and Jie Chen
Investigation of the Flow Properties of CBM Based on Stochastic Fracture Network Modeling
Reprinted from: *Materials* **2019**, *12*, 2387, doi:10.3390/ma12152387 27

Michele Bacciocchi, Raimondo Luciano, Carmelo Majorana and Angelo Marcello Tarantino
Free Vibrations of Sandwich Plates with Damaged Soft-Core and Non-Uniform Mechanical Properties: Modeling and Finite Element Analysis
Reprinted from: *Materials* **2019**, *12*, 2444, doi:10.3390/ma12152444 47

Dong Li and Demin Wei
Rate-Dependent Cohesive Zone Model for Fracture Simulation of Soda-Lime Glass Plate
Reprinted from: *Materials* **2020**, *13*, 749, doi:10.3390/ma13030749 77

Wojciech Gilewski and Jan Pełczyński
Material-Oriented Shape Functions for FGM Plate Finite Element Formulation
Reprinted from: *Materials* **2020**, *13*, 803, doi:10.3390/ma13030803 91

Federico Oyedeji Falope, Luca Lanzoni and Angelo Marcello Tarantino
FE Analyses of Hyperelastic Solids under Large Bending: The Role of the Searle Parameter and Eulerian Slenderness
Reprinted from: *Materials* **2020**, *13*, 1597, doi:10.3390/ma13071597 105

Daniel Gnoli, Sajjad Babamohammadi and Nicholas Fantuzzi
Homogenization and Equivalent Beam Model for Fiber-Reinforced Tubular Profiles
Reprinted from: *Materials* **2020**, *13*, 2069, doi:10.3390/ma13092069 125

Paweł Dunaj, Krzysztof Marchelek, Stefan Berczyński and Berkay Mizrak
Rigid Finite Element Method in Modeling Composite Steel-Polymer Concrete Machine Tool Frames
Reprinted from: *Materials* **2020**, *13*, 3151, doi:10.3390/ma13143151 155

Piotr Bilko and Leszek Małyszko
An Orthotropic Elastic-Plastic Constitutive Model for Masonry Walls
Reprinted from: *Materials* **2020**, *13*, 4064, doi:10.3390/ma13184064 171

William Hideki Ito, Anna Maria Ferrero and Paulo Ivo Braga de Queiroz
Numerical Analysis of Bowing Phenomenon Due to Thermal Stresses in Marble Slabs
Reprinted from: *Materials* **2020**, *13*, 4367, doi:10.3390/ma13194367 195

Sebastian Rothe, Christopher Blech, Hagen Watschke, Thomas Vietor and Sabine C. Langer
Material Parameter Identification for Acoustic Simulation of Additively Manufactured Structures
Reprinted from: *Materials* **2021**, *14*, 168, doi:10.3390/ma14010168 . . . . **213**

Alexandre Mathern and Jincheng Yang
A Practical Finite Element Modeling Strategy to Capture Cracking and Crushing Behavior of Reinforced Concrete Structures
Reprinted from: *Materials* **2021**, *14*, 506, doi:10.3390/ma14030506 . . . . **233**

About the Editors

Angelo Marcello Tarantino is full professor of structural engineering at the University of Modena and Reggio Emilia (Italy). Expert in the theory of elasticity and structural engineering, he is the author of textbooks and 120 scientific publications, of which 90 have been published in international journals, mainly as the sole author. He has carried out studies and research activities in France, England, the United States, Switzerland, and China. He is the principal investigator of several research projects.

Carmelo Majorana is full professor of solid and structural mechanics and engineering at the University of Padova (Italy). Expert in computational mechanics and structural engineering, he is the author of textbooks and more than 200 scientific papers, of which 100 have been published in international journals. He has carried out studies and research activities in Europe and the United States. He has been the principal investigator of several research projects.

Raimondo Luciano is full professor in structural mechanics at the University of Naples Parthenope and he is a member of the Comitato Nazionale dei Garanti per la Ricerca (CNGR). He was the coordinator of the faculty activities of civil and environmental engineering of the University of Cassino and Southern Lazio. He has been involved in many research activities since 1991 as assistant professor at the University of Cassino. He has been the local coordinator of several research projects. He has more than 170 publications in national and international journals and conferences. His h-index is equal to 43. He has collaborated with distinguished scientists, such as Prof. John R. Willis of the University of Cambridge, UK, Prof. Sia Nemat Nasser of the University of San Diego (USA), Prof. Ever J. Barbero of the University of West Virginia (USA), Prof. Hota Gangarao and Prof. Roberto Lopez Anido of the University of Maine (USA), and many others. He was the President of the National Committee of the MIUR for the PRIN Grants 2015 (http://prin.miur.it/). He was one of the four professors in the National Committee for Italian national qualification (ASN of MIUR) for full and associate professors in the 08/B2 competitive sector. He collaborates on the realization of the Italian Design Guidelines CNR DT/200 and many others for the use of composite materials in civil engineering. He has taken part in over 20 consultancy projects on structures in Italy (mainly) and diagnosis and strengthening of the design of bridges and buildings.

Michele Bacciocchi is adjunct professor at the University of the Republic of San Marino. He obtained his PhD at the University of Bologna (Italy). He was a research visiting scholar at Texas A&M University (United States). His research interests are the mechanics of solids and the development of numerical techniques. He is the co-author of more than 50 international peer-reviewed journal papers, six books, and more than 15 abstracts in international and national conferences.

materials

MDPI

Editorial

Special Issue: "Advances in Structural Mechanics Modeled with FEM"

Angelo Marcello Tarantino [1,*], Carmelo Majorana [2,*], Raimondo Luciano [3,*] and Michele Bacciocchi [4,*]

1 Department of Engineering "Enzo Ferrari" (DIEF), University of Modena and Reggio Emilia, Via Vivarelli, 41125 Modena, Italy
2 Department of Civil, Environmental and Architectural Engineering (DICEA), University of Padova, Via F. Marzolo, 35131 Padova, Italy
3 Engineering Department, University of Napoli Parthenope, Via Ammiraglio Ferdinando Acton, 80133 Napoli, Italy
4 Dipartimento di Economia, Scienze e Diritto (DESD), University of San Marino, Via Consiglio dei Sessanta, 47891 Dogana, San Marino
* Correspondence: angelomarcello.tarantino@unimore.it (A.M.T.); carmelo.maiorana@unipd.it (C.M.); raimondo.luciano@uniparthenope.it (R.L.); michele.bacciocchi@unirsm.sm (M.B.)

Citation: Tarantino, A.M.; Majorana, C.; Luciano, R.; Bacciocchi, M. Special Issue: "Advances in Structural Mechanics Modeled with FEM". *Materials* **2021**, *14*, 780. https://doi.org/10.3390/ma14040780

Academic Editor: Gabriele Milani
Received: 1 February 2021
Accepted: 4 February 2021
Published: 7 February 2021

The current Special Issue entitled "Advances in Structural Mechanics Modeled with FEM" aims to collect several numerical investigations and analyses focused on the use of the Finite Element Method (FEM). The undeniable spread of this methodology in the recent decades is due to the recent improvements of high computational resources and the large-scale deployment of increasingly powerful computers, which have allowed to develop numerical approaches related to various applications. This situation has represented the key aspect that led to the birth of this Special Issue.

It is well known that many structural and physical problems are not suitable to be solved analytically. Therefore, appropriate numerical methods have to be developed to get approximate but accurate solutions. Precisely, one of the most typical approaches that can be employed to this aim is the Finite Element Method, which is characterized by very good accuracy and reliability. In addition, it can be easily implemented to deal with several applications in many engineering fields. In fact, it is easy to find numerical analyses solved by means of the FEM in civil, mechanical, and aerospace engineering.

The Guest Editors would like to emphasize their enormous gratitude to the Editors-in-Chief of Materials for the fantastic opportunity to manage this Special Issue. Likewise, all the authors coming from many different countries who contributed to the success of the present Special Issue are gratefully acknowledged for their papers characterized by very high quality. The reviewers should be also mentioned for their support, providing acute observations which undoubtedly improved the submitted manuscripts. Finally, the Section Managing Editor, Ms. Floria Liu, is acknowledged for accurately and constantly managing the editorial process.

The success of this venture is proved by the twelve papers collected and published in just over a year and a half. A brief review of these papers is presented below to highlight the multidisciplinary and the quality of these researches.

A two-node beam element was developed to analyze the static bending of functionally graded (FG) beams by Nguyen et al. [1]. The quasi-3D beam theory which the formulation is based on allowed to avoid the shear-looking issue without selective or reduced integration. A power-law expression was introduced to characterize the graded mechanical properties of the structures. Many parametric tests were illustrated to emphasize the effect of the graded features on the bending response.

A finite element model was developed by Zhang et al. [2] to investigate the gas flow properties in the coal seam. The starting point of their research was the fact that coal is characterized by a large number of fractures, whose spatial distribution patterns may follow some macroscopic statistical laws. Therefore, they described the coal seam as a two-dimensional stochastic fracture network introducing various fracture geometric parameters to represent a fracture. The paper proved that the permeability of the coal seam is emphasized with the increase of fracture density, length, aperture.

The natural frequencies of sandwich plates were analyzed in the paper by Bacciocchi et al. [3] by using a finite element formulation based on the Reissner-Mindlin theory including the zig-zag effect. The plates were made of a damaged isotropic soft-core and two external orthotropic skins reinforced by randomly oriented Carbon nanotubes. A three-phase method combining the Eshelby–Mori–Tanaka scheme and the Halpin–Tsai approach was developed to evaluate the overall mechanical features of the composites. Various parametric investigations were carried out to highlight several effects on the dynamic response.

The fracture process of soda-lime glass under impact loading was investigated numerically in the paper by Li and Wei [4], establishing a rate-dependent cohesive zone model implemented in a commercial finite element code. The accuracy of the fracture model was enhanced by taking into account the strain rate effect and the tensile-shear mixed mode fracture. The paper proved that the rate-dependent cohesive zone model was more able to describe the impact failure features of a monolithic glass plate if compared to a cohesive zone model without the inclusion of the strain rate. The effects of strain rate sensitivity coefficient, mesh size and impact velocity were also discussed in the research.

Gilewski and Pełczyński developed a four-noded finite element for the analysis of thick plates made of (FG) materials [5]. A coordinate change based on NURBS functions was used to describe arbitrarily shaped domains. Full coupling of the membrane and bending states was included in the formulation, which was embedded in a commercial finite element code. A good convergence characterized their finite element in dealing with the analysis of nonhomogeneous FG plates.

The finite bending of prismatic hyperelastic solids was investigated in the paper by Falope et al. [6] taking into account the anticlastic effect. The corresponding finite element formulation was developed and the numerical results were compared with the ones obtained analytically. The theoretical framework was fully nonlinear, assuming finite displacements and deformations. Analogously, the constitutive law was based on the Mooney–Rivlin nonlinear model for rubberlike materials. The analyses aimed at discussing the Eulerian slenderness and the compactness index of the beam cross section, which were involved in the definition of the Searle parameter for bent solids.

Gnoli et al. [7] discussed the validity of commercial finite element codes that implement the classical isotropic beam model for studying some equivalent composite configurations. A simple approach to deal with composite beams as isotropic configurations was developed by removing those complexities related to the modelling of composite materials when 2D and 3D finite element approaches were used. To this aim, the stiffness matrix of the equivalent isotropic beam was computed by means of an analytical approach valid for various laminated configurations used in typical engineering applications.

Dunaj et al. [8] focused their research on the development of a methodology able to model the natural frequencies, mode shapes, and receptance functions of machine tool steel welded frames filled with strongly heterogeneous polymer concrete, by employing low-dimensional models based on the rigid finite elements method (RigFEM). The numerical solutions were validated with respect to the experimental results, highlighting the accuracy of the proposed approach in terms of relative error.

Masonry structures in the plane stress state were the main topic of the paper by Bilko and Małyszko [9]. In particular, their work dealt with orthotropic masonry at the material level and presented the implementation of a continuum structural model for the analysis in a finite element code. The mathematical elastoplasticity theory represented the theoretical framework for developing the constitutive relations, assuming small displacements and deformations. The authors proposed a generalization of the Hoffman orthotropic failure criterion in the plane stress state.

The study by Ito et al. [10] aimed at evaluating the internal stresses in marble slabs experiencing bowing phenomenon due to thermal loading, taking into account the true deformed shape in continuum media and defining the stress–strain relationship in the hypothesis of finite displacements. The results obtained by means of a finite element based procedure proved that that transient heat flux should induce higher stresses than just applying greater gradients of temperature in steady flux, which could clarify the larger decohesion through width occurring in bowing tests. They emphasized that previous published results concerning the mechanical behavior of bowing based on classical beam theories could turn out to be inadequate when the internal stresses were computed.

The paper by Rothe et al. [11] highlighted the enormous potentialities of additive manufacturing, especially as far as the advanced acoustic design measures like Acoustic Black Holes are concerned. They specified that the continuous alignment of the mechanical impedance could be accomplished by means of the layer-wise material deposition by setting different filling patterns and levels of filling which characterized the technique. To this aim, experimental tests were carried out considering several configurations in order to identify the material parameters in dependency on the frequency and the thickness. A 3D finite element model was developed for the parameter fitting, proving the reliability of their results.

Finally, Mathern and Yang presented a modeling strategy and many practical recommendations for the finite element analysis in nonlinear regime of reinforced concrete (RC) structures based on parametric studies of critical modeling choices [12]. The aim of the paper was the reliable prediction of flexural responses of RC elements, focusing on cracking behavior and crushing failure. These aspects should be taken into account when more complex cases are modeled, such as RC beams bonded with fiber reinforced polymer laminates. Their results were validated by means of experimental tests, as well.

Institutional Review Board Statement: Not applicable.

Informed Consent Statement: Not applicable.

Data Availability Statement: Data sharing not applicable.

Conflicts of Interest: The authors declare no conflict of interest.

References

1. Nguyen, H.N.; Hong, T.T.; Vinh, P.V.; Thom, D.V. An Efficient Beam Element Based on Quasi-3D Theory for Static Bending Analysis of Functionally Graded Beams. *Materials* **2019**, *12*, 2198. [CrossRef] [PubMed]
2. Zhang, B.; Li, Y.; Fantuzzi, N.; Zhao, Y.; Liu, Y.B.; Peng, B.; Chen, J. Investigation of the flow properties of CBM based on stochastic fracture network modeling. *Materials* **2019**, *12*, 2387. [CrossRef] [PubMed]
3. Bacciocchi, M.; Luciano, R.; Majorana, C.; Tarantino, A.M. Free vibrations of sandwich plates with damaged soft-core and non-uniform mechanical properties: modeling and finite element analysis. *Materials* **2019**, *12*, 2444. [CrossRef] [PubMed]
4. Li, D.; Wei, D. Rate-Dependent Cohesive Zone Model for Fracture Simulation of Soda-Lime Glass Plate. *Materials* **2020**, *13*, 749. [CrossRef] [PubMed]
5. Gilewski, W.; Pełczyński, J. Material-Oriented Shape Functions for FGM Plate Finite Element Formulation. *Materials* **2020**, *13*, 803. [CrossRef] [PubMed]
6. Oyedeji Falope, F.; Lanzoni, L.; Tarantino, A.M. FE Analyses of Hyperelastic Solids under Large Bending: The Role of the Searle Parameter and Eulerian Slenderness. *Materials* **2020**, *13*, 1597. [CrossRef] [PubMed]
7. Gnoli, D.; Babamohammadi, S.; Fantuzzi, N. Homogenization and Equivalent Beam Model for Fiber-Reinforced Tubular Profiles. *Materials* **2020**, *13*, 2069. [CrossRef] [PubMed]
8. Dunaj, P.; Marchelek, K.; Berczyński, S.; Mizrak, B. Rigid Finite Element Method in Modeling Composite Steel-Polymer Concrete Machine Tool Frames. *Materials* **2020**, *13*, 3151. [CrossRef] [PubMed]
9. Bilko, P.; Małyszko, L. An Orthotropic Elastic-Plastic Constitutive Model for Masonry Walls. *Materials* **2020**, *13*, 4064. [CrossRef]
10. Ito, W.H.; Ferrero, A.M.; de Queiroz, P.I.B. Numerical Analysis of Bowing Phenomenon Due to Thermal Stresses in Marble Slabs. *Materials* **2020**, *13*, 4367. [CrossRef] [PubMed]
11. Rothe, S.; Blech, C.; Watschke, H.; Vietor, T.; Langer, S.C. Material Parameter Identification for Acoustic Simulation of Additively Manufactured Structures. *Materials* **2021**, *14*, 168. [CrossRef] [PubMed]
12. Mathern, A.; Yang, J. A Practical Finite Element Modeling Strategy to Capture Cracking and Crushing Behavior of Reinforced Concrete Structures. *Materials* **2021**, *14*, 506. [CrossRef] [PubMed]

materials

MDPI

Article

An Efficient Beam Element Based on Quasi-3D Theory for Static Bending Analysis of Functionally Graded Beams

Hoang Nam Nguyen [1], Tran Thi Hong [2], Pham Van Vinh [3] and Do Van Thom [3,*]

1 Modeling Evolutionary Algorithms Simulation and Artificial Intelligence, Faculty of Electrical & Electronics Engineering, Ton Duc Thang University, Ho Chi Minh City 700000, Vietnam; nguyenhoangnam@tdtu.edu.vn

2 Center of Excellence for Automation and Precision Mechanical Engineering, Nguyen Tat Thanh University, Ho Chi Minh City 700000, Vietnam; hongtt@ntt.edu.vn

3 Department of Mechanics, Le Quy Don Technical University, Hanoi City 100000, Vietnam; phamvanvinh@lqdtu.edu.vn

* Correspondence: thom.dovan.mta@gmail.com

Received: 13 May 2019; Accepted: 21 June 2019; Published: 8 July 2019

Abstract: In this paper, a 2-node beam element is developed based on Quasi-3D beam theory and mixed formulation for static bending of functionally graded (FG) beams. The transverse shear strains and stresses of the proposed beam element are parabolic distributions through the thickness of the beam and the transverse shear stresses on the top and bottom surfaces of the beam vanish. The proposed beam element is free of shear-looking without selective or reduced integration. The material properties of the functionally graded beam are assumed to vary according to the power-law index of the volume fraction of the constituents through the thickness of the beam. The numerical results of this study are compared with published results to illustrate the accuracy and convenience rate of the new beam element. The influence of some parametrics on the bending behavior of FGM beams is investigated.

Keywords: beam element; Quasi-3D; static bending; functionally graded beam

1. Introduction

Functionally graded (FG) materials (FGM) are one of the advanced composites. In which the material properties of FGM vary continuously through one or more directions. A typical, the material properties of an FGM beam, plate and shell varies continuously through the thickness direction. Due to their advantages, the FGMs have used widely in many fields such as civil, aerospace, automobile, engineering, nuclear power plants and so on [1,2]. Since then, many scientists have been focused on the mechanical analysis of FG beams, plates and shells. In which, they used several theories and methods, for instance, analytical and numerical methods based on Euler-Bernoulli theory, Timoshenko theory or first-order shear deformation theory (FSDT), higher-order shear deformation theory (HSDT), Quasi-3D theory and Carrera Unified Formulation (CUF).

Sankar [3] developed an elasticity solution to analyze a simply supported FG beam subjected to transverse distribution loading. In his work, Sankar developed new beam theory which was similar to the Euler-Bernoulli beam theory. Zenkour [4] analyzed an exponentially graded thick rectangular plate using both 2-D and 3-D elasticity solutions. Zhong et al. [5] analyzed a cantilever FG beam using a general two-dimensional solution. The free vibration and buckling analysis of FG beams under mechanical and thermal loads were investigated by Trinh et al. [6] using the analytical method based on the state space approach and higher-order beam theory.

The Euler-Bernoulli beam theory ignores the shear deformation so that it provides an acceptable solution for thin beams only; this model requires a C^1 continuity for a compatible displacement formulation. Kien [7] used the Euler-Bernoulli beam model to analyze large displacement behaviors of tapered cantilever FG beams by the finite element method. Lee et al. [8] applied Euler-Bernoulli beam theory for free vibration analysis of FG beam using an exact transfer matrix expression.

Due to the shear deformation effects are more obvious in thick beams and plates, so the FSDT can be used in these cases. On the other hand, it requires only C^0 continuity for the deflection. One of the shortcomings of the FSDT is the transverse shear strain distributes in an inaccurate and it does not satisfy the stress-free boundary conditions at the top and bottom surfaces of the structures, so this model requires a shear correction factor. Menaa et al. [9] used the energy equivalence principle to derive a general expression for the static shear correction coefficients in FG beams. Modal analysis of FG beams with shear correction function was studied by Murin et al. [10]. Nguyen et al. [11] employed FSDT for static and free vibration of axially loaded FG beams. Nam et al. [12] investigated the mechanical behaviors of variable thickness FG beam using modified FSDT. Due to the simplicity and effectiveness of FSDT, many scientists have applied FSDT to analyze plates and beams, and it is still being improved to achieve higher accuracy.

The shear correction coefficients can be removed by using HSDT, which have been developed by many scientists. In this model, the transverse shear strain varies parabolically through the height of the structures, and the transverse shear stresses at the top and bottom surfaces of the structures are neglected so that it need not any shear correction coefficients. Shi [13] proposed a new simple third-order shear deformation theory (TSDT) to analyze static bending of rectangular plates. Kadoli et al. [14] used HSDT for static bending analysis of FG beams. Benatta et al. [15] studied static bending of FG short beam involving the effects of warping and shear deformation. Li et al. [16] investigated static bending and dynamic response of FG beams using HSDT. Thai et al. [17] applied various HSDT to analyze static bending and free vibration of FG beams. Refined shear deformation was applied for static bending and vibration analysis of FG beams by Vo et al. [18]. Tinh et al. [19] used finite element method (FEM) and a new TSDT for mechanical response analysis of heated FGM plates.

Both FSDT and HSDT ignore the effect of the thickness stretching, which is noticeable in thick beams and plates. A number of Quasi-3D theories have been developed, in which the effects of shear deformation and thickness stretching were included. Vo et al. [20] used a Quasi-3D theory with only four unknown components to investigate the static behavior of FG beams and FG sandwich beams. Neves et al. [21,22] and Hebali [23] developed a Quasi-3D theory with sinusoidal shear function and hyperbolic shear deformation theory to analyze the static bending and free vibration of FG plates. Mantari et al. [24] studied static bending of advanced composite plates using a generalized hybrid Quasi-3D shear deformation theory. Mantari et al. [25] used a four-unknown Quasi-3D shear deformation theory for analysis of advanced composite plates. Thai et al. [26] presented a Quasi-3D hyperbolic shear deformation theory for analysis FG plates. Fang et al. [27] applied Quasi-3D theory and isogeometric analysis to study thick porous beams. Nguyen et al. [28] and Yu et al. [29] used Quasi-3D theory and isogeometric analysis to investigate FG microplates and two-directional FG microbeams. Farzam-Rad et al. [30] applied Quasi-3D theory and isogeometric analysis to study FG plates based on the physical neutral surface. Tran et al. [31] employed a Quasi-3D model with six-variable for static analysis of laminated composite plate using isogeometric analysis. The most outstanding of Quasi-3D theory is applicable to analyze thick plates and beams where the normal deformation effect is significant.

Carrera [32] developed Unified Formulation (CUF), which produces any refined theories for many structures such as beams, plates, and shells. Cerrera et al. [33] applied CUF for free vibration finite element analysis of beams with a uniform section. Cerrera et al. [34] employed CUF for studying micropolar beams using an analytical method. The 1D CUF theories were applied to analyze FG beams using FEM by Giunta et al. [35] and Filippi et al. [36].

However, in HSDT and Quasi-3D theory, the displacement field is considered by the existence of the higher order derivative of the deflection of transverse. So that it involves the development of a C^1 continuity element, which can cause difficulty to originate the second derivative of deformation in

FEM. To overcome these continuity issues, Hermite interpolation functions with C^1 elements and some C^0 approximations have been adopted. Chakraborty et al. [37] developed a new beam element based on the FSDT and an exact solution of the static part of the governing differential equations for analysis of FGM structures. Nguyen et al. [38] applied the Timoshenko beam model and FEM for dynamic response of bi-directional FG beams subjected to moving load. Khan et al. [39] investigated the static bending and free vibration of FG beams using FEM. Heyliger [40] developed a higher order beam finite element for bending and vibration of beams. Kapuria et al. [41] studied bending behavior and free vibration response of layered FG beams using a third order zigzag theory and FEM. Based on refined shear deformation theory, Vo et al. [42] developed a finite element model to analyze free vibration and buckling of FG sandwich beams. Moallemi-Oreh et al. [43] used FEM for stability and free vibration analysis of the Timoshenko beam. Pascon [44] analyzed FG beams with variable Poisson's ratio using FEM. Yarasca et al. [45] studied FG sandwich beams using Hermite-Lagrangian finite element formulation. The use of higher-order shape function will cost much computation effort in comparison with linear shape function. Furthermore, the linear shape function is simpler in formulation and transformation than higher-order shape function. However, plate and beam element using linear shape function are mainly developed based on FSDT and HSDT. To author's knowledge, there is currently no beam element using linear shape function based on a Quasi-3D theory. Therefore, the development of a beam element using linear shape function based on a Quasi-3D theory is necessary.

This paper presents a new beam element based on Quasi-3D theory, which only requires C^0 shape functions. The organization of this study is as follows. Firstly, Section 2 defines the model and material properties of FG beams. In Section 3, the governing equations of FG beams based on Quasi-3D theory are given. The finite element formulations of the proposed beam element are presented in Section 4. In Section 5, some example problems are carried out to show the convergence and accurate rate of new beam element in comparison with published data. Then, the static bending behaviors of FG beams using the proposed beam element are studied. The influences of the distribution of materials properties, length-to-thickness ratio, boundary conditions and effect of normal strain are investigated. Finally, in the conclusion section, some remarks on the proposed beam element are given.

2. Functionally Graded Material

Consider an FG beam as shown in Figure 1, the length of the beam is L, the width of the beam is b, and the thickness of the beam is h. The Young's modulus varies continuously through the thickness of the beams with a power law distribution [19,20]:

$$E(z) = E_m + (E_c - E_m)\left(\frac{z}{h} + \frac{1}{2}\right)^p \tag{1}$$

in which subscript m denotes the metallic component and c denotes the ceramic component, E_m, E_c are respectively Young's modulus of the metal and ceramic, p is the power-law index. In this study, the Poisson's ratios ν of both components are assumed to be constant and equal.

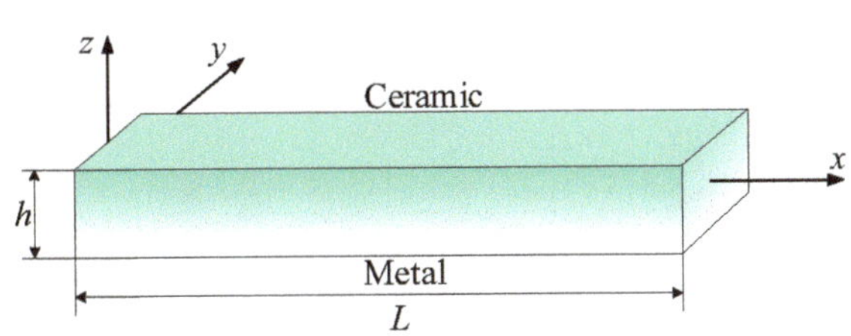

Figure 1. The FG beam model.

3. Governing Equations

The displacements of a point of the beam are expressed by

$$\begin{cases} u(x,z) = u(x) + f_1(z)\beta(x) + f_2(z)\alpha(x) \\ w(x,z) = w(x) + g(z)\varphi(x) \end{cases} \tag{2}$$

The functions $f_1(z)$, $f_2(z)$ are given by Shi [13]

$$f_1(z) = \frac{5}{4}\left(z - \frac{4z^3}{3h^2}\right);\ f_2(z) = -\frac{z}{4} + \frac{5z^3}{3h^2};\ g(z) = f'_1(z) \tag{3}$$

The strain field is obtained as follows

$$\begin{cases} \varepsilon_x = u_{,x} + f_1(z)\beta_{,x} + f_2(z)\alpha_{,x} \\ \varepsilon_z = g'(z)\varphi \\ \gamma_{xz} = w_{,x} + f_1'\varphi_{,x} + f_1'(z)\beta + f_2{\prime}(z)\alpha \end{cases} \tag{4}$$

in which the symbol (,) means the derivatives with respect to the quantity following it and the symbol $(')$ means the derivatives with respect to z direction.

Rewrite the strain components in the short form as follows

$$\varepsilon = \varepsilon^0 + f_1\varepsilon^1 + f_2\varepsilon^2 + g'\varepsilon^3 \tag{5}$$

where

$$\varepsilon^0 = \begin{Bmatrix} u_{,x} \\ 0 \end{Bmatrix},\ \varepsilon^1 = \begin{Bmatrix} \beta_{,x} \\ 0 \end{Bmatrix},\ \varepsilon^2 = \begin{Bmatrix} \alpha_{,x} \\ 0 \end{Bmatrix},\ \varepsilon^3 = \begin{Bmatrix} 0 \\ \varphi \end{Bmatrix} \tag{6}$$

Rewrite the transverse shear strain γ_{xz} as follows

$$\gamma_{xz} = f_1'(\gamma_0 + \gamma_1) + f_2'\gamma_2,\ \gamma_0 = \varphi_{,x},\ \gamma_1 = w_{,x} + \beta,\ \gamma_2 = w_{,x} + \alpha \tag{7}$$

The transverse shear strain is assumed to have a quadratic distribution across the thickness of the beam. In addition, the transverse shear strain equals to zero at the top and bottom surfaces of the beam. These conditions lead to

$$\begin{gathered} \gamma_2 = w_{,x} + \alpha = 0 \\ \gamma_{xz} = f_1'(z)(\gamma_0 + \gamma_1) \end{gathered} \tag{8}$$

The constitutive relations between the stress field and the strain field are expressed as follows

$$\begin{Bmatrix} \sigma_x \\ \sigma_z \\ \tau_{xz} \end{Bmatrix} = \begin{bmatrix} C_{11} & C_{13} & 0 \\ C_{13} & C_{33} & 0 \\ 0 & 0 & C_{55} \end{bmatrix} \begin{Bmatrix} \varepsilon_x \\ \varepsilon_z \\ \gamma_{xz} \end{Bmatrix} \tag{9}$$

In this study, Young's modulus E of FGM is a function of the coordinate, whereas, the Poisson's ratio is assumed to be constant and equal, the coefficients C_{ij} vary with the position according to the following formulas [17]

$$C_{11} = C_{33} = \frac{E(z)}{1-\nu^2},\ C_{13} = \nu C_{11},\ C_{55} = \frac{E(z)}{2(1+\nu)} \tag{10}$$

Equation (9) may be rewritten in the short form as

$$\sigma = D\varepsilon = D\left(\varepsilon^0 + f_1\varepsilon^1 + f_2\varepsilon^2 + g'\varepsilon^3\right),\ \tau_{xz} = f'_1 G\gamma_{xz} \tag{11}$$

where

$$D = \frac{E(z)}{1-\nu^2}\begin{bmatrix} 1 & \nu \\ \nu & 1 \end{bmatrix},\ G = C_{55} = \frac{E(z)}{2(1+\nu)} \tag{12}$$

4. Finite Element Formulation

The expression of the strain energy of the beam is

$$\Pi = \frac{1}{2}\int_V \left(\varepsilon^T.\sigma + \gamma_{xz}.\tau_{xz}\right)dV \tag{13}$$

The expression of the variation of strain energy can be calculated as follows

$$\delta\Pi = \int_V \left\{ \begin{array}{l} \left[\delta\varepsilon^0 + f_1\delta\varepsilon^1 + f_2\delta\varepsilon^2 + g'\delta\varepsilon^3\right]^T.\mathrm{D}\left(\varepsilon^0 + f_1\varepsilon^1 + f_2\varepsilon^2 + g'\varepsilon^3\right) \\ +\delta(\gamma_0+\gamma_1).f_1'.G.f'_1.(\gamma_0+\gamma_1) \end{array} \right\}dV \tag{14}$$

After integrating Equation (14) over the beam section and rewriting it in the matrix form, the variation of the strain energy can be computed as

$$\delta\Pi = \int_L \left(\delta\omega^T\mathrm{R} + \delta\gamma_{01}^T T_{01}\right)dx \tag{15}$$

where R and T_{01} are given by

$$\mathrm{R} = b\int_z \begin{bmatrix} \mathrm{D} & f_1\mathrm{D} & f_2\mathrm{D} & g'\mathrm{D} \\ f_1\mathrm{D} & f_1^2\mathrm{D} & f_1f_2\mathrm{D} & f_1g'\mathrm{D} \\ f_2\mathrm{D} & f_1f_2\mathrm{D} & f_2^2\mathrm{D} & f_2g'\mathrm{D} \\ g'\mathrm{D} & g'f_1\mathrm{D} & g'f_2\mathrm{D} & g'g'\mathrm{D} \end{bmatrix}\begin{bmatrix} \varepsilon^0 \\ \varepsilon^1 \\ \varepsilon^2 \\ \varepsilon^3 \end{bmatrix}dz\ , \quad T_{01} = b\int_z f_1'\tau dz = b\int_z \left(f_1'\right)^2 G\gamma_{01}dz \tag{16}$$

and

$$\delta\omega = \begin{bmatrix} \delta\varepsilon^0 \\ \delta\varepsilon^1 \\ \delta\varepsilon^2 \\ \delta\varepsilon^3 \end{bmatrix}, \quad \delta\gamma_{01} = \delta\varphi_{,x} + \delta w_{,x} + \delta\beta \tag{17}$$

where

$$\delta\varepsilon^0 = \left\{ \begin{array}{c} \delta u_{,x} \\ 0 \end{array} \right\}, \delta\varepsilon^1 = \left\{ \begin{array}{c} \delta\beta_{,x} \\ 0 \end{array} \right\}, \delta\varepsilon^2 = \left\{ \begin{array}{c} \delta\alpha_{,x} \\ 0 \end{array} \right\}, \delta\varepsilon^3 = \left\{ \begin{array}{c} 0 \\ \delta\varphi \end{array} \right\} \tag{18}$$

As consequence, Equation (16) can be rewritten as

$$\mathrm{R} = \mathrm{H}\omega\ ;\ T_{01} = H_s\gamma_{01} \tag{19}$$

where

$$\mathrm{H} = b\int_z \begin{bmatrix} \mathrm{D} & f_1\mathrm{D} & f_2\mathrm{D} & g'\mathrm{D} \\ f_1\mathrm{D} & f_1^2\mathrm{D} & f_1f_2\mathrm{D} & f_1g'\mathrm{D} \\ f_2\mathrm{D} & f_1f_2\mathrm{D} & f_2^2\mathrm{D} & f_2g'\mathrm{D} \\ g'\mathrm{D} & g'f_1\mathrm{D} & g'f_2\mathrm{D} & g'g'\mathrm{D} \end{bmatrix}dz \tag{20}$$

$$H_s = b\int_z \left(f_1'\right)^2 G dz \tag{21}$$

In the current work, a two-node beam element is considered, each node includes five degrees of freedom. The vector of displacement of node *i-th* is

$$\{d_i\} = \begin{bmatrix} u_i & w_i & \varphi_i & \beta_i & \alpha_i \end{bmatrix}^T \tag{22}$$

The nodal displacement vector of the proposed beam element, U, is defined by

$$\mathrm{U} = \begin{bmatrix} u_1 & w_1 & \varphi_1 & \beta_1 & \alpha_1 & u_2 & w_2 & \varphi_2 & \beta_2 & \alpha_2 \end{bmatrix}^T \tag{23}$$

The isoparametric geometry on two nodes of new beam element and the nodal variables are given by

$$\begin{cases} x = N_1 x_1 + N_2 x_2 & u = N_1 u_1 + N_2 u_2 \\ w = N_1 w_1 + N_2 w_2 & \varphi = N_1 \varphi_1 + N_2 \varphi_2 \\ \beta = N_1 \beta_1 + N_2 \beta_2 & \alpha = N_1 \alpha_1 + N_2 \alpha_2 \end{cases} \tag{24}$$

For the mixed finite element formulation, the authors approve a quadratic interpolation for α, β with parameters α_m, β_m and a constant shear resultant T_{01}, which will be eliminated later.

$$\begin{aligned} \alpha &= N_1 \alpha_1 + N_2 \alpha_2 + N_m \alpha_m \\ \beta &= N_1 \beta_1 + N_2 \beta_2 + N_m \beta_m \end{aligned} \tag{25}$$

$$T_{01} = T_0 - F(x), \quad \delta T_{01} = \delta T_0, \; F(x) = \int_0^x q(s)ds \tag{26}$$

in which, the shape functions N_1, N_2 and N_m are defined as follows

$$N_1 = \tfrac{1-\xi}{2}, \quad N_2 = \tfrac{1+\xi}{2}, \quad N_m = N_1 N_2 = \tfrac{1-\xi^2}{4} \tag{27}$$

where

$$\xi = \frac{2x - L}{L}, \; d\xi = \frac{2}{L}dx, \; dx = \frac{L}{2}d\xi \tag{28}$$

The first equation in Equation (8) is imposed in integral form as following

$$\int_0^L \gamma_2 dx = \int_0^L (w_{,x} + \alpha)dx = 0 \tag{29}$$

Substitute α and w from Equations (24) and (25) into Equation (29), the parameter α_m can be deduced as

$$\alpha_m = \frac{6}{L}\Big(w_1 - w_2 - \frac{L}{2}(\alpha_1 + \alpha_2)\Big) = \mathrm{B}_{f0}\mathrm{U} \tag{30}$$

where

$$\mathrm{B}_{f0} = \begin{bmatrix} 0 & \frac{6}{L} & 0 & 0 & -3 & 0 & \frac{-6}{L} & 0 & 0 & -3 \end{bmatrix} \tag{31}$$

Substitute Equations (24) and (30), into Equation (17), the strain variation vector, $\delta\omega$, can be obtained as

$$\delta\omega = \begin{bmatrix} \delta\varepsilon^0 \\ \delta\varepsilon^1 \\ \delta\varepsilon^2 \\ \delta\varepsilon^3 \end{bmatrix} \tag{32}$$

$$\delta\varepsilon^0 = \begin{Bmatrix} \delta u_{,x} \\ 0 \end{Bmatrix}, \delta\varepsilon^1 = \begin{Bmatrix} \delta\beta_{,x} \\ 0 \end{Bmatrix}, \delta\varepsilon^2 = \begin{Bmatrix} \delta\alpha_{,x} \\ 0 \end{Bmatrix}, \delta\varepsilon^3 = \begin{Bmatrix} 0 \\ \delta\varphi \end{Bmatrix} \tag{33}$$

Using Equation (30), the variable α becomes

$$\alpha = \sum N_i \alpha_i + N_m \alpha_m = \sum N_i \alpha_i + N_m\big(\mathrm{B}_{f0}\mathrm{U}\big) \tag{34}$$

Then

$$\delta\varepsilon^2 = \left\{ \begin{array}{c} \delta\alpha_{,x} \\ 0 \end{array} \right\} = \left\{ \begin{array}{c} \sum N_{i,x}\delta\alpha_i \\ 0 \end{array} \right\} + \left\{ \begin{array}{c} N_{m,x}\mathrm{B}_{f0}\delta\mathrm{U} \\ 0 \end{array} \right\} \tag{35}$$

Using Equations (33) and (35), the strain variation Equation (32) can be expressed as follows

$$\delta\omega = \mathrm{B}\delta\mathrm{U} = \left(\mathrm{B}_{mf} + \mathrm{B}_f\right)\delta\mathrm{U} \tag{36}$$

where

$$\mathrm{B}_{mf} = \left[\begin{array}{cc} \mathrm{B}^1_{mf} & \mathrm{B}^2_{mf} \end{array} \right] \tag{37}$$

$$\mathrm{B}^1_{mf} = \begin{bmatrix} N_{1,x} & 0 & 0 & 0 & 0 \\ 0 & 0 & 0 & 0 & 0 \\ 0 & 0 & 0 & N_{1,x} & 0 \\ 0 & 0 & 0 & 0 & 0 \\ 0 & 0 & 0 & 0 & N_{1,x} \\ 0 & 0 & 0 & 0 & 0 \\ 0 & 0 & 0 & 0 & 0 \\ 0 & 0 & N_1 & 0 & 0 \end{bmatrix}, \ \mathrm{B}_f = \begin{bmatrix} 0_{1\times10} \\ 0_{1\times10} \\ 0_{1\times10} \\ 0_{1\times10} \\ N_{m,x}\mathrm{B}_{f0} \\ 0_{1\times10} \\ 0_{1\times10} \\ 0_{1\times10} \end{bmatrix} \tag{38}$$

The shear strain variation $\delta\gamma_{01}$ can be obtained as

$$\delta\gamma_{01} = \delta\varphi_{,x} + \delta w_{,x} + \delta\beta \tag{39}$$

Rewrite Equation (39) into the matrix form as below

$$\delta\gamma_{01} = \mathrm{B_s}\delta\mathrm{U} \tag{40}$$

where

$$\mathrm{B_s} = \left[\begin{array}{cccccccccc} 0 & N_{1,x} & N_{1,x} & N_1 & 0 & 0 & N_{2,x} & N_{2,x} & N_2 & 0 \end{array} \right] \tag{41}$$

Using Equation (25) the strain fields $\delta\varepsilon$, $\delta\gamma_{01}$ become

$$\delta\omega = \mathrm{B}\delta\mathrm{U} + \mathrm{B}_m\delta\beta_m, \ \mathrm{B}_m = \left[\begin{array}{cccccccc} 0 & 0 & N_{m,x} & 0 & 0 & 0 & 0 & 0 \end{array} \right]^T \tag{42}$$

$$\delta\gamma_{01} = \mathrm{B_s}\delta\mathrm{U} + N_m\delta\beta_m \tag{43}$$

The variation of strain energy Equation (15) becomes

$$\delta\Pi = \int_0^L \left(\delta\omega^T\mathrm{R} + \delta\gamma_{01}T_{01} + \delta T_{01}\left(\gamma_{01} - \frac{T_{01}}{H_s}\right)\right)dx \tag{44}$$

Substituting $T_{01} = T_0 - F$, $\delta T_{01} = \delta T_0$ into Equation (44) we get

$$\delta\Pi = \int_0^L \left(\delta\omega^T\mathrm{R} + \delta\gamma_{01}(T_0 - F) + \delta T_0\left(\gamma_{01} - \frac{T_0 - F}{H_s}\right)\right)dx \tag{45}$$

$$\begin{aligned} \delta\Pi = \int_0^L \Big(& \delta\mathrm{U}^T\mathrm{B}^T\mathrm{HBU} + \delta\mathrm{U}^T\mathrm{B}^T\mathrm{HB}_m\beta_m + \\ & \delta\beta_m\mathrm{B}^T_{\beta m}\mathrm{HBU} + \delta\beta_m\mathrm{B}^T_{\beta m}\mathrm{HB}_{\beta m}\beta_m + \\ & \delta\mathrm{U}^T\mathrm{B_s}^T T_0 - \delta\mathrm{U}^T\mathrm{B_s}^T F + \delta\beta_m{}^T N_m T_0 - \delta\beta_m{}^T N_m F + \\ & \delta T_0\mathrm{B_s U} + \delta T_0 N_m\beta_m - \delta T_0\tfrac{1}{H_s}T_0 - \delta T_0\tfrac{-1}{H_s}F\Big)dx \end{aligned} \tag{46}$$

Rewritten Equation (46) in the matrix form

$$\delta\Pi = \int_0^L \left\{ \begin{bmatrix} \delta U^T & \delta T_0 & \delta\beta_m \end{bmatrix} \left(\begin{bmatrix} K_{uu} & K_{uT} & K_{u\beta} \\ K_{uT}^T & K_{TT} & K_{T\beta} \\ K_{u\beta}^T & K_{T\beta} & K_{\beta\beta} \end{bmatrix} \begin{bmatrix} U \\ T_0 \\ \beta_m \end{bmatrix} - \begin{bmatrix} f_u \\ f_T \\ f_\beta \end{bmatrix} \right) \right\} dx \tag{47}$$

where K_{uu}, K_{uT}, $K_{u\beta}$, K_{TT}, $K_{T\beta}$, $K_{\beta\beta}$, f_u, f_T and f_β are calculated as follows

$$K_{uu} = \int_{-1}^{1} B^T H B \frac{L}{2} d\xi = \int_{-1}^{1} \left(B_{mf} + B_f\right)^T H \left(B_{mf} + B_f\right) \frac{L}{2} d\xi \tag{48}$$

$$K_{uT} = \int_{-1}^{1} B_s^T \frac{L}{2} d\xi \tag{49}$$

$$K_{u\beta} = \int_{-1}^{1} B^T H B_{\beta m} \frac{L}{2} d\xi \tag{50}$$

$$K_{TT} = -\int_{-1}^{1} \frac{1}{H_s} \frac{L}{2} d\xi \tag{51}$$

$$K_{T\beta} = \int_{-1}^{1} N_m \frac{L}{2} d\xi \tag{52}$$

$$K_{\beta\beta} = \int_{-1}^{1} B_{\beta m}^T H B_{\beta m} \frac{L}{2} d\xi \tag{53}$$

$$f_u = \int_0^L N_w^T q dx + \int_0^L B_s^T F dx \tag{54}$$

$$f_T = \frac{-1}{H_s} \int_0^L F(x) dx \tag{55}$$

$$f_\beta = \int_0^L N_m(x) F(x) dx \tag{56}$$

where

$$N_w = \begin{bmatrix} 0 & N_1 & 0 & 0 & 0 & 0 & N_2 & 0 & 0 & 0 \end{bmatrix} \tag{57}$$

The parameter T_0 and rotation β_m are then removed in two steps as below.

Step 1: eliminate rotation parameter β_m

$$\beta_m = \frac{1}{K_{\beta\beta}} \left(f_b - K_{u\beta}^T U - K_{T\beta} T_0 \right) \tag{58}$$

Substitute Equation (58) into Equation (47), we get

$$\delta\Pi = \int_0^L \left\{ \begin{bmatrix} \delta \mathrm{U}^T & \delta T_0 \end{bmatrix} \left(\begin{bmatrix} \mathrm{K}_{11} & \mathrm{K}_{12} \\ \mathrm{K}_{12}^T & K_{22} \end{bmatrix} \begin{bmatrix} \mathrm{U} \\ T_0 \end{bmatrix} - \begin{bmatrix} \mathrm{f}_1 \\ f_2 \end{bmatrix} \right) \right\} dx \tag{59}$$

where K_{11}, K_{12}, K_{22}, f_1 and f_2 are expressed as

$$\mathrm{K}_{11} = \mathrm{K}_{uu} - \mathrm{K}_{u\beta} K_{\beta\beta}^{-1} \mathrm{K}_{u\beta}^T \tag{60}$$

$$\mathrm{K}_{12} = \mathrm{K}_{uT} - \mathrm{K}_{u\beta} K_{\beta\beta}^{-1} K_{T\beta} \tag{61}$$

$$K_{22} = K_{TT} - K_{T\beta} K_{\beta\beta}^{-1} K_{T\beta} \tag{62}$$

$$\mathrm{f}_1 = \mathrm{f}_u - \mathrm{K}_{u\beta} K_{\beta\beta}^{-1} f_\beta \tag{63}$$

$$f_2 = f_T - K_{T\beta} K_{\beta\beta}^{-1} f_\beta \tag{64}$$

Step 2: eliminate the stress resultant constant T_0

$$T_0 = K_{22}^{-1}\left(f_2 - \mathrm{K}_{12}^T \mathrm{U}\right) \tag{65}$$

Substitute Equation (65) into Equation (59), the expression of the variation of strain energy is manifested as

$$\delta\Pi = \delta \mathrm{U}^T\left(\mathrm{K}_{11} - \mathrm{K}_{12} K_{22}^{-1} \mathrm{K}_{12}^T\right)\mathrm{U} - \delta \mathrm{U}^T\left(\mathrm{f}_1 - \mathrm{K}_{12} K_{22}^{-1} f_2\right) \tag{66}$$

The stiffness matrix of new beam element K is now calculated as

$$\mathrm{K} = \mathrm{K}_{11} - \mathrm{K}_{12} K_{22}^{-1} \mathrm{K}_{12}^T \tag{67}$$

The nodal load vector of the beam element is expressed as

$$\mathrm{f} = \mathrm{f}_1 - \mathrm{K}_{12} K_{22}^{-1} f_2 \tag{68}$$

5. Numerical Results and Discussion

5.1. Convergence Study

In this subsection, some examples are given to determine the convergence rate of the proposed beam element. A homogeneous beam with the width $b = 1$ m, Young's modulus $E = 29000$ Pa, Poisson's ratio $\nu = 0.3$ as in the work of Heyliger [40] is considered here.

Firstly, a cantilever beam subject to a concentrated load $P = 100$ N at the free end side is considered. Secondly, a simply supported beam under a uniform load $q = 10$ N/m is investigated. The numerical results for some cases of L/h ratios and number of elements are shown in Tables 1 and 2. The numerical results are compared with results of Heyliger [40] using two-node beam C^1 continuous formulation element based on HSDT. The comparison shows that the new beam element has an excellent convergence rate. Although the new proposed beam element uses linear shape functions, it has a better convergence rate than that of the beam element using higher-order shape function, thus it costs less effort and time of computation.

Table 1. Comparison of maximum displacement for cantilever beams.

Length	Height	Source	Number of Elements			
			N = 2	N = 4	N = 8	N = 16
160	12	[40]	30.838	32.368	32.742	32.823
		Present	31.566	32.191	32.509	32.666
80		[40]	3.9234	4.1105	4.1506	4.1567
		Present	3.9987	4.0772	4.1160	4.1317
40		[40]	0.52266	0.54249	0.54540	0.54588
		Present	0.52608	0.53570	0.53907	0.53880
12		[40]	0.023551	0.023741	0.023874	0.023931
		Present	0.022902	0.022347	0.021967	0.021840
160	1	[40]	52968.0	55616.0	56278.0	56444.0
		Present	54302.8	55380.8	55931.1	56206.2
80		[40]	6621.8	6952.9	7035.6	7056.3
		Present	6788.5	6923.3	6992.0	7026.4
40		[40]	828.15	869.53	879.87	882.44
		Present	848.88	865.73	874.33	878.62
12		[40]	22.513	23.627	23.897	23.953
		Present	23.036	23.492	23.723	23.836

Table 2. Comparison of maximum displacement for simply supported beams.

Length	Height	Source	Number of Elements			
			N = 2	N = 4	N = 8	N = 16
160	12	[40]	19.779	20.529	20.691	20.717
		Present	20.692	20.690	20.690	20.690
80		[40]	1.3011	1.3415	1.3478	1.3486
		Present	1.3425	1.3422	1.3421	1.3421
40		[40]	0.096033	0.097481	0.097670	0.097703
		Present	0.096067	0.096060	0.096022	0.096018
12		[40]	0.0022234	0.0022206	0.0022204	0.0022204
		Present	0.0018828	0.0019482	0.0019524	0.0019523
160	1	[40]	33549.0	34873.0	35205.0	35287.0
		Present	35302.9	35302.9	35302.9	35302.9
80		[40]	2097.7	2180.4	2201.1	2206.3
		Present	2207.0	2207.0	2207.0	2207.0
40		[40]	131.31	136.49	137.77	138.08
		Present	138.09	138.09	138.09	138.09
12		[40]	1.0860	1.1267	1.1351	1.1364
		Present	1.1347	1.1346	1.1346	1.1346

5.2. Validation Study

Continuously, to confirm the accuracy of the proposed beam element, the static bending of an FG beam subjected to a uniform load q is investigated. The FG beams made of two components, which are Aluminum and Alumina. The material properties of Aluminum and Alumina are $E_m = 70\,\text{GPa}$, $E_c = 380\,\text{GPa}$, $v_m = 0.3$, $v_c = 0.3$. Two cases of slenderness ratios $L/h = 5$ and $L/h = 20$ of the FG beams are considered. The displacements and stresses are calculated in the normalized form as.

For simply-simply (SS) and clamped-clamped (CC) supported beams

$$w^* = \frac{100E_m h^3}{qL^4} w\left(\frac{L}{2}, 0\right) \quad (69)$$

For clamped-free (CF) supported beams

$$w^{*} = \frac{100E_m h^3}{qL^4} w(L,0) \tag{70}$$

For axial, normal and shear stresses

$$\sigma_x^{*} = \frac{h}{qL}\sigma_x\left(\frac{L}{2},\frac{h}{2}\right),\ \sigma_z^{*} = \frac{h}{qL}\sigma_z\left(\frac{L}{2},\frac{h}{2}\right),\ \sigma_{xz}^{*} = \frac{h}{qL}\sigma_{xz}(0,0) \tag{71}$$

The numerical results of dimensionless vertical displacement, normal stress and axial stress of FG beam using proposed beam are compared with the results of Li et al. [16] and Vo et al. [20] using analytical and finite element methods, which are given in Tables 3–7. The results in these tables show that the solutions from the present theory are very close with the results from HSDT of Li et al. [16] and Quasi-3D solutions of Vo [20] for different values of the power-law index, aspect ratio, and boundary conditions.

Table 3. The maximum nondimensional vertical displacement of FG SS beams.

L/h	Source	σ_z	p=0	p=1	p=2	p=5	p=10
5	Li et al. [16]	$=0$	3.1657	6.2599	8.0602	9.7802	10.8979
	Vo [20] (Navier)	$\neq 0$	3.1397	6.1338	7.8606	9.6037	10.7578
	Vo [20] (FEM)	$\neq 0$	3.1397	6.1334	7.8598	9.6030	10.7572
	Present	$\neq 0$	3.1388	6.1316	7.8570	9.5992	10.7526
20	Li et al. [16]	$=0$	2.8962	5.8049	7.4415	8.8151	9.6879
	Vo [20] (Navier)	$\neq 0$	2.8947	5.7201	7.2805	8.6479	9.5749
	Vo [20] (FEM)	$\neq 0$	2.8947	5.7197	7.2797	8.6471	9.5743
	Present	$\neq 0$	2.8938	5.7179	7.2770	8.6435	9.5698

Table 4. The nondimensional normal stress of FG SS beams.

L/h	Source	σ_z	p=0	p=1	p=2	p=5	p=10
5	Vo [20] (Navier)	$\neq 0$	0.1352	0.0670	0.0925	0.0180	−0.0181
	Vo [20] (FEM)	$\neq 0$	0.1352	0.0672	0.0927	0.0183	−0.0179
	Present	$\neq 0$	0.1351	0.0669	0.0924	0.0179	−0.0183
20	Vo [20] (Navier)	$\neq 0$	0.0337	−0.5880	−0.6269	−1.1698	−1.5572
	Vo [20] (FEM)	$\neq 0$	0.0338	−0.5874	−0.6261	−1.1690	−1.5560
	Present	$\neq 0$	0.0337	−0.5880	−0.6270	−1.1696	−1.5570

Table 5. The nondimensional axial stress of FG SS beams.

L/h	Source	σ_z	p=0	p=1	p=2	p=5	p=10
5	Li et al. [16]	$=0$	3.8020	5.8837	6.8812	8.1030	9.7063
	Vo [20] (Navier)	$\neq 0$	3.8005	5.8812	6.8818	8.1140	9.7164
	Vo [20] (FEM)	$\neq 0$	3.8020	5.8840	6.8860	8.1190	9.7220
	Present	$\neq 0$	3.7994	5.8793	6.8792	8.1101	9.7108
20	Li et al. [16]	$=0$	15.0130	23.2054	27.0989	31.8112	38.1372
	Vo [20] (Navier)	$\neq 0$	15.0125	23.2046	27.0988	31.8137	38.1395
	Vo [20] (FEM)	$\neq 0$	15.0200	23.2200	27.1100	31.8300	38.1600
	Present	$\neq 0$	15.0079	23.1970	27.0884	31.7987	38.1176

Table 6. The nondimensional shear stress of FG SS beams.

L/h	Source	σ_z	p=0	p=1	p=2	p=5	p=10
5	Li et al. [16]	$=0$	0.7500	0.7500	0.6787	0.5790	0.6436
	Vo [20] (Navier)	$\neq 0$	0.7233	0.7233	0.6622	0.5840	0.6396
	Vo [20] (FEM)	$\neq 0$	0.7291	0.7291	0.6661	0.5873	0.6439
	Present	$\neq 0$	0.7233	0.7233	0.6622	0.5839	0.6396
20	Li et al. [16]	$=0$	0.7500	0.7500	0.6787	0.5790	0.6436
	Vo [20] (Navier)	$\neq 0$	0.7432	0.7432	0.6809	0.6010	0.6583
	Vo [20] (FEM)	$\neq 0$	0.7466	0.7466	0.6776	0.6036	0.6675
	Present	$\neq 0$	0.7454	0.7457	0.6828	0.6022	0.6595

Table 7. The maximum nondimensional vertical displacement of FG CC and CF beams.

L/h	Boundary Condition	Source	p=0	p=1	p=2	p=5	p=10
5	CC	Vo [20]	0.8327	1.5722	2.0489	2.6929	3.1058
		Present	0.8367	1.5787	2.0568	2.7039	3.1193
	CF	Vo [20]	28.5524	56.2002	71.7295	86.1201	95.7582
		Present	28.5743	56.2359	71.7607	86.1492	95.7903
20	CC	Vo [20]	0.5894	1.1613	1.4811	1.7731	1.9694
		Present	0.5894	1.1612	1.4806	1.7726	1.9689
	CF	Vo [20]	27.6217	54.6285	69.5266	82.4836	91.2606
		Present	27.6087	54.6051	69.4911	82.4327	91.1965

Figure 2 displays the comparison of the distribution of vertical displacement along with the depth of the FG beam with different values of the power-law index. It can be observed that the vertical displacements are variable across the thickness of the beam and they are in good agreement with published results of Li et al. [16] and Vo [20].

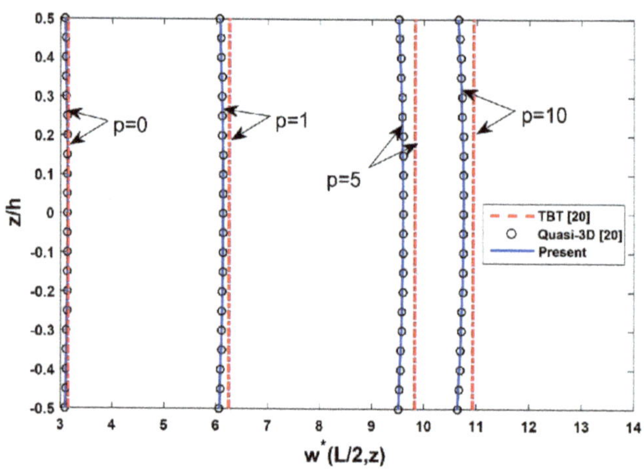

Figure 2. The comparison of the nondimensional transverse displacement $w^*(L/2,z)$ across the depth of FG SS beams subjected to a uniform load with $L/h = 5$.

The distributions of shear stress and axial stress along the depth of FG beam are compared with results of Li et al. [16] and Vo et al. [20] as in Figures 3 and 4. According to Figure 3, the shear stress distribution is parabolic along with the thickness and asymmetric for the FG beams. From Figure 4, the axial stress variation is not linear across the thickness of the FG beam, and its variation is linear across the thickness for isotropic (full ceramic or full metal) beams only. In general, the values of the axial stress do not equal to zeros at the mid-plane of the FG beams. Both shear stress variation and axial stress variation through the thickness of the FG beam using the proposed beam element are in remarkable agreement with those of Vo [20] using Quasi-3D theory.

According to the comparison, the results of the proposed beam element are very close actual adjacent to the Li et al. [16] and Vo et al. [20] solutions. Therefore, the new beam element can be applied to analyze FG beams.

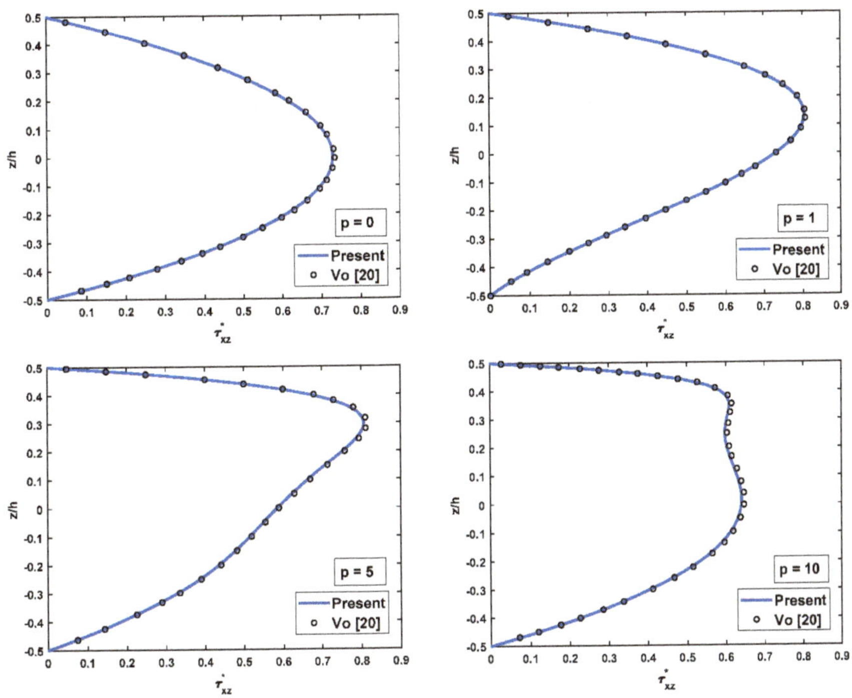

Figure 3. The comparison of the nondimensional shear stress τ_{xz}^{*} across the depth of FG SS beams subjected to a uniform load for different values of p with $L/h = 5$.

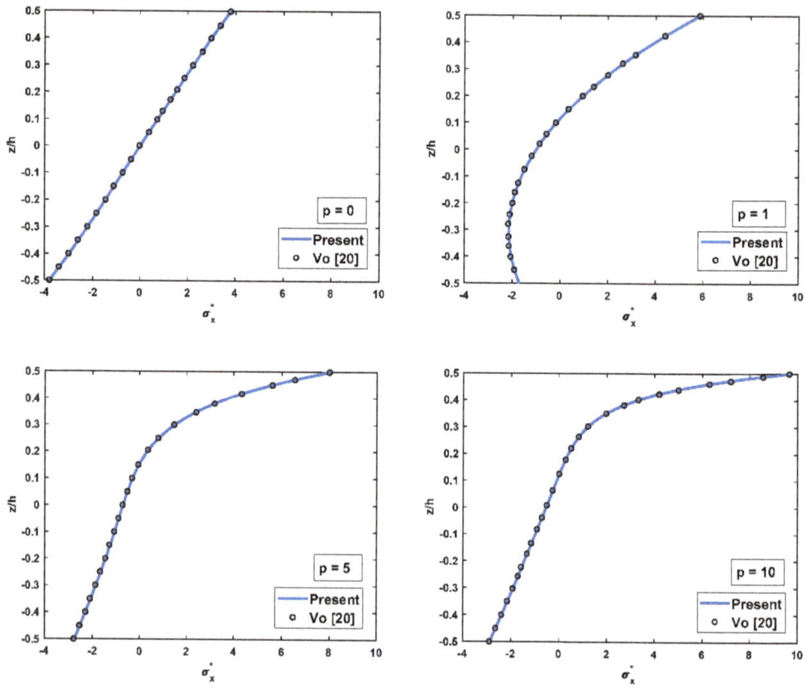

Figure 4. The comparison of the nondimensional axial stress σ_x^* across the depth of FG SS beams subjected to a uniform load for different values of p with $L/h = 5$.

5.3. Static Behaviour of FG Beams

In this subsection, an FG beam which is produced of Aluminum and Zirconium dioxide (Al/ZrO_2) under uniform distribution load q is investigated using proposed beam element. Various power-law indexes, slenderness ratios and boundary conditions are considered. The material properties of Al are $E_m = 70$ GPa, $\nu_m = 0.3$, and the material properties of ZrO_2 are $E_c = 200$ GPa, $\nu_c = 0.3$. The non-dimensional formulas are applied as follows

$$w* = \frac{100h^3E_m}{qL^4}w, \sigma_x* = \frac{h}{qL}\sigma_x, \sigma_z* = \frac{h}{qL}\sigma_z, \sigma_{xz}* = \frac{h}{qL}\sigma_{xz}. \tag{72}$$

In this study, some cases of boundary conditions are considered.

For the simply-simply supported (SS) beam: $u = w = 0$ at $x = 0,\ L$;

For the clamped-clamped supported (CC) beam: $u = w = \alpha = \beta = 0$ at $x = 0,\ L$;

For the clamped-simple supported (CS) beam: $u = w = \alpha = \beta = 0$ at $x = 0$, and $u = w = 0$ at $x = L$;

For the clamped-free supported (CF) beam: $u = w = \alpha = \beta = 0$ at $x = 0$.

The numerical results for bending behaviors of FG beams under uniform load are shown in Tables 8–11 and Figures 5 and 6. Table 8 and Figure 5 shows the nondimensional maximum vertical displacement of FG beams for some cases of boundary conditions, power-law index and different values of the length-to-height ratio. To show more clearly the effect of the power-law index and slenderness ratio, Figure 6 shows the dependence of nondimensional maximum vertical displacement of FG beams on the continuous transformation of the power-law index and slenderness ratio. It shows that the deflection of FG beams increases when increasing the power-law index. This can be explained that increasing the value of the power-law index leads to an increase in the component of metal in FGM, so that the FG beam becomes more flexible. Furthermore, it can be observed that the nondimensional maximum vertical displacement depends not only on the power law index, slenderness ratio but also boundary conditions, which is more pronounced for CC and CS beams than SS and CF beams Furthermore, it can be observed that the nondimensional maximum vertical displacement depends on power-law distribution index, boundary conditions and the length-to-thickness ratio. In addition, boundary conditions have more strongly effects on the deflection of CC and CS beams than those of SS and CF beams.

Table 8. Nondimensional maximum vertical displacement of FG beams subjected to a uniform load.

Boundary Condition	p	L/h = 5	L/h = 10	L/h = 20	L/h = 100
SS	0	5.9637	5.5917	5.4983	5.4684
	1	9.4520	8.9008	8.7625	8.7182
	2	10.8090	10.1178	9.9444	9.8888
	5	12.1559	11.2427	11.0136	10.9402
	10	13.1998	12.1936	11.9411	11.8602
CC	0	1.5898	1.2166	1.1200	1.0877
	1	2.4783	1.9253	1.7823	1.7344
	2	2.8903	2.2046	2.0270	1.9676
	5	3.3827	2.4869	2.2545	2.1770
	10	3.6885	2.7014	2.4453	2.3599
CS	0	2.8431	2.4121	2.3013	2.2650
	1	4.4681	3.8293	3.6651	3.6114
	2	5.1625	4.3680	4.1635	4.0966
	5	5.9291	4.8883	4.6201	4.5323
	10	6.4522	5.3055	5.0099	4.9132
CF	0	54.2912	52.8297	52.4566	52.3199
	1	86.3563	84.1912	83.6384	83.4359
	2	98.2871	95.5870	94.8959	94.6463
	5	109.4815	105.9308	105.0199	104.6959
	10	118.7523	114.8445	113.8429	113.4858

Table 9. Nondimensional axial stress $\sigma_x^*(L/2, h/2)$ of FG beams subjected to a uniform load.

Boundary Condition	p	L/h = 5	L/h = 10	L/h = 20	L/h = 100
SS	0	3.7994	7.5229	15.0080	74.9791
	1	5.1277	10.1431	20.2300	101.0602
	2	5.6251	11.1138	22.1592	110.6870
	5	6.3879	12.6061	25.1275	125.5017
	10	7.2947	14.4105	28.7315	143.5145
CC	0	1.3158	2.5271	5.0069	24.9844
	1	1.7824	3.4099	6.7496	33.6740
	2	1.9604	3.7389	7.3948	36.8820
	5	2.2314	4.2440	8.3871	41.8189
	10	2.5409	4.8479	9.5884	47.8209
CS	0	1.9797	3.7881	7.4945	37.3702
	1	2.6694	5.1069	10.1027	50.3710
	2	2.9383	5.6014	11.0697	55.1718
	5	3.3561	6.3638	12.5573	62.5539
	10	3.8300	7.2728	14.3568	71.5291
CF	0	−3.7207	−7.5172	−15.0722	−75.4217
	1	−5.0080	−10.1282	−20.3126	−101.6530
	2	−5.4757	−11.0879	−22.2441	−111.3295
	5	−6.1993	−12.5682	−25.2212	−126.2416
	10	−7.0995	−14.3779	−28.8453	−144.3695

Table 10. Nondimensional shear stress $\sigma^{*}_{xz}(0,0)$ of the FG beams subjected to a uniform load.

Boundary Condition	p	L/h = 5	L/h = 10	L/h = 20	L/h = 100
SS	0	0.7233	0.7370	0.7454	0.8112
	1	0.7233	0.7370	0.7455	0.8143
	2	0.6857	0.6989	0.7068	0.7660
	5	0.6513	0.6640	0.6714	0.7199
	10	0.6821	0.6954	0.7031	0.7536
CC	0	0.3330	0.1316	−0.2769	−4.4717
	1	0.3322	0.1283	−0.2900	−4.7991
	2	0.3060	0.1198	−0.2567	−4.1445
	5	0.2809	0.1137	−0.2160	−3.3285
	10	0.2937	0.1179	−0.2275	−3.4354
CS	0	0.3727	0.0596	−0.5617	−6.8939
	1	0.3710	0.0544	−0.5814	−7.3858
	2	0.3419	0.0525	−0.5203	−6.3910
	5	0.3148	0.0542	−0.4483	−5.1543
	10	0.3290	0.0550	−0.4717	−5.3222
CF	0	−0.2182	−1.5237	−4.0538	−29.8114
	1	−0.2231	−1.5435	−4.1329	−31.7882
	2	−0.2098	−1.4184	−3.7546	−27.6572
	5	−0.1921	−1.2813	−3.3351	−22.5588
	10	−0.2071	−1.3524	−3.5048	−23.3207

Table 11. Nondimensional normal stress $\sigma^{*}_{z}(L/2, h/2)$ of FG beams subjected to a uniform load.

Boundary Condition	p	L/h = 5	L/h = 10	L/h = 20	L/h = 100
SS	0	0.1351	0.0675	0.0337	0.0065
	1	0.0499	−0.2005	−0.5512	−2.9963
	2	0.0389	−0.2557	−0.6781	−3.6571
	5	0.0375	−0.3064	−0.8035	−4.3225
	10	0.0788	−0.2659	−0.7435	−4.0561
CC	0	0.1351	0.0675	0.0337	0.0065
	1	0.1501	−0.0001	−0.1504	−0.9923
	2	0.1611	−0.0112	−0.1890	−1.2118
	5	0.1820	−0.0174	−0.2255	−1.4326
	10	0.2145	0.0055	−0.2008	−1.3429
CS	0	0.1342	0.0657	0.0300	−0.0121
	1	0.1217	−0.0535	−0.2554	−1.5158
	2	0.1265	−0.0760	−0.3165	−1.8471
	5	0.1405	−0.0942	−0.3763	−2.1831
	10	0.1752	−0.0674	−0.3438	−2.0545
CF	0	0.1314	0.0601	0.0189	−0.0678
	1	0.3459	0.3915	0.6329	2.9244
	2	0.4008	0.4682	0.7696	3.5814
	5	0.4653	0.5492	0.9077	4.2336
	10	0.4791	0.5346	0.8575	3.9489

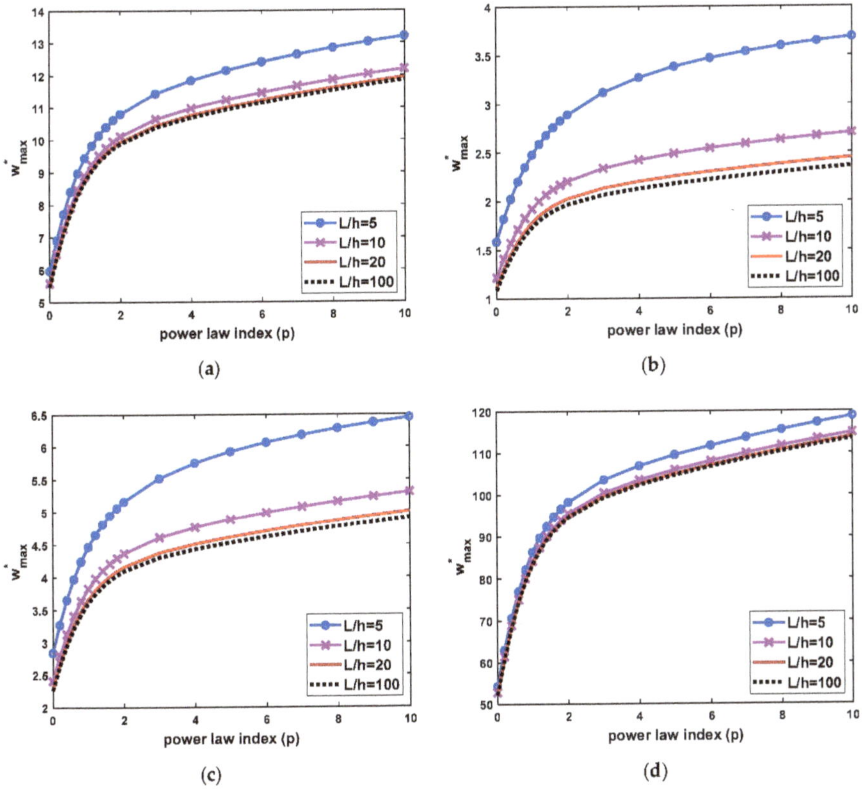

Figure 5. Nondimensional maximum transverse deflection w^*_{max} depends on the power-law index and length-to-thickness ratio of FG beams subjected to a uniform load, (**a**) SS beams, (**b**) CC beams, (**c**) CS beams and (**d**) CF beams.

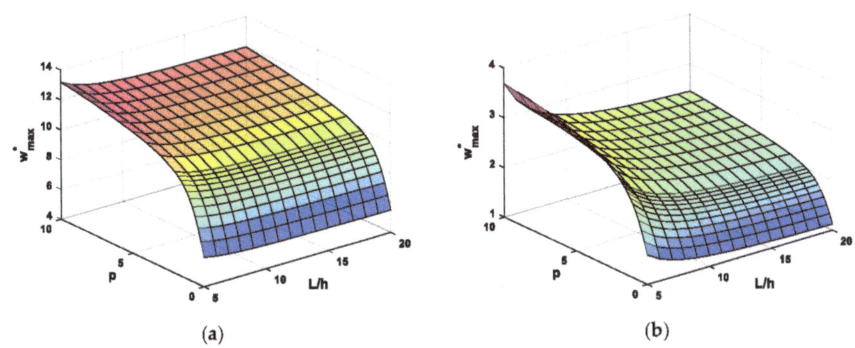

Figure 6. *Cont.*

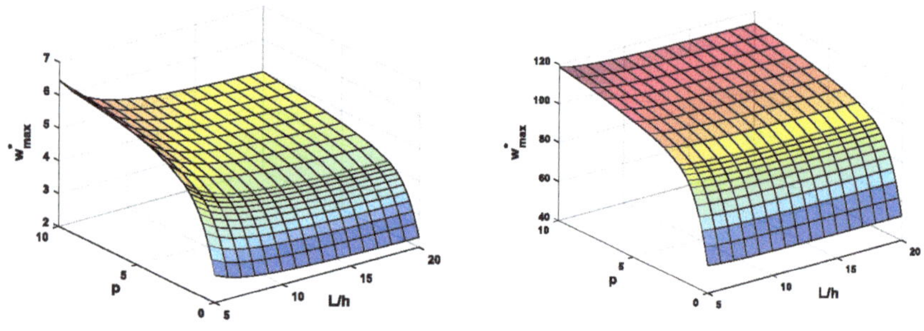

(c) (d)

Figure 6. Nondimensional maximum transverse deflection w^*_{max} as a function of the power-law index and length-to-thickness ratio of FG beams subjected to a uniform load, (**a**) SS beams, (**b**) CC beams, (**c**) CS beams and (**d**) CF beams.

Table 9 shows the nondimensional axial stress $\sigma^*_x(L/2, h/2)$ of FG beams subjected to a uniform load depends on some parameters and boundary conditions. Table 7 and Figure 7 present the distributions of nondimensional axial stress along with the depth of the SS and CC FG beams for different values of the power-law distribution index. The most significant aspect of this figure is that the axial stress distribution of FG beams is much more different from those of isotropic beams. As seen from Table 7 and Figure 7, the axial stress variation is not linear along with the thickness of the FG beams, the tensile stresses at the top are maximum. The values of the axial stresses do not equal to zeros at the mid-plane of the FG beams. This indicates that the neutral plane of the FG beams does not appear at the mid-plane, it is near the top face of the FG beams. This can be explained by the variation of the modulus of elasticity across the depth of the FG beams.

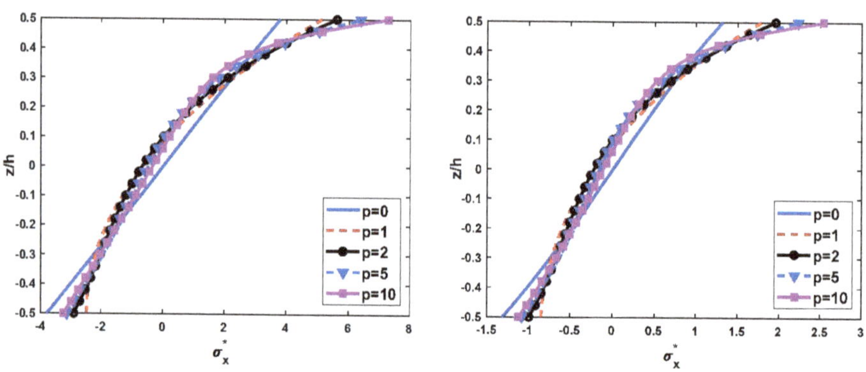

(a) (b)

Figure 7. Nondimensional axial stress $\sigma^*_x(L/2, z)$ through the thickness of FG beams subjected to a uniform load with $L/h = 5$, (**a**) SS beams and (**b**) CC beams.

The non-dimensional shear stress distributions across the thickness of the beams made of FGM with different values of the power-law distribution index and some cases of boundary conditions are presented in Figure 8. The shear stresses of the full ceramic (isotropic) beams are symmetric about the mid-plane of the beams. In addition, the shear stress distributions are greatly influenced by the power-law index. In addition, Figure 8 shows the great dependence of the shear stress distribution on the power-law index.

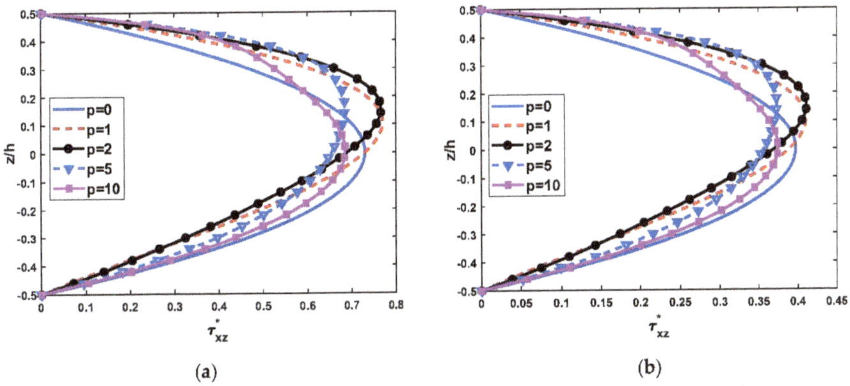

Figure 8. Nondimensional shear stress $\tau^*_{xz}(0,z)$ across the depth of FG beams subjected to a uniform load with $L/h = 5$, (**a**) SS beams and (**b**) CC beams.

The non-dimensional normal stresses of the FG beams under uniform distribution load are shown in Table 11, which highlight the effect of thickness stretching on bending behaviors of FG beam from Quasi-3D theory. Due to the thickness stretching effect, the vertical displacement obtained from present Quasi-3D theory is smaller than those of HSDT and FSDT.

The variation of the vertical displacement through the thickness of the FG beam for SS and CC boundary conditions are shown in Figure 9. According to Figure 9, the difference among the present Quasi-3D theory and other HSDT or FSDT is meaningful for thickness stretching. In this present Quasi-3D theory, the vertical displacement is not constant through the thickness of the beams as in HSDT and FSDT.

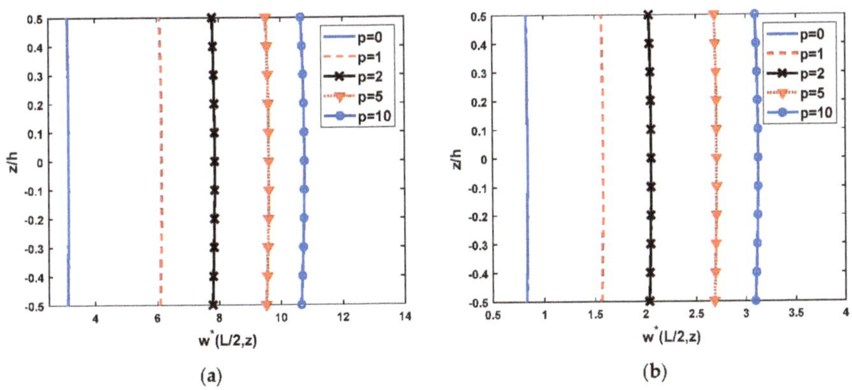

Figure 9. The distribution across the thickness of nondimensional vertical displacement of FG beam subjected to a uniform load with $L/h = 5$, (**a**) SS beams and (**b**) CC beams.

Finally, to show more obviously the influence of normal deformation on the deflection of FG beams, we suggest the deflection ratio which is well-defined as the fraction of transverse displacement obtained by present Quasi-3D beam theory to that calculated by HSDT. The effect of normal deformation on the deflection of SS and CC supported FG beams is exhibited in Figure 10 for different values of power-law distribution index and slenderness ratio. Figure 10 shows that the deflection ratio is almost smaller than unity. It shows that the deflection will be decreased when the normal deformation effect is included. In the case of CC beams, there is a range of power-law index and slenderness ratio that causes the deflection ratio to be greater than unity, which represents that the normal deformation has more affect strongly than bending and shear deformation in this case.

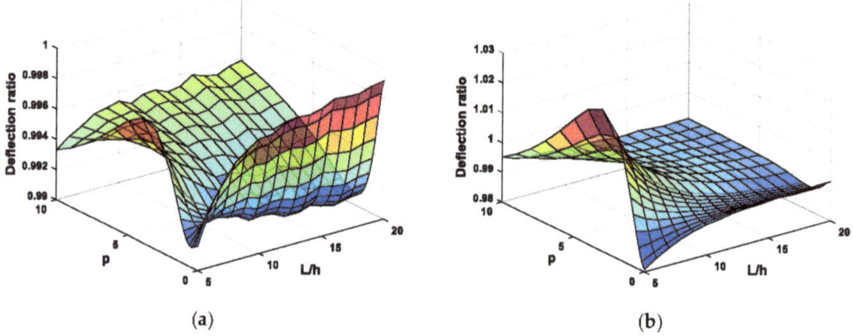

Figure 10. The deflection ratio of FG beams subjected to uniform load, (**a**) SS beams and (**b**) CC beams.

6. Conclusions

In this paper, a new efficient Quasi-3D beam element was developed for static bending analysis of FG beams. Using mixed formulation, only C^0 continuous shape functions are required for finite element formulation of the new beam element. In addition, the new beam element presents the excellent results of displacement and stress even for a coarse mesh. Therefore, the proposed beam element costs less effort and time of computation than those using higher order shape functions, consequently, it can be widely applied for complex structural analysis. The shear stresses vary parabolically across the thickness of the FG beam, and equal to zeros at two free surfaces of beams, so it does not need any shear correction factors. The new beam element includes shear deformation and normal deformation. Effect of normal deformation is significant, and it should be considered in the static bending analysis of FG beams, especially for medium and very thick FG beams. The numerical results of the FG beams using the proposed beam element are in good agreement with other published results. The new beam element is accurate and efficient for bending behavior of FG beams. The influences of some parameters such as the power-law distribution index and length-to-thickness ratio are investigated.

Author Contributions: Investigation, H.N.N.; Methodology, H.N.N., T.T.H.; Software, T.T.H.; Formal Analysis, H.N.N., P.V.V. an T.T.H.; Writing—original draft, D.V.T., P.V.V.; Writing—review & editing, D.V.T., P.V.V.; Project Administration, D.V.T.; Funding Acquisition, D.V.T.

Funding: This research and the APC was funded by Vietnam National Foundation for Science and Technology Development (NAFOSTED) grant number 107.02-2018.30.

Acknowledgments: DVT gratefully acknowledges the support of Vietnam National Foundation for Science and Technology Development (NAFOSTED) under grant number 107.02-2018.30.

Conflicts of Interest: The authors declare no conflict of interest.

References

1. Anandakumar, G.; Kim, J.H. On the modal behavior of a three-dimensional functionally graded cantilever beam: Poisson's ratio and material sampling effects. *Compos. Struct.* **2010**, *92*, 1358–1371. [CrossRef]
2. Ebrahimi, F.; Mokhtari, M. Free vibration analysis of a rotating Mori–Tanaka-based functionally graded beam via differential transformation method. *Arab. J. Sci. Eng.* **2016**, *41*, 577–590. [CrossRef]
3. Sankar, B.V. An elasticity solution for functionally graded beams. *Compos. Sci. Technol.* **2001**, *61*, 689–696. [CrossRef]
4. Zenkour, A.M. Benchmark trigonometric and 3-D elasticity solutions for an exponentially graded thick rectangular plate. *Arch. Appl. Mech.* **2007**, *77*, 197–214. [CrossRef]

5. Zhong, Z.; Yu, T. Analytical solution of a cantilever functionally graded beam. *Compos. Sci. Technol.* **2007**, *67*, 481–488. [CrossRef]
6. Trinh, L.C.; Vo, T.P.; Thai, H.T.; Nguyen, T.K. An analytical method for the vibration and buckling of functionally graded beams under mechanical and thermal loads. *Compos. Part B* **2016**, *100*, 152–163. [CrossRef]
7. Kien, N.D. Large displacement behaviour of tapered cantilever Euler–Bernoulli beams made of functionally graded material. *Appl. Math. Comput.* **2014**, *237*, 340–355. [CrossRef]
8. Lee, J.W.; Lee, Y.J. Free vibration analysis of functionally graded Bernoulli-Euler beams using an exact transfer matrix expression. *Int. J. Mech. Sci.* **2016**, *122*, 1–17. [CrossRef]
9. Menaa, R.; Tounsi, A.; Mouaici, F.; Mechab, I.; Zidi, M.; Bedia, E.I. Analytical solutions for static shear correction factor of functionally graded rectangular beams. *Mech. Adv. Mater. Structu.* **2012**, *19*, 641–652. [CrossRef]
10. Murin, J.; Aminbaghai, M.; Hrabovsky, J.; Kutis, V.; Kugler, S. Modal analysis of the FGM beams with effect of the shear correction function. *Compos. Part B* **2013**, *45*, 1575–1582. [CrossRef]
11. Nguyen, T.K.; Vo, T.P.; Thai, H.T. Static and free vibration of axially loaded functionally graded beams based on the first-order shear deformation theory. *Compos. Part B* **2013**, *55*, 147–157. [CrossRef]
12. Nam, V.H.; Vinh, P.V.; Chinh, N.V.; Thom, D.V.; Hong, T.T. A new beam model for simulation of the mechanical behaviour of variable thickness functionally graded material beams based on modified first order shear deformation theory. *Materials* **2019**, *12*, 404. [CrossRef] [PubMed]
13. Shi, G. A new simple third-order shear deformation theory of plates. *Int. J. Solids Struct.* **2007**, *44*, 4399–4417. [CrossRef]
14. Kadoli, R.; Akhtar, K.; Ganesan, N. Static analysis of functionally graded beams using higher order shear deformation theory. *Appl. Math. Model.* **2008**, *32*, 2509–2525. [CrossRef]
15. Benatta, M.A.; Mechab, I.; Tounsi, A.; Adda Bedia, E.A. Static analysis of functionally graded short beams including warping and shear deformation effects. *Comput. Mater. Sci.* **2008**, *44*, 765–773. [CrossRef]
16. Li, X.F.; Wang, B.L.; Han, J.C. A higher-order theory for static and dynamic analyses of functionally graded beams. *Arch. Appl. Mech.* **2010**, *80*, 1197–1212. [CrossRef]
17. Thai, H.T.; Vo, T.P. Bending and free vibration of functionally graded beams using various higher-order shear deformation beam theories. *Int. J. Mech. Sci.* **2012**, *62*, 57–66. [CrossRef]
18. Vo, T.P.; Thai, H.T.; Nguyen, T.K.; Iman, F. Static and vibration analysis of functionally graded beams using refined shear deformation theory. *Meccanica* **2014**, *49*, 155–168. [CrossRef]
19. Tinh, Q.B.; Thom, V.D.; Lan, H.T.T.; Duc, H.D.; Satoyuki, T.; Dat, T.P.; Thien-An, N.V.; Yu, T.T.; Sohichi, H. On the high temperature mechanical behaviors analysis of heated functionally graded plates using FEM and a new third-order shear deformation plate theory. *Compos. Part B* **2016**, *92*, 218–241. [CrossRef]
20. Vo, T.P.; Thai, H.T.; Nguyen, T.K.; Iman, F.; Lee, J. Static behaviour of functionally graded sandwich beams using a Quasi-3D theory. *Compos. Part B* **2015**, *68*, 59–74. [CrossRef]
21. Neves, A.M.A.; Ferreira, A.J.M.; Carrera, E.; Roque, C.M.C.; Cinefra, M.; Jorge, R.M.N.; Soares, C.M.M. A Quasi-3D sinusoidal shear deformation theory for the static and free vibration analysis of functionally graded plates. *Compos. Part B* **2012**, *43*, 711–725. [CrossRef]
22. Neves, A.M.A.; Ferreira, A.J.A.; Carrera, E.; Cinefra, M.; Roque, C.M.C.; Jorge, R.M.N.; Soares, C.M.M. A Quasi-3D hyperbolic shear deformation theory for the static and free vibration analysis of functionally graded plates. *Compos. Struct.* **2012**, *94*, 1814–1825. [CrossRef]
23. Hebali, H.; Tounsi, A.; Houari, M.S.A.; Bessain, A.; Adda Bedia, E.A. New Quasi-3D hyperbolic shear deformation theory for the static and free vibration analysis of functionally graded plates. *J. Eng. Mech.* **2014**, *140*, 374–383. [CrossRef]
24. Mantari, J.L.; Soares, C.G. Generalized hybrid Quasi-3D shear deformation theory for the static analysis of advanced composite plates. *Compos. Struct.* **2012**, *94*, 2561–2575. [CrossRef]
25. Mantari, J.L.; Soares, C.G. Four-unknown Quasi-3D shear deformation theory for advanced composite plates. *Compos. Struct.* **2014**, *109*, 231–239. [CrossRef]
26. Thai, H.T.; Vo, T.P.; Bui, T.Q.; Nguyen, T.K. A Quasi-3D hyperbolic shear deformation theory for functionally graded plates. *Acta Mech.* **2014**, *225*, 951–964. [CrossRef]
27. Fang, W.; Yu, T.; Lich, L.V.; Bui, T.Q. Analysis of thick porous beams by a Quasi-3D theory and isogeometric analysis. *Compos. Struct.* **2019**, *221*, 110890. [CrossRef]

28. Nguyen, H.X.; Nguyen, T.N.; Abdel-Wahab, M.; Bordas, S.P.A.; Hung, N.X.; Vo, T.P. A refined Quasi-3D isogeometric analysis for functionally graded microplates based on the modifed couple stress theory. *Comput. Methods Appl. Mech. Eng.* **2017**, *313*, 904–940. [CrossRef]
29. Yu, T.; Zhang, J.; Hu, H.; Bui, T.Q. A novel size-dependent Quasi-3D isogeometric beam model for two-directional FG microbeams analysis. *Compos. Struct.* **2019**, *221*, 76–88. [CrossRef]
30. Farzam-Rad, S.A.; Hassani, B.; Karamodin, H. Isogeometric analysis of functionally graded plates using a new Quasi-3D shear deformation theory based on physical neutral surface. *Compos. Part B* **2017**, *108*, 174–189. [CrossRef]
31. Tran, L.V.; Wahab, M.A.; Niiranen, J. A six-variable Quasi-3D model for static analysis of laminated composite plates using isogeometric analysis. *Int. Conf. Numer. Model. Eng.* **2018**, *20*, 135–142. [CrossRef]
32. Carrera, E. Theories and finite elements for multilayered plates and shells: A unified compact formulation with numerical assessment and benchmarking. *Arch. Comput. Methods Eng.* **2003**, *10*, 215–296. [CrossRef]
33. Carrera, E.; Petrolo, M.; Nali, P. Unified formulation applied to free vibrations finite element analysis of beams with arbitrary section. *Shock Vib.* **2011**, *18*, 485–502. [CrossRef]
34. Cerrera, E.; Zozulya, V.V. Carrera unified formulation (CUF) for the micropolar beams: Analytical solutions. *Mech. Adv. Mater. Struct.* **2019**. [CrossRef]
35. Giunta, G.; Belouettar, S.; Carrera, E. Analysis of FGM beams by means of classical and advanced theories. *Mech. Adv. Mater. Struct.* **2010**, *17*, 622–635. [CrossRef]
36. Filippi, M.; Carrera, E.; Zenkour, A.M. Static analyses of FGM beams by various theories and finite elements. *Compos. Part B* **2015**, *72*, 1–9. [CrossRef]
37. Chakraborty, A.; Gopalakrishnan, S.; Reddy, J.N. A new beam finite element for the analysis of functionally graded materials. *Int. J. Mech. Sci.* **2003**, *45*, 519–539. [CrossRef]
38. Nguyen, D.K.; Nguyen, Q.H.; Tran, T.T.; Bui, V.T. Vibration of bi-dimensional functionally graded Timoshenko beams excited by a moving load. *Acta Mech.* **2017**, *228*, 141–155. [CrossRef]
39. Khan, A.A.; Alam, M.N.; Rahman, N.; Wajid, M. Finite element modelling for static and free vibration response of functionally graded beam. *Lat. Am. J. Solids Struct.* **2016**, *13*, 690–714. [CrossRef]
40. Heyliger, P.R. A higher order beam finite element for bending and vibration problems. *J. Sound Vib.* **1988**, *126*, 309–326. [CrossRef]
41. Kapuria, S.; Bhattacharyya, M.; Kumar, A.N. Bending and free vibration response of layered functionally graded beams: A theoretical model and its experimental validation. *Compos. Struct.* **2008**, *82*, 390–402. [CrossRef]
42. Vo, T.P.; Thai, H.T.; Nguyen, T.K.; Maheri, A.; Lee, J. Finite element model for vibration and buckling of functionally graded sandwich beams based on a refined shear deformation theory. *Eng. Struct.* **2014**, *64*, 12–22. [CrossRef]
43. Moallemi-Oreh, A.; Karkon, M. Finite element formulation for stability and free vibration analysis of Timoshenko beam. *Adv. Acoust. Vib.* **2013**, *2013*, 841215. [CrossRef]
44. Pascon, J.P. Finite element analysis of flexible functionally graded beams with variable Poisson's ratio. *Eng. Comput.* **2016**, *33*, 2421–2447. [CrossRef]
45. Yarasca, J.; Mantari, J.L.; Arciniega, R.A. Hermite-Lagrangian finite element formulation to study functionally graded sandwich beams. *Compos. Struct.* **2016**, *140*, 567–581. [CrossRef]

materials

MDPI

Article

Investigation of the Flow Properties of CBM Based on Stochastic Fracture Network Modeling

Bo Zhang [1,2], Yong Li [1,2,3,*], Nicholas Fantuzzi [1,4], Yuan Zhao [1,5], Yan-Bao Liu [6,*], Bo Peng [3] and Jie Chen [1,2]

1 State Key Laboratory of Coal Mine Disaster Dynamics and Control, Chongqing University, Chongqing 400044, China
2 School of Resources & Safety Engineering, Chongqing University, Chongqing 400044, China
3 Technology Center of Sichuan Coal Industry Group, Chengdu 610091, China
4 DICAM, University of Bologna, 40126 Bologna, Italy
5 Sinohydro Bureau 8 Co. LTD., POWERCHINA, Changsha 410004, China
6 CCTEG Chongqing Engineering Co. LTD., Chongqing 400016, China
* Correspondence: yong.li@cqu.edu.cn (Y.L.); yanbao_liu@163.com (Y.-B.L.); Tel.: +86-188-8337-0816 (Y.L.)

Received: 22 June 2019; Accepted: 24 July 2019; Published: 26 July 2019

Abstract: Coal contains a large number of fractures, whose characteristics are difficult to describe in detail, while their spatial distribution patterns may follow some macroscopic statistical laws. In this paper, several fracture geometric parameters (FGPs) were used to describe a fracture, and the coal seam was represented by a two-dimensional stochastic fracture network (SFN) which was generated and processed through a series of methods in MATLAB. Then, the processed SFN image was able to be imported into COMSOL Multiphysics and converted to a computational domain through the image function. In this way, the influences of different FGPs and their distribution patterns on the permeability of the coal seam were studied, and a finite element model to investigate gas flow properties in the coal seam was carried out. The results show that the permeability of the coal seam increased with the rising of fracture density, length, aperture, and with the decrease of the angle between the fracture orientation and the gas pressure gradient. It has also been found that large-sized fractures have a more significant contribution to coal reservoir permeability. Additionally, a numerical simulation of CBM extraction was carried out to show the potential of the proposed approach in the application of tackling practical engineering problems. According to the results, not only the connectivity of fractures but also variations of gas pressure and velocity can be displayed explicitly, which is consistent well with the actual situation.

Keywords: Monte Carlo method; coalbed methane; stochastic fracture network; fracture geometric parameters; dual-porosity and dual-permeability media; finite element method

1. Introduction

Coal is a kind of porous medium with many fractures formed in it after a long-term geological process. The existence of these weak structures has a great influence on the flow of coalbed methane (CBM) that can not only lead to mine hazards but also provide a substantial source of energy in both industry and households [1–3]. Therefore, it is of great significance to investigate the characteristics of fractures and their internal gas flow properties for both CBM exploitation and gas outburst prevention [4,5].

In terms of the investigation of flow in porous media, Darcy's law was the earliest linear seepage model to emerge [6,7]. Since its emergence, numerous experiments and theoretical investigations on gas flow and transport properties in various porous media have been performed and reported for single-phase, multiphase, saturated, and partially saturated domains [8–10]. However, because of the complex structure of porous media and the limitations of considering real geological conditions, conventional experiments and theoretical achievements fail in describing the uncertainties of flow and transport properties in porous media. For example, the continuous equivalent model averages the permeability of reservoirs and overlooks the influence stemming from the discontinuity caused by fractures [11,12]. This model is similar to the "black box" and its central problem is to solve the permeability tensor. Obviously, it is not suitable for capturing some important mechanical behaviors of fractured rocks, especially when some large-sized fractures exist. Barenblatt et al. have proposed the dual medium model, though it is not suitable to describe the flow behavior of fractures [13].

With the increasingly used computer and simulation methods, many researchers have modeled porous media using a discrete fracture network (DFN), which generally contains three categories (geological-mapping-based DFN, geomechanically grown DFN, and stochastic fracture network (SFN)) [14]. A DFN has advantages in the discretization of fracture networks and preservation of the relationships between fractures and fracture sets compared to the extensively used lattice Boltzmann method (LBM) in which fractures are discretized into cells or edges [15]. Geological-mapping-based DFN illustrates fracture patterns from limited exposure of outcrops, boreholes, or tunnels [16]. Geomechanically grown DFN can reproduce natural fractures through a DFN simulator based on paleostress conditions [17]. SFN is a simple and convenient DFN model generated from statistical data of fracture network characteristics which applies various scale fractures [18]. Fracture network characteristics can be described by FGPs, which basically comprise fracture density, length, aperture, and orientation. These FGPs all have an obvious impact on the porosity and permeability of the whole fracture structure [19–21]. However, due to the complexity of the stress environment, formation history, and lithotype, the distribution patterns of FGPs in various areas and buried depths are different and cannot be characterized by in situ observations or some conventional laboratory approaches including nuclear magnetic resonance and mercury intrusion porosimetry. Micron computed tomography scanning technology and scanning electron microscopy have limitations in the description of large-scale fracture systems and the determination of fractures' connectivity features [2,22]. Seismological surveys may be able to assess and image large-scale structures but the current technology can hardly detect widely-spreading medium and small fractures due to the resolution limit [23]. In contrast, SFN modelling, as a probabilistic realization of a fracture network based on the theory of random processes like the Monte Carlo method [24,25], provides an efficient way to generate fracture networks containing differently distributed FGPs derived from field observations and measurements [26–32]. To date, SFN modelling has been used in many fields such as civil, mining, enhanced geothermal systems, and water resource engineering [33–40].

In the present work, even though the coal matrix is regarded as a homogeneous body with fixed porosity and permeability, SFN is used to describe the fracture system of the coal seam, which can reflect the relationships between fractures and describe the heterogeneity of the coal seam in a statistical sense. Coming to the topic of this paper, although previous researchers have made great contributions to solve engineering problems using the SFN model, few studies have directly combined it with numerical simulation software to investigate engineering problems. The intention of this paper is twofold. Firstly, a coal seam is represented by a two-dimensional SFN image. The pixels in the picture represent fractures or the matrix and the image resolution determines the accuracy of the fracture size (the higher the resolution is, the smaller the size of the fracture which can be generated). Secondly, a finite element analysis of gas flow in the coal seam is carried out in COMSOL Multiphysics 5.4 on the basis that an SFN image has been transformed to a computational domain. In this way, the temporal and spatial distribution of gas velocity and pressure is able to be reflected more intuitively, which provides a more efficient method to study the flow characteristics of CBM.

2. The CBM Industry with a Worldwide View

As a kind of unconventional natural gas, CBM has been exploited in about 30 countries, and among them, the U.S. is the first and most successful country in CBM exploitation, with the largest production in the world throughout the given period shown in Figure 1b and the second largest estimated CBM resources, which are at 49.2 Tcm, as shown in Figure 1a. According to the line graph in Figure 1b, the U.S. saw a considerable increase in CBM production within the period 1989 to 2008. However, after its production peaked at 55.67 Bcm in 2008, the figure had fallen to 28.88 Bcm by 2016. This phenomenon may stem from three areas: the government provided effective policy support at first, which greatly stimulated the commercial production of CBM; later, the average daily production of single wells continued to decrease, which resulted in the economic benefits declining; and falling natural gas prices led to a sharp drop in investment. The development history of the U.S. CBM industry has important reference significance for other countries, including China.

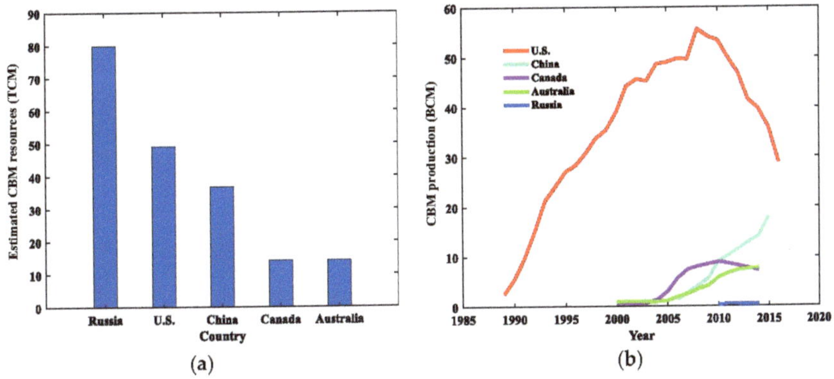

Figure 1. (**a**) The bar chart compares the five highest ranking countries in terms of estimated coalbed methane (CBM) resources. (**b**) The line graph shows CBM production of these five countries from 1989 to 2016, 2000 to 2015, 2000 to 2014, 2000 to 2014, and 2010 to 2014, respectively. Source: International Energy Agency https://www.iea.org/ugforum/ugd/, U.S. Energy Information Administration, Washington, DC https://www.eia.gov/dnav/ng/ng_prod_coalbed_s1_a.htm and National Development and Reform Commission http://www.ndrc.gov.cn/zcfb/zcfbghwb/201112/t20111231_585486.html.

With considerable CBM resources estimated at almost 37 Tcm, as shown in Figure 1a, China has great potential for CBM exploitation. Although it has started to exploit CBM commercially relatively late, production statistical data shows that Chinese CBM production has increased significantly from 1 Bcm in 2000 to 20 Bcm in 2015, which reflects a similar trend to the U.S.'s figures in the early years, as shown in Figure 1b. Compared to the large integrated network of CBM pipelines in the U.S., Australia, and Canada, commercial utilization of CBM in China is localized with most production coming from high-rank coals in the Ordos or Qinshui basins [41]. In terms of the geological conditions, Chinese CBM reservoirs generally reflect low permeability (under 1×10^{-3} μm^2), low gas pressure, low resource abundance (under 1.3×10^8 m^3/km^2) and great buried depth (over 600 m) compared with the U.S.'s CBM reservoir conditions (permeability over 2×10^{-3} μm^2, resource abundance over 2×10^8 m^3/km^2, and buried depth under 500 m) [42]. With these challenges faced with the development of the Chinese CBM industry, it is of great importance to have a good understanding of CBM occurrence conditions (especially pores and fractures within reservoirs, which have fundamental impacts on gas transport).

3. SFN Reconstruction and Processing

In this section, the basic theory for SFN generation, SFN image processing, and the techniques adopted for transforming an SFN image into a computational domain are presented.

3.1. Basic Theory and Method of SFN Reconstruction

The Monte Carlo method, which is known as a statistical simulation method, is based on the large number theorem and the central limit theorem. The basic idea is that when the problem is the probability of a random event, the probability of the random event is estimated by the frequency of the occurrence of this event by some "experimental" method or some digital features of the random variable. The main means of the Monte Carlo method is to use random numbers to carry out statistical tests and produce random numbers that follow a certain distribution function, which basically contains two steps:

(1) The linear congruence method is used to generate uniformly distributed random numbers in [0,1] interval [22], i.e.,

$$\begin{cases} x_{i+1} = (ax_i + c)mod(m) \\ \xi_{i+1} = \frac{x_{i+1}}{m} \end{cases} \tag{1}$$

where x_{i+1} is a random variable corresponding to a random number ξ_{i+1}; a is a multiplier; c is the increment; m is a modulus; *mod(m)* represents the remainder of the modulus; the subscript i is an integer; and the initial value is zero.

(2) The obtained uniformly distributed random numbers are used to generate other random numbers that are subject to different distributions based on statistical data (averages and standard variances of FGPs).

Taking normal distribution as an example, the probability density function is expressed as

$$f(x) = \frac{1}{\sqrt{2\pi}\sigma} e^{-\frac{(x-\mu)^2}{2\sigma^2}}, -\infty < x < +\infty \tag{2}$$

Furthermore, the probability distribution function can be derived as

$$F(x) = \int_{-\infty}^{x} f(t)dt = \frac{1}{\sqrt{2\pi}\sigma} \int_{-\infty}^{x} e^{\frac{1}{2}} \left(\frac{t - t^{-2}}{\sigma}\right) dt \tag{3}$$

The random number of normal distribution can be obtained as

$$x = \mu_x + \sigma_x \sqrt{-2 \ln \xi} \cos(2\pi\xi) \tag{4}$$

where x is a random number which is subject to normal distribution; and ξ is a random number of uniform distribution in the [0,1] interval.

Four distribution functions in Table 1 and four FGPs of SFN (density (ρ), fracture length (l), fracture aperture (d), and fracture direction (θ)) have been investigated in this paper. A single fracture is represented by a straight line in SFN. The center coordinates of a fracture are (x_0, y_0) and center points of all fractures are uniformly distributed in an SFN. By using Equation (5), the starting point and endpoint coordinates of a fracture can be obtained. In addition, the number of fractures is determined by fracture density, and the fracture orientation is defined by the angle from the X-axis rotated along counter clockwise to the fracture. In this way, SFNs can be reconstructed.

$$\begin{cases} x = x_0 \pm \left(\frac{l}{2}\right) \cos\theta \\ y = y_0 \pm \left(\frac{l}{2}\right) \sin\theta \end{cases} \tag{5}$$

Table 1. The probability density functions and corresponding random variables used in this work.

Distribution	Probability Density Function	Random Variable
Uniform	$f(x) = \frac{1}{b-a}(a \le x \le b)$	$x_r = (b-a)\xi + a$
Normal	$f(x) = \frac{1}{\sqrt{2\pi}\sigma}e^{-\frac{(x-\mu)^2}{2\sigma^2}}$	$x_r = \sqrt{\frac{12}{n}}\left(\sum\limits_{i=1}^{n}\xi_i - \frac{n}{2}\right)\sigma + \mu$
Exponential	$f(x) = \frac{1}{\mu}e^{-(x/\mu)}(x > 0)$	$x_r = -\mu\ln(1-\xi)$
Lognormal	$f(x) = \frac{1}{\sqrt{2\pi}\sigma x}e^{-\frac{1}{2}\left(\frac{\ln x-\mu}{\sigma}\right)^2}(x > 0)$	$x_r = e^{\sqrt{\frac{12}{n}}(\sum_{i=1}^{n}\xi_i - \frac{n}{2})\sigma+\mu}$

3.2. SFN Image Processing

For the sake of simplicity and to show the efficient use of such an SNF model in a numerical simulation, the simulations were restricted to two-dimensional images in the present study. A series of images representing fracture structures were obtained and processed through MATLAB 2016b for reconstructing the coal reservoir. Figure 2a represents an SFN which has been processed for frame removing, grey processing, binarization, and color reversion. The white regions depict fractures and the black regions depict the coal matrix. Figure 2b processed by image function defines a continuous computational domain for CBM flow simulation in COMSOL. Similarly, based on these steps, other SFN images were processed to obtain different computational domains.

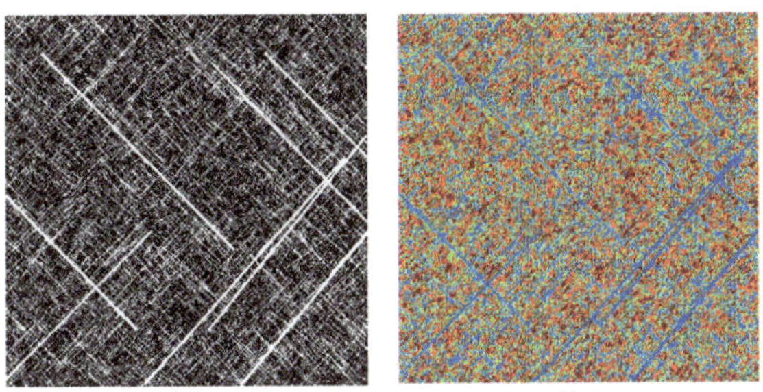

(a) (b)

Figure 2. The process of transforming a stochastic fracture network (SFN) image to a computational domain: **(a)** a stochastic fracture image generated and processed in MATLAB; **(b)** the same image which has been imported into COMSOL through the image function.

3.3. Image function

The image function makes it possible to import an image to COMSOL and map the image's RGB data to a scalar (single channel) function output value. By default the function's output uses the mapping (R+G+B)/3. An image is defined on a two-dimensional domain, and we typically describe the image function using spatial coordinates: *im(x,y)*. According to Section 3.2, we made all the images binarized, so *im(x,y)* = 0 or 1. If the area on the image represents the coal matrix (the red area in Figure 2b), *im(x,y)* = 0, and where *im(x, y)* = 1 this represents fractures (the blue area in Figure 2b). Therefore, the porosity and permeability of an SFN can be divided into two parts by the image function using

$$\varphi = \left(\varphi_f - \varphi_m\right) \times im(x,y) + \varphi_m \quad (6)$$

$$K = \left(K_f - K_m\right) \times im(x,y) + K_m \quad (7)$$

where φ is the porosity of the SFN; φ_m is the porosity of the coal matrix; φ_f is the fracture porosity; and K is the permeability of the SFN. K_m is the permeability of the coal matrix and K_f is the permeability of fracture.

4. Numerical Simulation of Gas Flow in the Coal Seam

The basic assumptions, computational geometry, governing equations, and numerical techniques adopted for investigating the influence of different FGPs on the permeability of the SFN are presented in this section.

4.1. Basic Assumptions

In this study, the following basic assumptions were made:

(1) The coal seam is represented by an SFN and treated as a dual-porosity reservoir that is composed of fractures and the coal matrix.
(2) FGPs consist of density, length, aperture, and orientation.
(3) The flow in the coal seam is a single phase and saturated Darcy flow.
(4) Gas absorption is described by the Langmuir law.
(5) Coupling effects of multiple physical fields are ignored.

4.2. Fracture Flow Model

Numerous investigations including theoretical analyses and numerical modeling about the flow behavior in fractured rocks have been conducted with various rocks. Researchers have normally assumed laminar flow in a single fracture with two fracture surfaces. According to the Navier-Stokes equation, the average flow rate through a plane void can be calculated. It has been found that flow transmissivity is proportional to the cube of the fracture aperture, which also known as the "cubic flow equation" [43], i.e.,

$$q = -\frac{b^3}{12\mu}\frac{dP}{dx} \tag{8}$$

where q is the flow rate of a fracture at a unit height in the Z direction (Figure 3); b is the fracture aperture; and μ is the fluid dynamic viscosity.

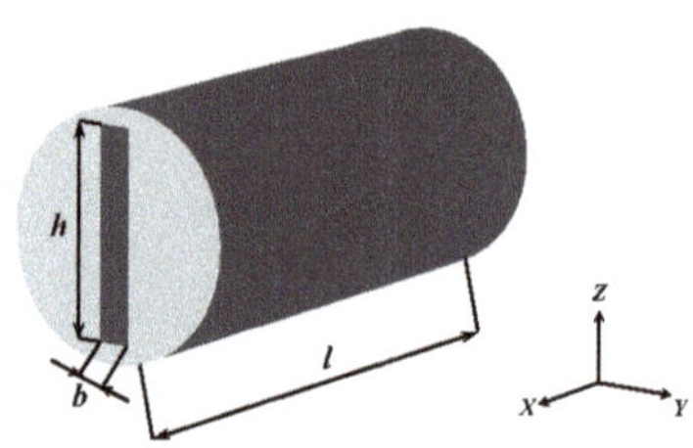

Figure 3. Diagram of a 3D fracture in coal.

As shown in Figure 3, when the height of the fracture is h, the flow rate Q of the total rock cross section can be described as

$$Q = hq = -\frac{A\varphi_f b^2}{12\mu}\frac{dP}{dx} = -\frac{K_f A}{\mu}\frac{dP}{dx} \tag{9}$$

where A is the area of the rock cross section.

Then the permeability of the fracture becomes:

$$K_f = \frac{\varphi_f b^2}{12} \tag{10}$$

In this work, b is the average value of the fracture aperture in the SFN and φ_f is defined by the ratio of pixels representing fractures to total pixels.

4.3. Governing Equations and Boundary Conditions

Based on the basic assumptions of the SFN model and ignoring gas adsorption, the continuity equation of gas flow in the coal seam can be expressed as

$$\begin{cases} \nabla\cdot\left(\rho_g V\right) = 0 \\ V = -\frac{K}{\mu}\nabla P \\ \rho_g = \frac{M_g P}{RT} \end{cases} \tag{11}$$

where ρ_g is the gas density; V is the gas velocity; μ is the dynamic viscosity of gas; P is the gas pressure; M_g is molar mass of gas [kg/mol]; R is the gas constant [J/(mol·K)]; and T is the temperature of the coal seam.

By substituting Equations (7) and (10) into Equation (11), the gas flow equation can be derived as:

$$\nabla\cdot\left[\frac{M_g P}{RT}\left(-\frac{K_m\left(\frac{\varphi_f b^2}{12K_m}-1\right)im(x,y)+K_m}{\mu}\nabla P\right)\right]=0 \tag{12}$$

The geometry of the fracture network and boundary conditions are shown in Figure 4. This geometry was imported from the processed image, with dimensions along the X and Y axes being both 15 m. The gas flow is pressure driven with a constant pressure gradient maintained from the inlet to the outlet.

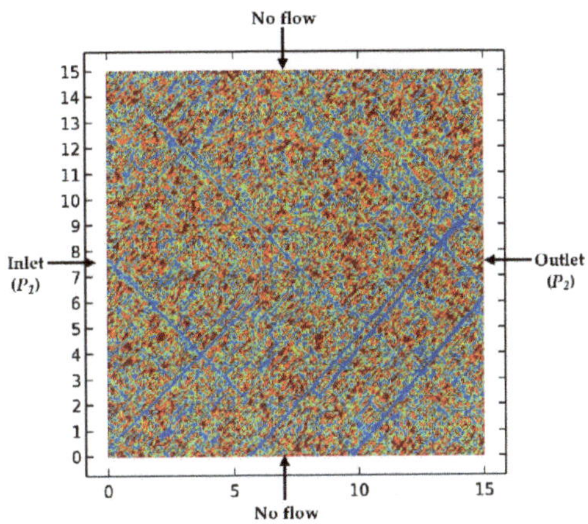

Figure 4. The representative computational geometries used for the numerical study. All dimensions are in meters.

4.4. Numerical Simulation Results

Based on the boundary conditions (Figure 4), combined with the parameters in Table 2, the distribution features of gas pressure and velocity in the SFN were obtained through steady flow computation in COMSOL, as shown in Figure 5. The gas pressure was seen to gradually decrease from the inlet to the outlet, and the gas velocity in the fractures was much greater than that in the matrix. Subsequently, the average value of the gas velocity at the outlet of the fracture network was able to be obtained by integral and averaging (Figure 5b).

Table 2. Parameters used for simulation.

Parameter	Value
Gas dynamic viscosity (μ)	11.067 [Pa.s]
Matrix permeability (k_m)	0.02 [mD]
Matrix porosity (φ_m)	0.01
Gas pressure of inlet (P_1)	1 [MPa]
Gas pressure of outlet (P_2)	0 [MPa]
Coal seam temperature (T)	310 [K]

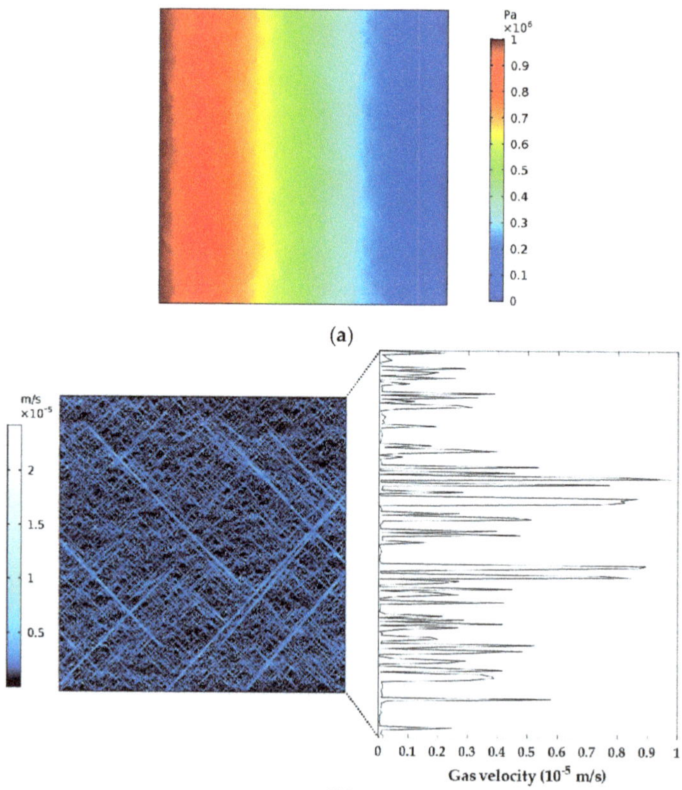

Figure 5. Numerical simulation results: (**a**) the spatial distribution features of gas pressure in the SFN; (**b**) the distribution of the gas velocity in the whole network attached with gas velocity at the outlet of the SFN.

5. Parametric Study

In this section, the purpose was to compare the permeability of different fracture networks without obtaining their values. According to Darcy's law of gas seepage in porous media (Equation (14)), the permeability (K) is proportional to gas velocity (V) when all the other parameters are fixed. Hence, the permeability was able to be compared by the variation of mean gas velocities obtained from velocity curves like Figure 5b.

$$K = \frac{2QP_{\alpha}\mu L}{A(P_1^2 - P_2^2)} = \frac{2VP_{\alpha}\mu L}{P_1^2 - P_2^2} \tag{13}$$

where P_α is the standard atmospheric pressure and L represents the distance between the gas inlet and the outlet in the direction of the gas pressure gradient.

To investigate the influences of FGPs on the permeability of the coal seam, a parametric study was performed by varying every parameter and calculating the corresponding mean gas velocity at the outlet of the SFN. The reference values of the FGPs are listed in Table 3. In this part, several groups of SFN images are described in each subsection and the influences of fracture density, length, aperture, and orientation on the permeability of the SFN are discussed respectively on the basis of making all the FGPs subject to normal distribution. Additionally, impacts caused by the distribution of FGPs and the combination of differently scaled fractures are studied. Every group of SFNs contains three SFN images generated with all the same FGPs, and the average value of three simulation outcomes is taken as a reference quantity.

Table 3. Reference values of fracture geometric parameters (FGPs).

Density (m^{-2})	Orientation (°)		Length (m)		Aperture (m)	
	Mean	Standard Deviation	Mean	Standard Deviation	Mean	Standard Deviation
3.2	10/100	3°	2.3	0.1	0.03	0.002

5.1. *Influence of Fracture Density on CBM Flow*

To find out the influence of fracture density on the permeability of the coal seam, other FGPs have been kept the same as in Table 3. The mean gas velocity at the outlet of the SFNs generated with different densities has been compared (Figure 6). It is easy to see that the connected region increased obviously with the rising of the fracture density, which is more conducive to the flow of gas. The increasing gas velocity indicates that the permeability of the coal seam increased with greater fracture density.

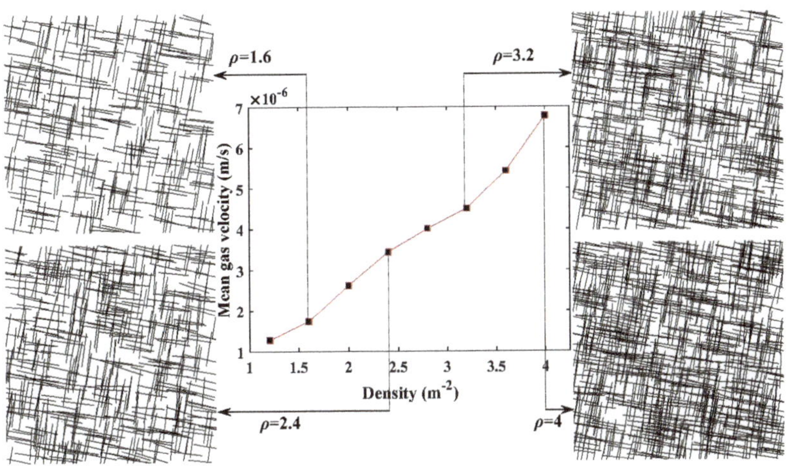

Figure 6. Relationship between the density of fractures and the mean velocity of gas.

5.2. *Influence of Fracture Length on CBM Flow*

To investigate the influence of fracture length on the permeability of the SFN, the mean velocity of gas at the outlet of the SFNs was compared by making the fracture length vary from 1.5 m to 3 m, as shown in Figure 7, and keeping the other FGPs the same as that in Table 3. As shown in the graph, gas velocity increased gradually with the rising of the fracture length, which indicates that longer fracture length can result in the increase of a reservoir's permeability.

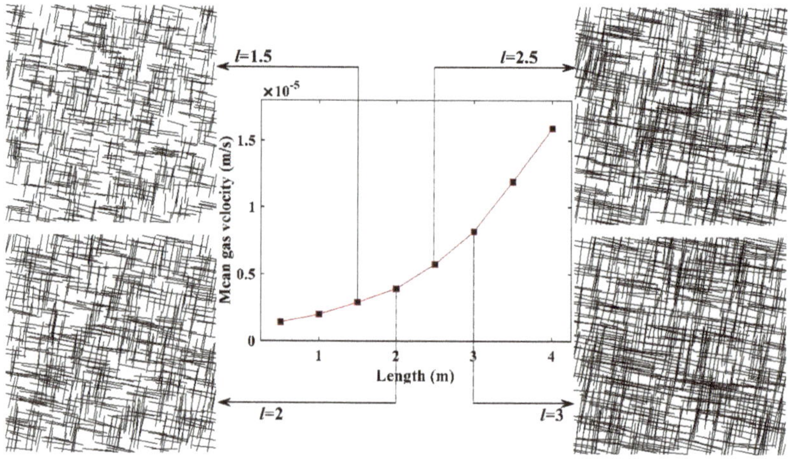

Figure 7. Relationship between the fracture length and the mean velocity of methane at the outlet.

5.3. *Influence of Fracture Aperture on CBM Flow*

By choosing different values of fracture aperture and the same other parameters shown in Table 3, the influence of fracture aperture on the permeability of coal seam was investigated. According to Figure 8, gas velocity rose considerably with the increase of the fracture aperture. This result from a side illustrates that the permeability of the coal seam will increase as the fracture aperture becomes larger.

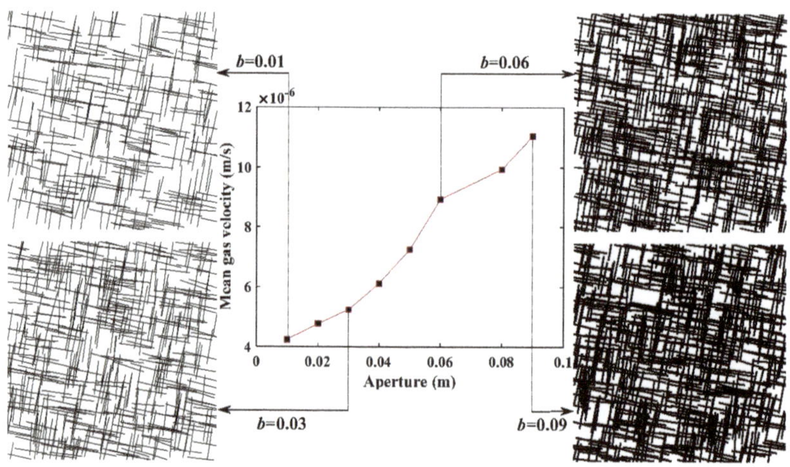

Figure 8. Relationship between the fracture aperture and the mean velocity of methane at the outlet.

As Figures 7 and 8 depict, gas velocity rose by approximately 5×10^{-6} m/s as the fracture length increased from 1.5 m to 3 m. However, gas velocity rose by almost 7×10^{-6} m/s as the fracture aperture increased from 0.01 m to 0.09 m. Thus, it can be concluded that the fracture aperture has a more obvious effect on permeability than fracture length.

5.4. *Influence of Fracture Orientation on CBM Flow*

Considering the symmetry of the SFN surface, it was feasible to choose fracture direction angles of 10°, 30°, 50°, 70°, and 90° for the investigation. By making FGPs the same as that in Table 3 except for the orientation, the mean velocity of gas at the outlet of SFNs with different fracture orientations was compared (Figure 9).

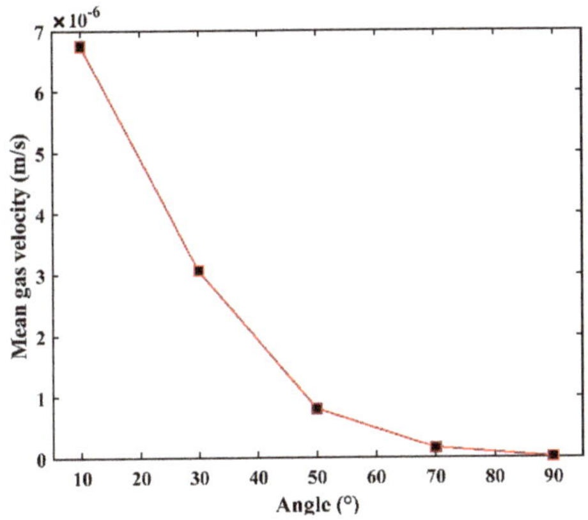

Figure 9. Relationship between fracture orientation and the mean velocity of methane at the outlet.

The change of fracture orientation can be seen to not affect the flow capacity of the fracture network, but it does change the direction of the gas flow. In the context of the present work, the fracture orientation means the angle between the gas flow direction and the pressure gradient. As shown in Figure 9, gas velocity decreased gradually with the increase in fracture direction angle, which illustrates that the permeability of SFN became smaller.

5.5. *Influence of Fracture Size on CBM Flow*

According to long-term field and laboratory measurements and statistics of a newly exposed coal face and collected coal samples, Fu et al. [44] proposed a comprehensive classification method of coal seam fractures through statistical analysis of the fractures' morphological characteristics. Fracture size is divided into four grades: large fractures, middle fractures, small fractures, and micro fracture (Table 4).

Table 4. Classification of coal fracture size.

Fracture Size	Fracture Aperture (mm)	Length (m)	Density (m^{-1})
Large (L)	>100	>10	1~10
Middle (M)	10~100	1~10	10~100
Small (S)	5~15	0.01~1	10~200
Micro (Mi)	<10	0.01~0.1	20~500

In this work, seven groups of SFNs with different combinations of differently scaled fractures have been researched. The fracture densities of the SFNs were all 1.6 m^{-2} and the other FGPs were determined according to Table 4. Through numerical computation, gas velocities of each group SFN were obtained (Figure 10). It is clearly shown that the larger proportion of larger scale fractures in the SFN corresponded to greater gas flow velocity, which indicates that large scale fractures make a dominating contribution to the permeability of the coal seam.

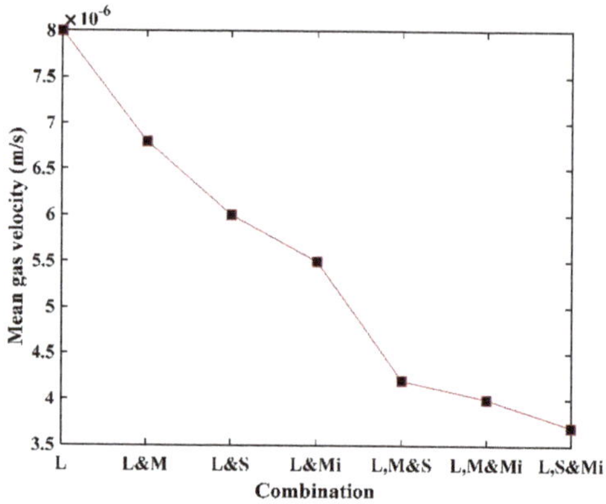

Figure 10. Relationship between the combination of different fractures and the mean gas velocities at the outlets of the SFNs.

5.6. *Influence of Distribution Patterns of FGPs on CBM Flow*

The distribution patterns of FGPs are reflected by distribution functions. Generally, distribution patterns of different FGPs are not the same in the real case. In this paper, uniform distribution, normal distribution, lognormal distribution, and exponential distribution were selected to study the influence of different distribution patterns of FGPs on the permeability of the SFN, which correspond to four groups of SFNs, with all the FGPs being set to follow one distribution function (Figure 11). The fracture density SFNs were all 3.2 m^{-2}, with the other FGPs kept the same as that in Table 3.

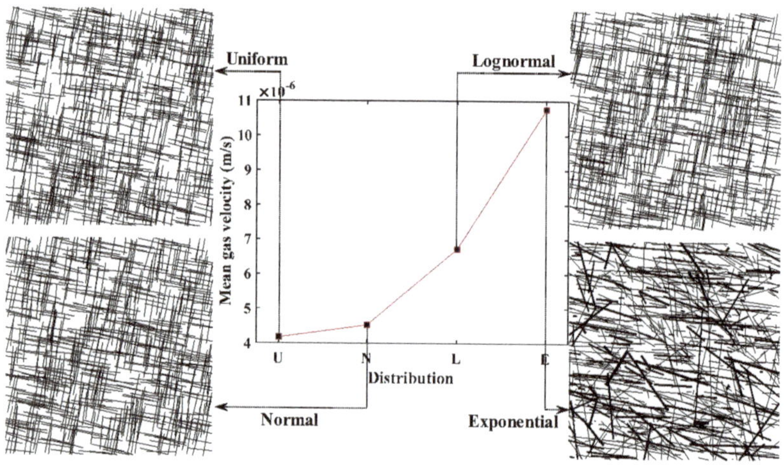

Figure 11. Comparison of gas velocities when FGPs are differently distributed.

The gas velocity curve (Figure 11) shows that when the FGPs were exponentially distributed, the gas velocity was the largest, with lognormal distribution followed by normal distribution, and with gas velocity being the smallest when the value of the FGPs was uniformly distributed.

In order to illustrate this result, a comparison of four probability density plots with the same mean and variance of fracture length is taken as an example. As shown in Figure 12, normal distribution is symmetric around the point $x = \mu$, which is at the same time the mean of the distribution. Lognormal distribution is a positive skew distribution with its peak shifted to the left and a long tail to the right side of the mean. When the standard deviation is small, lognormal distribution is shown to be very close to normal distribution in the short term; however, lognormal distribution has more values of fracture length distributed upward in the long run. For uniform distribution, the probability density is constant within two boundaries and the value range of the fracture length is smaller compared with the normal distribution. Relatively speaking, the probability density change of the exponentially distributed fracture length is small, and the variable values are more widely distributed than that in the lognormal distribution. Therefore, combined with the conclusion of the previous subsection, the result of Figure 11 can be well supported.

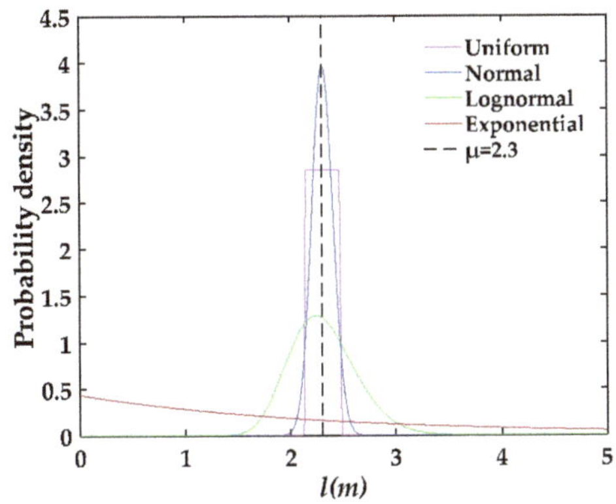

Figure 12. The probability density distributions of fracture length with μ = 2.3 m; σ = 0.1 m.

6. CBM Extraction Simulation Based on SFN Modeling

In order to test the application in solving practical engineering problems through the proposed approach, an SFN image with four scales of fractures was generated to represent the coal seam (Figure 13a). Based on Section 4, a CBM extraction numerical simulation with consideration of gas adsorption and desorption was carried out in this part. Boundary conditions and initial conditions adopted in the simulation are shown in Figure 13b.

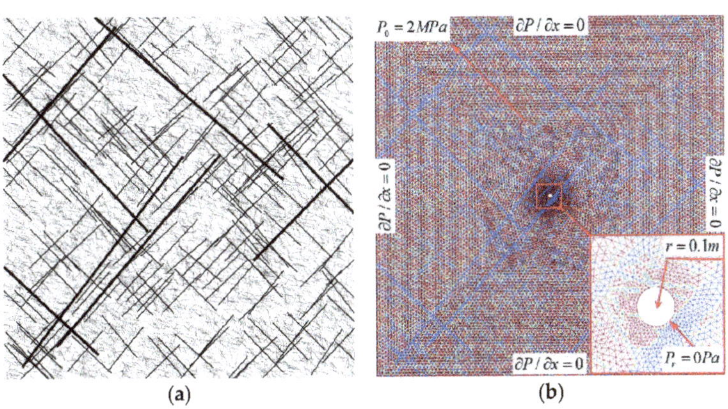

Figure 13. **(a)** The SFN generated for the gas extraction simulation; **(b)** the same image which has been imported into COMSOL and for which a mesh containing 28,676 elements has been generated in the computational domain. Boundary conditions and initial conditions have also been indicated in the figure.

In practical engineering applications, the data of FGPs are collected by field and laboratory measurements and then the probability density functions are determined according to the fitting of the data. For the sake of simplicity, we determined that the fracture lengths all obey a lognormal distribution; the direction angles of the large and middle fractures obey a lognormal distribution; the direction angles of the small and micro fractures obey an exponential distribution; and fracture apertures at all scales follow a normal distribution. The specific values of the FGPs are shown in Table 5. The numerical simulation parameters in COMSOL have been derived from Table 6.

Table 5. FGPs for coalbed reconstruction.

Fractures	Density (m^{-2})	Orientation (°)		Length (m)		Aperture (m)	
		Mean	Standard Deviation	Mean	Standard Deviation	Mean	Standard Deviation
Large	0.03	45/135	3	11	0.5	0.11	0.01
Middle	1	45/135	3	2.3	0.1	0.03	0.002
Small	30	45	-	0.4	0.05	0.005	0
Micro	100	45	-	0.06	0.01	0.001	0

Table 6. Computational parameters in simulation.

Parameter	Value
Gas density under standard conditions (ρ_{gs})	0.716 [kg/m^3]
Langmuir pressure constant (P_L)	3.034 [MPa]
Langmuir volume constant (V_L)	0.036 [m^3/kg]
Density of coal skeleton (ρ_s)	1370 [kg/ m^3]

6.1. *Governing Equations and Parameters*

CBM content in the reservoir consisted of absorbed gas and free gas, which is defined as [45]

$$m_{\mathrm{g}} = \frac{M_{\mathrm{g}}P}{RT}\varphi_f + \rho_{\mathrm{gs}}\rho_{\mathrm{s}}\frac{V_{\mathrm{L}}P}{P_{\mathrm{L}}+P} \tag{14}$$

where ρ_{gs} is the density of the gas under standard conditions; ρ_s is the coal skeleton density; V_L is the Langmuir volume constant; and P_L is the Langmuir pressure constant.

Under the influence of the gas concentration and pressure gradient, the gas in the matrix is shown to migrate into fractures. On the basis of mass conservation, Equation (15) is able to be obtained, i.e.,

$$\frac{\partial m_{\mathrm{g}}}{\partial t} + \nabla\cdot\left(\rho_g V\right) = 0 \tag{15}$$

Substituting Equations (7), (10), and (14) into Equation (15), the gas migration equation in the coal seam can be written as

$$\begin{gathered}\frac{\partial}{\partial t}\left\{\frac{V_{\mathrm{L}}P}{P_{\mathrm{L}}+P}\rho_{\mathrm{s}}\rho_{\mathrm{gs}}\right\} + \frac{\partial}{\partial t}\left(\varphi_f\frac{M_{\mathrm{g}}}{RT}P\right)\\ +\nabla\cdot\left(-\frac{M_{\mathrm{g}}\left(\frac{\varphi_f b^2}{12K_m}-1\right)im(x,y)K_m+M_{\mathrm{g}}K_m}{\mu RT}P\nabla P\right)=0\end{gathered} \tag{16}$$

6.2. *Simulation Results Analysis*

Figure 14 gives information about the spatial and temporal distributions of gas pressure in the coal seam at four points at the times 0, 10, 20, and 30 h. It is apparent that the gas pressure around the borehole gradually decreased with time. The pressure decreased quickly near the borehole and the decrease became slower as it moved away from the borehole. This phenomenon resulted in the pressure drop funnel forming around the borehole, which can also be observed in the line graph. Figure 15 illustrates the gas velocity distribution in the coalbed at different times. It is noticeable that the gas velocity in fractures was much larger than that in matrix. Gas velocity increased from T = 0 h to T = 10 h and then decreased with the gas pressure becoming small. Additionally, the area where the gas velocity changed obviously got larger first and then became smaller. The simulation results show the characteristics of CBM flow during the process of gas extraction, which indicates that SFN images combined with finite element analysis have great potential in the application of tackling engineering problems.

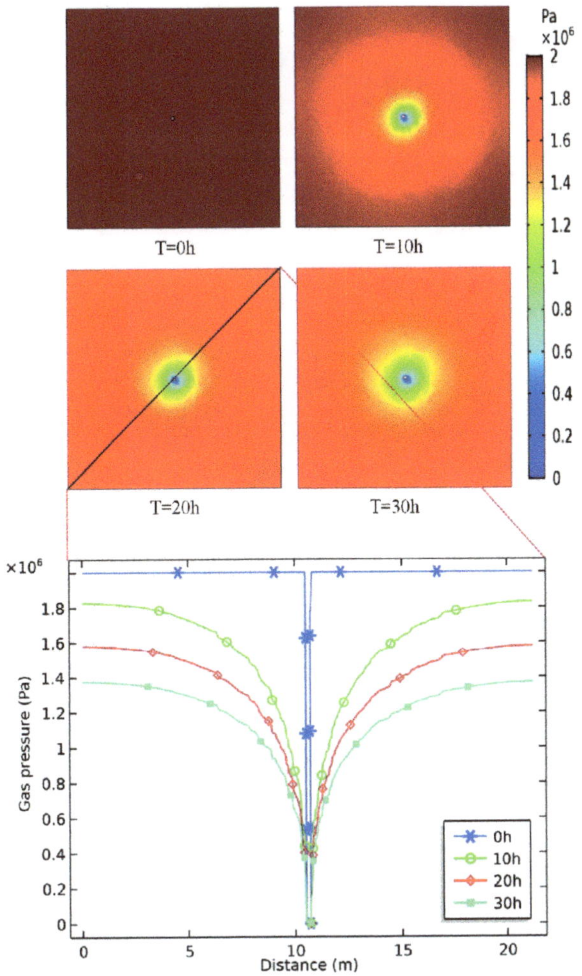

Figure 14. Gas pressure distribution around the borehole at different times.

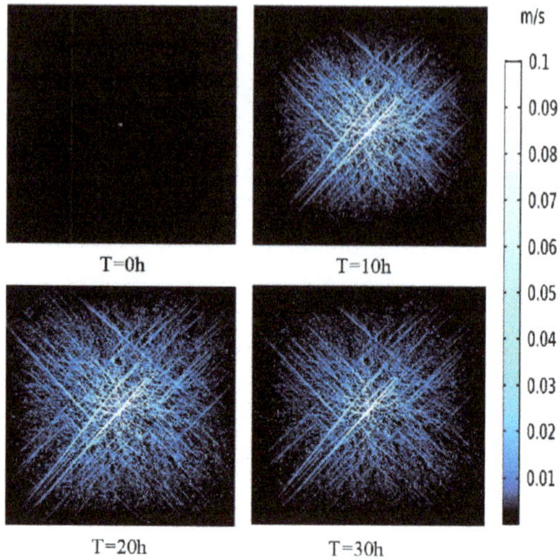

Figure 15. Gas velocity distribution around the borehole at different times.

7. Conclusions

Traditionally, pore-fracture scale simulations are conducted using the lattice Boltzmann method. However, in this work a relatively simple technique to show CBM migration through finite element analysis, which is based on 2D SFN modeling and image function, has been proposed, in which the dual-porosity medium coalbed is represented by an SFN image which can be generated by a self-built program in MATLAB. Influences of different FGPs and their distributions on the permeability of SFN were analyzed and a CBM extraction simulation in COMSOL was carried out. Although some limits, such as generating SFNs without taking other FGPs like tortuosity and roughness into consideration, difficulty in combining large scale fractures with nanoscale pores, and the absence of multi-field coupling effects analysis, still exist, the proposed method provides an efficient way to research CBM flow properties in the coal seam, which has great potential not only for gas-induced hazard prevention but also CBM industry development. According to the present study, the following conclusions can be drawn:

(1) Based on the Monte Carlo method, SFNs with different FGPs were able to be generated, with the simulation results showing that the permeability of SFN increases with larger values of density, length, and aperture, and smaller values of the angle between the fractures and gas pressure gradient. The fracture aperture has a larger influence on permeability than fracture length according to the variation range of gas velocity and the value of FGPs (Figures 7 and 8).

(2) The contribution order of different scales of fractures to the coal reservoir permeability from large to small is: large-size fractures, middle-size fractures, small-size fractures, and micro fractures, which also confirms the first conclusion because larger fracture size is shown to correspond to larger trace length and aperture.

(3) The impacts on reservoir permeability of FGP distribution were examined. On the condition that the values of all FGPs are kept the same, permeability ranking of SFNs from large to small is SFN with exponentially distributed FGPs, with lognormally distributed FGPs, with normally distributed FGPs, and with uniformly distributed FGPs.

(4) The gas extraction simulation can reflect CBM flow properties at each stage of the entire extraction process, including the temporal and spatial variations of gas velocity and pressure, the differences of gas velocity and pressure between fractures and coal matrix, and the gradually formed gas pressure drop cone.

Author Contributions: Conceptualization, J.C.; methodology, B.Z.; software, B.Z.; validation, B.Z.; formal analysis, Y.-B.L.; investigation, B.Z. and Y.L.; resources, Y.Z.; data curation, N.F. and Y.Z.; writing—original draft preparation, B.Z.; writing—review and editing, Y.L.; visualization, B.Z.; supervision, Y.L.; project administration, Y.L.; funding acquisition, B.P.

Funding: This research was supported by the Graduate Research and Innovation Foundation of Chongqing, China (CYB18032), the National Natural Science Foundation of China (51474039, 51404046), the Chongqing University Postgraduates' Innovation Project (CYS15007), and the Basic Research of Frontier and Application of Chongqing (cstc2015jcy jA90019, cstc2016jcyjA0085).

Conflicts of Interest: The authors declare no conflict of interest.

References

1. Xu, H.; Sang, S.; Yang, J.; Jin, J.; Hu, Y.; Liu, H.; Li, J.; Zhou, X.; Ren, B. Selection of suitable engineering modes for CBM development in zones with multiple coalbeds: A case study in western Guizhou Province, Southwest China. *J. Nat. Gas Sci. Eng.* **2016**, *36*, 1264–1275. [CrossRef]
2. Mostaghimi, P.; Armstrong, R.T.; Gerami, A.; Hu, Y.; Jing, Y.; Kamali, F.; Liu, M.; Liu, Z.; Lu, X.; Ramandi, H.L.; et al. Cleat-scale characterisation of coal: An overview. *J. Nat. Gas Sci. Eng.* **2017**, *39*, 143–160. [CrossRef]
3. Zhang, Z.G.; Cao, S.G.; Li, Y.; Guo, P.; Yang, H.; Yang, T. Effect of moisture content on methane adsorption- and desorption-induced deformation of tectonically deformed coal. *Adsorpt. Sci. Technol.* **2018**, *36*, 1648–1668. [CrossRef]
4. Li, H.; Ogawa, Y.; Shimada, S. Mechanism of methane flow through sheared coals and its role on methane recovery. *Fuel* **2003**, *82*, 1271–1279. [CrossRef]
5. Wang, H.; Cheng, Y.; Wang, W.; Xu, R. Research on comprehensive CBM extraction technology and its applications in China's coal mines. *J. Nat. Gas Sci. Eng.* **2014**, *20*, 200–207. [CrossRef]
6. Sharp, J.M.; Simmons, C.T. The Compleat Darcy: New lessons learned from the first english translation of Les Fontaines Publiques de la Ville de Dijon. *Groundwater* **2010**, *43*, 457–460. [CrossRef]
7. Ta, D.; Cao, S.; Steyl, G.; Yang, H.; Li, Y. Prediction of Groundwater Inflow into an Iron Mine: A Case Study of the Thach Khe Iron Mine, Vietnam. *Mine Water Environ.* **2019**, *38*, 310–324. [CrossRef]
8. Berkowitz, B. Characterizing flow and transport in fractured geological media: A review. *Adv. Water Resour.* **2002**, *25*, 861–884. [CrossRef]
9. Fu, Q.; Cloutier, A.; Laghdir, A.; Stevanovic, T. Surface Chemical Changes of Sugar Maple Wood Induced by Thermo-Hygromechanical (THM) Treatment. *Materials* **2019**, *12*, 1946. [CrossRef]
10. Bear, J. Dynamics of fluids in porous media. *Eng. Geol.* **1972**, *7*, 174–175. [CrossRef]
11. Snow, D.T. Anisotropie Permeability of Fractured Media. *Water Resour. Res.* **1969**, *5*, 1273–1289. [CrossRef]
12. Huang, Y.; Hu, S.; Gu, Z.; Sun, Y. Fracture Behavior and Energy Analysis of 3D Concrete Mesostructure under Uniaxial Compression. *Materials* **2019**, *12*, 1929. [CrossRef] [PubMed]
13. Barenblatt, G.; Zheltov, I.; Kochina, I. Basic concepts in the theory of seepage of homogeneous liquids in fissured rocks. *J. Appl. Math. Mech.* **1960**, *24*, 1286–1303. [CrossRef]
14. Lei, Q.; Latham, J.-P.; Tsang, C.F. The use of discrete fracture networks for modelling coupled geomechanical and hydrological behaviour of fractured rocks. *Comput. Geotech.* **2017**, *85*, 151–176. [CrossRef]
15. Ren, J.; Guo, P.; Guo, Z.; Wang, Z. A Lattice Boltzmann Model for Simulating Gas Flow in Kerogen Pores. *Transp. Porous Media* **2015**, *106*, 285–301. [CrossRef]
16. Belayneh, M.; Cosgrove, J.W. Fracture-pattern variations around a major fold and their implications regarding fracture prediction using limited data: An example from the Bristol Channel Basin. *Geol. Soc. Lond. Spéc. Publ.* **2004**, *231*, 89–102. [CrossRef]
17. Pollard, D.D.; Renshaw, C.E. Numerical simulation of fracture set formation: A fracture mechanics model consistent with experimental observations. *J. Geophys. Res. Space Phys.* **1994**, *99*, 9359–9372.
18. Davy, P.; Le Goc, R.; Darcel, C. A model of fracture nucleation, growth and arrest, and consequences for fracture density and scaling. *J. Geophys. Res. Solid Earth* **2013**, *118*, 1393–1407. [CrossRef]
19. Boadu, F.K. Fractured rock mass characterization parameters and seismic properties: Analytical studies. *J. Appl. Geophys.* **1997**, *37*, 1–19. [CrossRef]
20. Dershowitz, W.S.; Einstein, H.H. Characterizing rock joint geometry with joint system models. *Rock Mech. Rock Eng.* **1988**, *21*, 21–51. [CrossRef]

21. Neuman, S.P. Multiscale relationships between fracture length, aperture, density and permeability. *Geophys. Res. Lett.* **2015**, *35*, 1092–1104. [CrossRef]
22. Zhao, Y.; Liu, X.; Chen, B.; Yang, F.; Zhang, Y.; Wang, P.; Robinson, I. Three-Dimensional Characterization of Hardened Paste of Hydrated Tricalcium Silicate by Serial Block-Face Scanning Electron Microscopy. *Materials* **2019**, *12*, 1882. [CrossRef] [PubMed]
23. Boadu, F. Relating the hydraulic properties of a fractured rock mass to seismic attributes: Theory and numerical experiments. *Int. J. Rock Mech. Min. Sci.* **1997**, *34*, 885–895. [CrossRef]
24. Spanier, J.; Gelbard, E.M.; Bell, G. Monte Carlo Principles and Neutron Transport Problems. *Phys. Today* **1970**, *23*, 56–57. [CrossRef]
25. Giacobbo, F.; Patelli, E. Monte Carlo simulation of nonlinear reactive contaminant transport in unsaturated porous media. *Ann. Nucl. Energy* **2008**, *34*, 51–63. [CrossRef]
26. Adler, P.M.; Thovert, J.F. *Fractures and Fracture Networks*; Springer: Dordrecht, The Netherlands, 1999; p. 15.
27. Cacas, M.C.; Daniel, J.M.; Letouzey, J. Nested geological modelling of naturally fractured reservoirs. *Pet. Geosci.* **2001**, *7*, S43–S52. [CrossRef]
28. Hanano, M. Contribution of fractures to formation and production of geothermal resources. *Renew. Sustain. Energy Rev.* **2004**, *8*, 223–236. [CrossRef]
29. Alghalandis, Y.F.; Dowd, P.A.; Xu, C. The RANSAC Method for Generating Fracture Networks from Micro-seismic Event Data. *Math. Geol.* **2013**, *45*, 207–224.
30. Mardia, K.V.; Nyirongo, V.B.; Walder, A.N.; Xu, C.; Dowd, P.A.; Fowell, R.J.; Kent, J.T. Markov Chain Monte Carlo Implementation of Rock Fracture Modelling. *Math. Geol.* **2007**, *39*, 355–381. [CrossRef]
31. Xu, C.; Dowd, P. A new computer code for discrete fracture network modelling. *Comput. Geosci.* **2010**, *36*, 292–301. [CrossRef]
32. Pardo-Igúzquiza, E.; Dowd, P.A.; Xu, C.; Durán-Valsero, J.J. Stochastic simulation of karst conduit networks. *Adv. Water Resour.* **2012**, *35*, 141–150. [CrossRef]
33. Seifollahi, S.; Dowd, P.A.; Xu, C.; Fadakar, A.Y. A Spatial Clustering Approach for Stochastic Fracture Network Modelling. *Rock Mech. Rock Eng.* **2014**, *47*, 1225–1235. [CrossRef]
34. Nelson, R.A. *Geologic Analysis of Naturally Fractured Reservoirs*, 2nd ed.; Gulf Professional Publishing: Woburn, MA, USA, 2001; pp. 323–332.
35. Cacace, M.; Blöcher, G.; Watanabe, N.; Moeck, I.; Börsing, N.; Scheck-Wenderoth, M.; Kolditz, O.; Huenges, E. Modelling of fractured carbonate reservoirs: Outline of a novel technique via a case study from the Molasse Basin, southern Bavaria, Germany. *Environ. Earth Sci.* **2013**, *70*, 3585–3602. [CrossRef]
36. Yeh, H.D.; Chang, Y.C. Recent advances in modeling of well hydraulics. *Adv. Water Resour.* **2013**, *51*, 27–51. [CrossRef]
37. Choi, S.; Wold, M. Simulation of Fluid Flow in Coal Using a Discrete Fracture Network Model. In Proceedings of the SPE Asia Pacific Oil and Gas Conference, Melbourne, Australia, 7–10 November 1994.
38. Lorig, L.J.; Darcel, C.; Damjanac, B.; Pierce, M.; Billaux, D. Application of discrete fracture networks in mining and civil geomechanics. *Min. Technol.* **2015**, *124*, 239–254. [CrossRef]
39. Jing, Y.; Armstrong, R.T.; Mostaghimi, P. Rough-walled discrete fracture network modelling for coal characterisation. *Fuel* **2017**, *191*, 442–453. [CrossRef]
40. Maryška, J.; Severýn, O.; Vohralík, M. Numerical simulation of fracture flow with a mixed-hybrid FEM stochastic discrete fracture network model. *Comput. Geosci.* **2005**, *8*, 217–234. [CrossRef]
41. Moore, T.A. Coalbed methane: A review. *Int. J. Coal Geol.* **2012**, *101*, 36–81. [CrossRef]
42. Aminian, K.; Rodvelt, G. Chapter 4—Evaluation of Coalbed Methane Reservoirs. *Coal Bed Methane* **2014**, *43*, 63–91.
43. Koyama, T.; Neretnieks, I.; Jing, L. A numerical study on differences in using Navier–Stokes and Reynolds equations for modeling the fluid flow and particle transport in single rock fractures with shear. *Int. J. Rock Mech. Min. Sci.* **2008**, *45*, 1082–1101. [CrossRef]
44. Fu, X.H.; Qin, Y. *Theories and Techniques of Permeability Prediction of Multiphase Medium Coalbed-Methane Reservoirs*; China Mining University Press: Xuzhou, China, 2003.

45. Wang, J.; Kabir, A.; Liu, J.; Chen, Z. Effects of non-Darcy flow on the performance of coal seam gas wells. *Int. J. Coal Geol.* **2012**, *93*, 62–74. [CrossRef]

materials

MDPI

Article

Free Vibrations of Sandwich Plates with Damaged Soft-Core and Non-Uniform Mechanical Properties: Modeling and Finite Element Analysis

Michele Bacciocchi [1,2,*], Raimondo Luciano [3], Carmelo Majorana [4] and Angelo Marcello Tarantino [5]

1 Department of Civil, Chemical, Environmental, and Materials Engineering (DICAM), University of Bologna, Viale del Risorgimento, 40136 Bologna, Italy

2 Dipartimento di Economia, Scienze e Diritto (DESD), University of San Marino, Via Consiglio dei Sessanta, 47891 Dogana, San Marino

3 Engineering Department, University of Napoli Parthenope, Via Ammiraglio Ferdinando Acton, 80133 Napoli, Italy

4 Department of Civil, Environmental and Architectural Engineering (DICEA), University of Padova, Via F. Marzolo, 35131 Padova, Italy

5 Department of Engineering "Enzo Ferrari" (DIEF), University of Modena and Reggio Emilia, Via Vivarelli, 41125 Modena, Italy

* Correspondence: michele.bacciocchi@unibo.it; Tel.: +39-051-209-3494

Received: 27 June 2019; Accepted: 29 July 2019; Published: 31 July 2019

Abstract: The paper aims to investigate the natural frequencies of sandwich plates by means of a Finite Element (FE) formulation based on the Reissner-Mindlin Zig-zag (RMZ) theory. The structures are made of a damaged isotropic soft-core and two external stiffer orthotropic face-sheets. These skins are strengthened at the nanoscale level by randomly oriented Carbon nanotubes (CNTs) and are reinforced at the microscale stage by oriented straight fibers. These reinforcing phases are included in a polymer matrix and a three-phase approach based on the Eshelby-Mori-Tanaka scheme and on the Halpin-Tsai approach, which is developed to compute the overall mechanical properties of the composite material. A non-uniform distribution of the reinforcing fibers is assumed along the thickness of the skin and is modeled analytically by means of peculiar expressions given as a function of the thickness coordinate. Several parametric analyses are carried out to investigate the mechanical behavior of these multi-layered structures depending on the damage features, through-the-thickness distribution of the straight fibers, stacking sequence, and mass fraction of the constituents. Some final remarks are presented to provide useful observations and design criteria.

Keywords: three-phase composite materials; Finite Element modeling; sandwich plates; zig-zag theory; carbon nanotubes; free vibrations

1. Introduction

Since its early development and the publication of the first research papers [1–6], the Finite Element (FE) method has shown its potentiality in solving easily and accurately many structural problems which could not be solved analytically. Nowadays, this feature is even more emphasized by the great and continuous technological advancements reached in computer sciences in terms of available computational resources. As highlighted in the books which can be certainly considered as milestones in the development of the FE method [7–13], the easy implementation and the possibility to reduce complex continuous problems into simpler discrete ones has encouraged its rapid spread among many researchers and engineers [14–20].

In the current paper, the FE technique is implemented in a computational code to solve the free vibration analysis of three-layered sandwich plates with a damaged soft-core and non-uniform mechanical properties. In particular, the structures are made by an isotropic core that undergoes a progressive uniform damage, which is modeled as a decay of the mechanical properties expressed in terms of engineering constants. The damage model is the one illustrated in the book by Lemaitre and Chaboche [21], and it is representative of the formation of microcracks and discontinuities within the considered medium. Since these defects are uniformly distributed and affect the central layer of the plates independently from the direction, this phenomenon is known as "isotropic damage" and it is fully described by a scalar parameter. Further details concerning the phenomenological aspects related to the damage of materials can be found in the book by Reddy and Miravete [22], whereas some structural applications which investigate the effect of damage are illustrated in [23–33].

The soft-core of the structures is surrounded and strengthened by two external sinks (or face-sheets), which are stiffer and thinner than the central layer. These plies are made of a polymer matrix which contains randomly oriented Carbon nanotubes (CNTs) and keeps together straight Carbon fibers. The role of the matrix in composite materials is clearly illustrated in the books by Vinson [34], Jones [35], Reddy [36], and Barbero [37]. These works should be taken into account as complete references for the analysis and modelling of composite materials. As highlighted in the papers [38,39], the matrix is reinforced at different levels. At the nano-scale, the CNTs provide the matrix with additional stiffness [40–49]. The single fiber of Carbon nanotube (CNT) is modeled as a transversely isotropic cylinder as suggested by Odegard et al. [50] and it assumed to be randomly oriented in the matrix, following the approach developed in the paper by Shi et al. [51]. The Eshelby-Mori-Tanaka scheme is applied at this level to evaluate the mechanical properties of this enriched matrix [52,53], which has isotropic features due to the random orientation of CNTs [54]. On the other hand, the reinforcing phase at the micro-scale is given by oriented straight fibers. The semi-empirical Halpin-Tsai homogenization procedure [55–57], which is based on the use of the Hill's elastic moduli [58,59], is employed at this level to obtain an accurate estimation of the overall mechanical features of the composite skins. Therefore, a multi-phase approach has been developed to this aim [60–63].

The development of micromechanics theories able to evaluate the mechanical features of fiber-reinforced composites has always fascinated many researchers, as illustrated in a complete manner by Chamis and Sendeckyj [64]. For completeness purposes, it should be recalled that different methods can be found in the literature as alternatives to the Halpin-Tsai semi-empirical homogenization procedure, such as self- consistent models [65,66], variational methodologies [67,68], and techniques based on the mechanics of materials [69–71]. Further details can be found also in the papers [72,73].

With respect to previous contributions, the straight fibers are graded in the thickness direction. In other words, the volume fraction distribution of the reinforcing fibers is non-uniform along the thickness of the sinks. This approach is typically applied in the class of granular composites, also known as Functionally Graded (FG) materials, to characterize the gradual variation of isotropic constituents [74–80]. On the other hand, in this research the graded constituent is represented by the orthotropic reinforcing fibers, which can be also arbitrarily oriented in the planar direction. A power-law function is used to characterize the through-the-thickness variation of their volume fraction. FG materials are typically used as constituents in plates and shells [81–86], beams [87,88], micro- and nano-structures [89–91].

Thus, it should be noted that the plates are characterized by noticeable differences in terms of mechanical properties at the layer interfaces. For this reason, a suitable structural model must be introduced to describe accurately their mechanical behavior. As clearly highlighted in the papers by Carrera [92–96], laminated and sandwich structures are characterized by a piece-wise continuous displacement field along the thickness direction. The differences in terms of transverse deformability among the layers give rise to a change in the slope of the displacement components between two adjacent layers. This outcome is known as zig-zag effect. As specified in [97–99], this effect is well-captured by Layer-Wise (LW) models, in which the degrees of freedom are assumed as independent parameters along each layer. Nevertheless, this approach is quite onerous in terms of computational resources. Effective and accurate solutions can be obtained more easily by introducing the Murakami's function in the displacement fields of Equivalent Single Layers (ESL) models, in which all the degrees of freedom are defined within the reference surface of the structure independently from the number of layers. The Murakami's function, in fact, is able to capture the zig-zag effect in an accurate manner without increasing excessively the computations [100]. In general, classical models for laminated structures, such as the well-known Reissner-Mindlin (RM) theory, are not able to take into account this effect. Nevertheless, these theories could be also enhanced by adding the Murakami's function in their kinematic models. The results that can be obtained are accurate and the computational cost is reduced, if compared to the one that characterizes higher-order ESL or LW models [92]. Therefore, in the present paper the in-plane displacement field of the RM model is enriched by the Murakami's function in order to capture the effective behavior of sandwich plates with damaged soft-core and non-uniform mechanical properties. It should be mentioned that further details concerning Zig-zag theories can be found in the review paper by Carrera [94].

Finally, a brief description of the research outline is presented. The geometric and mechanical characterization of the plates are illustrated after this introduction in Section 2. In particular, the multi-phase approach including the Eshelby-Mori-Tanaka scheme and the Halpin-Tsai homogenization procedure is described to provide the mechanical properties of the skin. In the same Section, the damage model is introduced for the isotropic soft-core. On the other hand, Section 3 is focused on the theoretical aspects of the Reissner-Mindlin Zig-zag (RMZ) theory. The corresponding FE model is developed, and the fundamental system of equations is deducted by means of the Hamilton's principle [36,101]. The results of the numerical applications are presented in Section 4. Here, the effect of the Murakami's function is discussed by means of the comparison between the RM and RMZ approaches, and the proposed model is validated as well. Then, several applications are illustrated to investigate the effects of the progressive damage, the non-uniform distribution of the fiber volume fraction, the in-plane fiber orientation, and the material properties on the natural frequencies of the structures and on their corresponding mode shapes. Finally, the matrix form of the fundamental operators required in the governing equations are defined in Appendix A.

2. Geometric and Mechanical Characterization

The paper is focused on the vibrational behavior of laminated sandwich plates with an inner damaged soft-core. The plates under consideration are characterized by a planar size given by the lengths of their sides L_x, L_y, in which x, y denote the principal directions of the local reference system. The extension along the coordinate z is specified by the overall thickness h of the composite structures, which is given by $h = 2h_s + h_c$, where h_s stands for the thickness of the external face-sheets and h_c is the thickness of the core. Thus, the plates consist in three layers, in which the external ones are made of orthotropic materials and have the same thickness whereas the central one is isotropic. The upper and lower thickness coordinates of the generic k-th ply are denoted by z_{k+1}, z_k, respectively. The geometric features of a plate element are shown in Figure 1.

The mechanical characterization of the k-th layer is carried out in terms of the engineering constants of the orthotropic material, which are the Young's moduli $E_{11}^{(k)}, E_{22}^{(k)}$, the shear moduli $G_{12}^{(k)}, G_{13}^{(k)}, G_{23}^{(k)}$, and the Poisson's ratio $\nu_{12}^{(k)}$. As far as the isotropic core is concerned ($k = 2$), only two independents parameters are needed, which are the Young's modulus $E^{(k)} = E_{11}^{(k)} = E_{22}^{(k)}$ and the Poisson's ratio $\nu^{(k)} = \nu_{12}^{(k)}$, whereas its shear modulus $G^{(k)} = G_{12}^{(k)} = G_{13}^{(k)} = G_{23}^{(k)}$ is given by:

$$G^{(k)} = \frac{E^{(k)}}{2\left(1 + \nu^{(k)}\right)}. \tag{1}$$

In the following sections, the evaluation of these engineering constants is discussed for both the external layers and the soft-core. It should be specified that a perfect bonding is assumed between two adjacent layers in the proposed model.

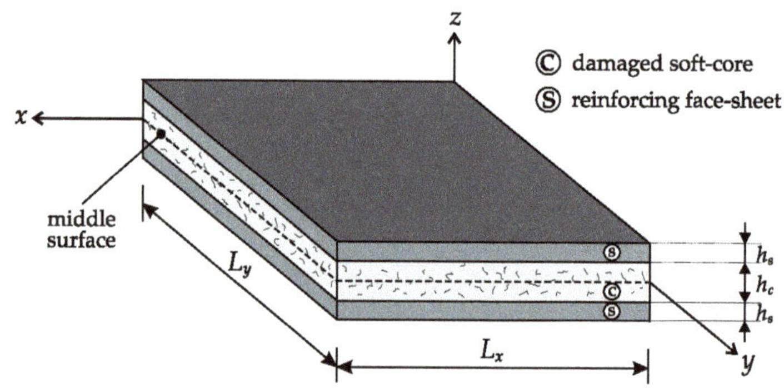

Figure 1. Geometric features of a laminated sandwich plate with an inner damaged soft-core.

2.1. Mechanical Properties of the Face-Sheets

A multiscale approach is employed to evaluate the overall mechanical properties of the three-phase composite face-sheets [38,39]. These layers are made of fiber-reinforced materials, which means that a polymer matrix (epoxy resin) is strengthened by oriented straight fibers (Carbon fibers). The third constituent is given by randomly oriented Carbon nanotubes (CNTs), which are inserted in the matrix and are used to further increase its mechanical features. The overall properties are computed by means of a two-step approach. Firstly, the Eshelby-Mori-Tanaka scheme is applied to obtain the properties of the epoxy resin including CNTs [52,53]. At this stage, the composite turns out to be isotropic due to the fact that the nanoparticles are randomly oriented, as illustrated in the paper by Shi et al. [51]. Then, the Halpin-Tsai approach is applied to combine the features of the enriched matrix with the properties of the reinforcing straight fibers [55–57].

At the nano-scale level, the single fiber of CNT is modeled as a linear-elastic, transversely isotropic and homogeneous cylindrical solid, as proposed in the paper by Odegard et al. [50]. Its mechanical properties are described by five parameters, which are the Hill's elastic moduli [58,59], defined as k_C, l_C, m_C, n_C, p_C. Its density ρ_C is required to compute the volume fraction of CNTs V_C as follows:

$$V_C = \left(\frac{\rho_C}{w_C \rho_M} - \frac{\rho_C}{\rho_M} + 1\right)^{-1}, \tag{2}$$

where ρ_M represents the density of the polymer matrix, whereas w_C stands for the mass fraction of CNTs. A uniform distribution of CNTs is assumed in each layer along the thickness direction. It should be recalled that the matrix volume fraction is given by $V_M = 1 - V_C$. The polymer matrix is isotropic, and it is fully characterized by its Young's modulus E_M and Poisson's ratio ν_M. In the following, its bulk modulus K_M and shear modulus G_M are required in order to apply the Eshelby-Mori-Tanaka approach [52,53]. The following definitions are needed for this purpose:

$$K_M = \frac{E_M}{3(1-2\nu_M)}, \qquad G_M = \frac{E_M}{2(1+\nu_M)}. \tag{3}$$

These quantities are noticeably affected by the presence of randomly oriented CNTs. As illustrated in the paper by Shi et al. [51], if the agglomeration of CNTs is neglected, the bulk modulus K_M^* and the shear modulus G_M^* of the enriched matrix are given by:

$$K_M^* = K_M + \frac{V_C(\delta_C - 3K_M\alpha_C)}{3(V_M + V_C\alpha_C)}, \qquad G_M^* = G_M + \frac{V_C(\eta_C - 2G_M\beta_C)}{2(V_M + V_C\beta_C)}, \tag{4}$$

where the following quantities, which can be computed by defining the Hill's elastic moduli of CNTs, are introduced:

$$\begin{aligned} \alpha_C &= \tfrac{3(K_M+G_M)+k_C+l_C}{3(G_M+k_C)}, \\ \beta_C &= \tfrac{1}{5}\left(\tfrac{4G_M+2k_C+l_C}{3(G_M+k_C)} + \tfrac{4G_M}{G_M+p_C} + \tfrac{2(G_M(6K_M+8G_M))}{G_M(3K_M+G_M)+m_C(3K_M+7G_M)}\right), \\ \delta_C &= \tfrac{1}{3}\left(n_C + 2l_C + \tfrac{(2k_C+l_C)(3K_M+G_M-l_C)}{G_M+k_C}\right), \\ \eta_C &= \tfrac{1}{5}\left(\tfrac{2}{3}(n_C - l_C) + \tfrac{8G_M p_C}{G_M+p_C} + \tfrac{2(k_C-l_C)(2G_M+l_C)}{3(G_M+k_C)} + \tfrac{8m_C G_M(3K_M+4G_M)}{3K_M(m_C+G_M)+G_{PM}(7m_C+G_M)}\right). \end{aligned} \tag{5}$$

Finally, the evaluation of K_M^* and G_M^* allows to compute the Young's modulus E_M^* and the Poisson's ratio ν_M^* of the polymer matrix enriched by CNTs:

$$E_M^* = \frac{9K_M^*G_M^*}{3K_M^* + G_M^*}, \qquad \nu_M^* = \frac{3K_M^* - 2G_M^*}{6K_M^* + 2G_M^*}, \tag{6}$$

whereas its density is given by:

$$\rho_M^* = (\rho_C - \rho_M)V_C + \rho_M. \tag{7}$$

The mechanical properties of a single CNT fiber in terms of its Hill's elastic moduli, as well as its density, are listed in Table 1. Such properties are valid for a single-walled Carbon nanotube with 10 as chiral index and armchair structure. Further details about the mechanical characterization of CNT can be found in [38].

Table 1. Mechanical properties of a single CNT fiber.

Hill's Elastic Moduli	Density
$k_C = 271$ GPa $l_C = 88$ GPa $m_C = 17$ GPa $n_C = 1089$ GPa $p_C = 442$ GPa	$\rho_C = 1400\,\text{kg/m}^3$

In order to apply the Halpin-Tsai approach that allows to compute the overall mechanical properties of the composite given by this enriched matrix including straight Carbon fibers, the Hill's elastic moduli of the matrix $k_M^*, l_M^*, m_M^*, n_M^*, p_M^*$ are needed. For this purpose, the following definitions are introduced:

$$\begin{aligned} k_M^* = \tfrac{E_M^*}{2(1+\nu_M^*)(1-2\nu_M^*)}, \quad l_M^* = 2\nu_M^* k_M^*, \quad m_M^* = \left(1 - 2\nu_M^*\right)k_M^*, \\ n_M^* = 2\left(1 - \nu_M^*\right)k_M^*, \quad p_M^* = \left(1 - 2\nu_M^*\right)k_M^*. \end{aligned} \tag{8}$$

Analogously, the Hill's elastic moduli of the Carbon fibers k_F, l_F, m_F, n_F, p_F are required. The reinforcing fibers are assumed as transversely isotropic and their mechanical properties are given by the corresponding Young's moduli E^F_{11}, E^F_{22}, shear modulus G^F_{12} and Poisson's ratios ν^F_{12}, ν^F_{23}. Once these quantities are known, the Hill's elastic moduli can be easily evaluated:

$$k_F = \frac{E^F_{22}}{2\left(1-\nu^F_{23}-2\nu^F_{21}\nu^F_{12}\right)}, \quad l_F = 2\nu^F_{12}k_F, \quad m_F = \frac{1-\nu^F_{23}-2\nu^F_{21}\nu^F_{12}}{1+\nu^F_{23}}k_F,$$
$$n_F = 2\left(1-\nu^F_{23}\right)\frac{E^F_{11}}{E^F_{22}}k_F, \quad p_F = G^F_{12}, \tag{9}$$

where $\nu^F_{21} = E^F_{22}\nu^F_{12}/E^F_{11}$. As shown in the previous step, the density of the fibers ρ_F is required to compute the reference value of the corresponding volume fraction $\widetilde{V}_F$, once their mass fraction w_F is defined

$$\widetilde{V}_F = \left(\frac{\rho_F}{w_F\rho^*_M} - \frac{\rho_F}{\rho^*_M} + 1\right)^{-1}. \tag{10}$$

Such constant quantity can be multiplied by a peculiar function $f^{(k)}(z)$ which depends on the thickness coordinate z. Consequently, the volume fraction distribution of the fibers V_F is given by $V_F = V_F(z) = \widetilde{V}_F f^{(k)}(z)$ and it is clearly non-uniform along the thickness of the face-sheets.

In the current paper, two different functions $f^{(k)}(z)$ are introduced to specify the volume fraction distribution V_F. It should be noted that they can be chosen arbitrarily in the two face-sheets of the sandwich structures under consideration. Power-law functions are used to this aim and the definitions of $f^{(k)}(z)$ are specified below:

$$f^{(k)} = \begin{cases} f^{(k)}_1(z) = \left(\frac{z-z_k}{z_{k+1}-z_k}\right)^\alpha \\ f^{(k)}_2(z) = \left(\frac{z_{k+1}-z}{z_{k+1}-z_k}\right)^\alpha \end{cases} \tag{11}$$

where α represents the arbitrary exponent of the distributions. As shown in Figure 2, several configurations can be obtained according to the value given to $\alpha \in [0,\infty]$. It should be noted that the extreme values of α, which correspond to the constant values $f^{(k)} = 0,\ f^{(k)} = 1$, are able to characterize a uniform distributions of the fiber along the thickness or their absence (as a consequence, the polymer matrix is the only constituent of the layer).

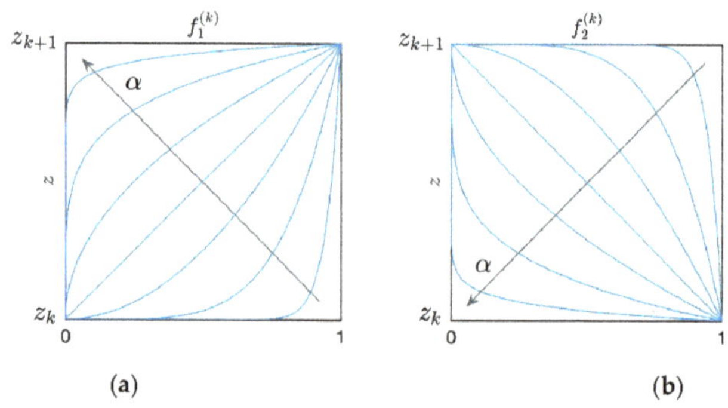

Figure 2. Through-the-thickness representation of the function $f^{(k)}$ employed to characterize the volume fraction distribution of the fibers in the k-th layer of the plate, for increasing values of the exponent α: (**a**) $f^{(k)}_1$; (**b**) $f^{(k)}_2$.

At this point, the Halpin-Tsai approach can be applied to obtain the Hill's elastic moduli of the composite layers, which are denoted by k, l, m, n, p:

$$
\begin{aligned}
k &= \frac{k_M^*(k_F+m_M^*)V_M^*+k_F(k_M^*+m_M^*)V_F}{(k_F+m_M^*)V_M^*+(k_M^*+m_M^*)V_F}, \\
l &= V_F l_F + V_M^* l_M^* + \frac{l_F-l_M^*}{k_F-k_M^*}\left(k - V_F k_F - V_M^* k_M^*\right), \\
m &= m_M^* \frac{2V_F m_F(k_M^*+m_M^*)+2V_M^* m_F m_M^*+V_M^* k_M^*(m_F+m_M^*)}{2V_F m_M^*(k_M^*+m_M^*)+2V_M^* m_F m_M^*+V_M^* k_M^*(m_F+m_M^*)}, \\
n &= V_F n_F + V_M^* n_M^* + \left(\frac{l_F-l_M^*}{k_F-k_M^*}\right)^2\left(k - V_F k_F - V_M^* k_M^*\right), \\
p &= \frac{(p_F+p_M^*)p_M^* V_M^*+2p_F p_M^* V_F}{(p_F+p_M^*)V_M^*+2p_M^* V_F},
\end{aligned}
\tag{12}
$$

being $V_M^* = 1 - V_F$. The engineering constants of these layers can be evaluated according to the following definitions:

$$
E_{11}^{(k)} = n - \frac{l^2}{k}, \quad E_{22}^{(k)} = \frac{4m\left(kn - l^2\right)}{kn - l^2 + mn}, \quad \nu_{12}^{(k)} = \frac{l}{2k}, \quad G_{12}^{(k)} = G_{13}^{(k)} = p, \quad G_{23}^{(k)} = m. \tag{13}
$$

The introduction of the function $f^{(k)}(z)$ in the definition of V_F causes the dependency on the thickness coordinate z of each engineering constant specified in Equation (13). Therefore, one gets $E_{11}^{(k)}(z)$, $E_{22}^{(k)}(z)$, $\nu_{12}^{(k)}(z)$, $G_{12}^{(k)}(z)$, $G_{13}^{(k)}(z)$ and $G_{23}^{(k)}(z)$, for $k = 1, 3$. Finally, the density of the composite face-sheets is given by:

$$
\rho^{(k)} = \left(\rho_F - \rho_M^*\right)V_F + \rho_M^*. \tag{14}
$$

In the following, the same constituents are used in the external layers assuming also the same values of the mass fractions of both CNTs and fibers. The relation $\nu_{21}^{(k)} = E_{22}^{(k)}\nu_{12}^{(k)}/E_{11}^{(k)}$ is also required to compute the Poisson's ratio $\nu_{21}^{(k)}$. As emphasized in the introduction and illustrated in the paper [38], different approaches and homogenization techniques could be used to the same aim. For instance, the rule of the mixture represents the most exploited methodology.

2.2. Mechanical Properties of the Damaged Matrix

The core of the sandwich structures considered in the paper is made of the same polymer matrix used in the face-sheets. Nevertheless, a damage model is introduced to provide an analytical description of an irreversible rheological process that causes the decay of the mechanical properties, in terms of engineering constants. An isotropic damage is considered in the following, which is fully characterized by a scalar D as illustrated in the book by Lemaitre and Chaboche [21]. The elastic modulus of the damaged material is given by:

$$
E^{(k)} = (1 - D)E_M \tag{15}
$$

for $k = 2$, in which E_M is the original value of matrix Young's modulus, for $0 \leq D < 1$ It is clear that $D = 0$ identifies a virgin material, whereas a fully damaged material is characterized by $D = 1$. Having in mind relation (1), the shear modulus is subjected to the same damage. On the contrary, the Poisson's ratio $\nu^{(k)} = \nu_M$ and the density $\rho^{(k)} = \rho_M$ of the core ($k = 2$) are kept constant. Finally, it should be specified that the damage does not depend on the spatial coordinates and affects uniformly the core. For conciseness purposes, the mechanical properties of the undamaged epoxy resin and the Carbon fibers are summarized in Table 2.

Table 2. Mechanical properties of the layer constituents.

Constituent	Young's Moduli	Shear Moduli	Poisson's Ratios	Density
Carbon fibers	$E_{11}^F = 230\,\text{GPa}$ $E_{22}^F = 15\,\text{GPa}$	$G_{12}^F = 50\,\text{GPa}$	$\nu_{12}^F = 0.20$ $\nu_{23}^F = 0.25$	$\rho_F = 1800\,\text{kg/m}^3$
Epoxy resin	$E_M = 3.27\,\text{GPa}$	–	$\nu_M = 0.38$	$\rho_M = 1200\,\text{kg/m}^3$

3. Finite Element Model Based on A First-Order Zig-Zag Plate Theory

A linear theory is used to model the mechanical behavior of sandwich plates with an inner soft-core. With respect to the well-known Reissner-Mindlin (RM) theory, the in-plane expansion is enriched by two more degrees of freedom, which are able to capture the zig-zag effect [92–96]. Here, the corresponding (FE) formulation is presented. The displacement field for the generic e-th element is given by:

$$\begin{aligned} U_x^{(e)}(x,y,z,t) &= u_x^{(e)}(x,y,t) + z\phi_x^{(e)}(x,y,t) + F_z(z)\psi_x^{(e)}(x,y,t) \\ U_y^{(e)}(x,y,z,t) &= u_y^{(e)}(x,y,t) + z\phi_y^{(e)}(x,y,t) + F_z(z)\psi_y^{(e)}(x,y,t) \\ U_z^{(e)}(x,y,z,t) &= u_z^{(e)}(x,y,t) \end{aligned} \tag{16}$$

in which the three-dimensional displacement components are denoted by $U_x^{(e)}, U_y^{(e)}, U_z^{(e)}$. The spatial coordinates of the plate are given by x, y, z, as shown in Figure 1, whereas t is the time variable. The zig-zag effect is modeled by means of the Murakami's function $F_z(z)$ defined below for a multilayered structure:

$$F_z = (-1)^k \frac{2z}{z_{k+1} - z_k} - \frac{z_{k+1} + z_k}{z_{k+1} - z_k} \tag{17}$$

where k identifies the generic layer. This function allows to introduce a discontinuity in the slope of the three-dimensional displacements $U_x{}^{(e)}, U_y{}^{(e)}$ along the thickness direction at each layer interface. Its meaning is well-described in the papers by Carrera [94]. It should be noted that the current model is characterized by seven degrees of freedom per node, two more than the classical RM approach. In particular, $u_x^{(e)}, u_y^{(e)}, u_z^{(e)}$ represent the translational displacements along x, y, z, $\phi_x^{(e)}, \phi_y^{(e)}$ are the rotations about the principal axes y, x respectively, whereas $\psi_x^{(e)}, \psi_y^{(e)}$ denote the magnitude of the zig-zag effect. A nine-node quadratic rectangular element is used to develop the FE formulation and each degrees of freedom is approximated by means of Lagrange interpolating functions N_i, for $i = 1, \ldots, 9$. The node numbering is illustrated in Figure 3.

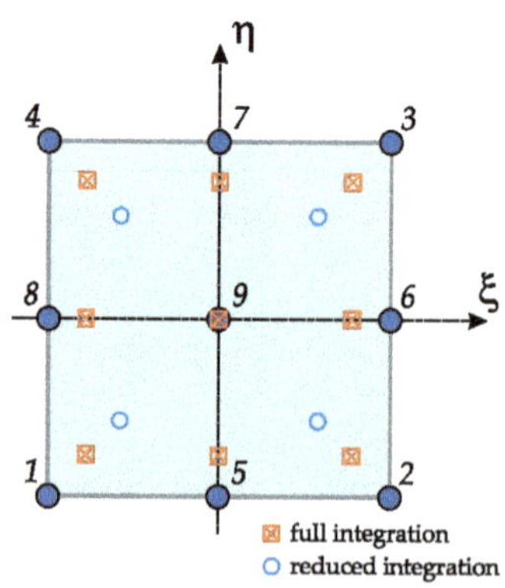

Figure 3. Discrete element: Node numbering and natural coordinates.

Due to this approximation, the degrees of freedom can be written as a function of the corresponding nodal displacements $u_{x,i}^{(e)}, u_{y,i}^{(e)}, u_{z,i}^{(e)}, \phi_{x,i}^{(e)}, \phi_{y,i}^{(e)}, \psi_{x,i}^{(e)}, \psi_{y,i}^{(e)}$:

$$
\begin{aligned}
u_x^{(e)}(x,y,t) &= \sum_{i=1}^{9} N_i(x,y) u_{x,i}^{(e)}(t) = \overline{\mathbf{N}}\mathbf{u}_x^{(e)} \\
u_y^{(e)}(x,y,t) &= \sum_{i=1}^{9} N_i(x,y) u_{y,i}^{(e)}(t) = \overline{\mathbf{N}}\mathbf{u}_y^{(e)} \\
u_z^{(e)}(x,y,t) &= \sum_{i=1}^{9} N_i(x,y) u_{z,i}^{(e)}(t) = \overline{\mathbf{N}}\mathbf{u}_z^{(e)} \\
\phi_x^{(e)}(x,y,t) &= \sum_{i=1}^{9} N_i(x,y) \phi_{x,i}^{(e)}(t) = \overline{\mathbf{N}}\boldsymbol{\phi}_x^{(e)} \\
\phi_y^{(e)}(x,y,t) &= \sum_{i=1}^{9} N_i(x,y) \phi_{y,i}^{(e)}(t) = \overline{\mathbf{N}}\boldsymbol{\phi}_y^{(e)} \\
\psi_x^{(e)}(x,y,t) &= \sum_{i=1}^{9} N_i(x,y) \psi_{x,i}^{(e)}(t) = \overline{\mathbf{N}}\boldsymbol{\psi}_x^{(e)} \\
\psi_y^{(e)}(x,y,t) &= \sum_{i=1}^{9} N_i(x,y) \psi_{y,i}^{(e)}(t) = \overline{\mathbf{N}}\boldsymbol{\psi}_x^{(e)}
\end{aligned}
\tag{18}
$$

where $\overline{\mathbf{N}} = \begin{bmatrix} N_1 & \cdots & N_9 \end{bmatrix}$ is the vector of the Lagrange interpolating functions. Analogously, the nodal displacements are defined in vector form as follows:

$$
\begin{aligned}
\mathbf{u}_x^{(e)} = \begin{bmatrix} u_{x,1}^{(e)} & \cdots & u_{x,9}^{(e)} \end{bmatrix}^T, \quad \mathbf{u}_y^{(e)} = \begin{bmatrix} u_{y,1}^{(e)} & \cdots & u_{y,9}^{(e)} \end{bmatrix}^T, \quad \mathbf{u}_z^{(e)} = \begin{bmatrix} u_{z,1}^{(e)} & \cdots & u_{z,9}^{(e)} \end{bmatrix}^T, \\
\boldsymbol{\phi}_x^{(e)} = \begin{bmatrix} \phi_{x,1}^{(e)} & \cdots & \phi_{x,9}^{(e)} \end{bmatrix}^T, \quad \boldsymbol{\phi}_y^{(e)} = \begin{bmatrix} \phi_{y,1}^{(e)} & \cdots & \phi_{y,9}^{(e)} \end{bmatrix}^T, \\
\boldsymbol{\psi}_x^{(e)} = \begin{bmatrix} \psi_{x,1}^{(e)} & \cdots & \psi_{x,9}^{(e)} \end{bmatrix}^T, \quad \boldsymbol{\psi}_x^{(e)} = \begin{bmatrix} \psi_{y,1}^{(e)} & \cdots & \psi_{y,9}^{(e)} \end{bmatrix}^T.
\end{aligned}
\tag{19}
$$

This notation is useful to define the vector $\overline{\mathbf{u}}^{(e)}$ which includes all the nodal degrees of freedom:

$$
\overline{\mathbf{u}}^{(e)} = \begin{bmatrix} \mathbf{u}_x^{(e)} & \mathbf{u}_y^{(e)} & \mathbf{u}_z^{(e)} & \boldsymbol{\phi}_x^{(e)} & \boldsymbol{\phi}_y^{(e)} & \boldsymbol{\psi}_x^{(e)} & \boldsymbol{\psi}_y^{(e)} \end{bmatrix}^T \tag{20}
$$

The interpolating functions N_i assume the well-known definitions presented in the book by Reddy [11]. These polynomials are conveniently expressed as functions of the natural coordinates ξ, η, with $\xi, \eta \in [-1, 1]$, introduced in the so-called master element (or parent element) depicted in Figure 3. Thus, the interpolating functions become $N_i = N_i(\xi, \eta)$.

The Lagrange functions N_i are also employed to define the geometric shape of each element. According to the principles of an isoparametric formulation, the coordinates $x^{(e)}, y^{(e)}$ within the generic e-th element can be defined as follows:

$$
x^{(e)} = \sum_{i=1}^{9} N_i(\xi, \eta) x_i^{(e)}, \quad y^{(e)} = \sum_{i=1}^{9} N_i(\xi, \eta) y_i^{(e)} \tag{21}
$$

in which the i-th node of the element under consideration is identified by the couple of nodal coordinates $x_i^{(e)}, y_i^{(e)}$. Such coordinates are included in the corresponding vectors $\mathbf{x}^{(e)}, \mathbf{y}^{(e)}$, which assume the following aspects:

$$
\mathbf{x}^e = \begin{bmatrix} x_1^{(e)} & \cdots & x_9^{(e)} \end{bmatrix}^T, \quad \mathbf{y}^e = \begin{bmatrix} y_1^{(e)} & \cdots & y_9^{(e)} \end{bmatrix}^T. \tag{22}
$$

The isoparametric formulation allows to move easily all the computations in the parent space. To this aim, the Jacobian matrix $\mathbf{J}$ is required to perform the coordinate change. In order to define this matrix, the derivates of the interpolating functions with respect to the natural coordinates ξ, η are needed and are collected in the corresponding vectors $\mathbf{B}_\xi, \mathbf{B}_\eta$ defined below:

$$\mathbf{B}_\xi = \begin{bmatrix} \frac{\partial N_1}{\partial \xi} & \cdots & \frac{N_9}{\partial \xi} \end{bmatrix}, \quad \mathbf{B}_\xi = \begin{bmatrix} \frac{\partial N_1}{\partial \eta} & \cdots & \frac{N_9}{\partial \eta} \end{bmatrix}. \tag{23}$$

At this point, the Jacobian matrix $\mathbf{J}$ can be introduced as specified in [11]:

$$\mathbf{J} = \begin{bmatrix} \frac{\partial x^{(e)}}{\partial \xi} & \frac{\partial y^{(e)}}{\partial \xi} \\ \frac{\partial x^{(e)}}{\partial \eta} & \frac{\partial y^{(e)}}{\partial \eta} \end{bmatrix} = \begin{bmatrix} \sum_{i=1}^{9} x_i^{(e)} \frac{\partial N_i}{\partial \xi} & \sum_{i=1}^{9} y_i^{(e)} \frac{\partial N_i}{\partial \xi} \\ \sum_{i=1}^{9} x_i^{(e)} \frac{\partial N_i}{\partial \eta} & \sum_{i=1}^{9} y_i^{(e)} \frac{\partial N_i}{\partial \eta} \end{bmatrix} = \begin{bmatrix} \mathbf{B}_\xi \mathbf{x}^{(e)} & \mathbf{B}_\xi \mathbf{y}^{(e)} \\ \mathbf{B}_\eta \mathbf{x}^{(e)} & \mathbf{B}_\eta \mathbf{y}^{(e)} \end{bmatrix}. \tag{24}$$

Assuming that the determinant of the Jacobian matrix is positive, the matrix $\mathbf{J}$ can be inverted. Since in the following only regular rectangular elements are considered, this assumption is always satisfied and the matrix $\mathbf{J}^{-1}$ is admissible and can be used to compute the derivatives of the interpolating functions in the physical domain defined by the coordinates x, y. The following relation is needed for this purpose:

$$\begin{bmatrix} \mathbf{B}_x \\ \mathbf{B}_y \end{bmatrix} = \mathbf{J}^{-1} \begin{bmatrix} \mathbf{B}_\xi \\ \mathbf{B}_\eta \end{bmatrix} \tag{25}$$

where $\mathbf{B}_x$ and $\mathbf{B}_x$ collect the derivatives of the shape functions with respect to x and y, respectively. These operators are required to define the compatibility equations of the RMZ model. In particular, the three-dimensional strain components for a rectangular plate can be obtained by means of the elasticity equations applied to the displacement fields (16) and assume the following definitions for the e-th element:

$$\begin{aligned}
\varepsilon_{xx}^{(e)} &= \frac{\partial U_x^{(e)}}{\partial x} = \frac{\partial u_x^{(e)}}{\partial x} + z\frac{\partial \phi_x^{(e)}}{\partial x} + F_z \frac{\partial \psi_x^{(e)}}{\partial x} = \varepsilon_x^{0(e)} + zk_x^{0(e)} + F_z \varepsilon_x^1 = \mathbf{B}_x \mathbf{u}_x^{(e)} + z\mathbf{B}_x \boldsymbol{\phi}_x^{(e)} + F_z \mathbf{B}_x \boldsymbol{\psi}_x^{(e)} \\
\varepsilon_{yy}^{(e)} &= \frac{\partial U_y^{(e)}}{\partial y} = \frac{\partial u_y^{(e)}}{\partial y} + z\frac{\partial \phi_y^{(e)}}{\partial y} + F_z \frac{\partial \psi_y^{(e)}}{\partial y} = \varepsilon_y^{0(e)} + zk_y^{0(e)} + F_z \varepsilon_y^1 = \mathbf{B}_y \mathbf{u}_y^{(e)} + z\mathbf{B}_y \boldsymbol{\phi}_y^{(e)} + F_z \mathbf{B}_y \boldsymbol{\psi}_y^{(e)} \\
\gamma_{xy}^{(e)} &= \frac{\partial U_y^{(e)}}{\partial x} + \frac{\partial U_x^{(e)}}{\partial y} = \frac{\partial u_y^{(e)}}{\partial x} + \frac{\partial u_x^{(e)}}{\partial y} + z\left(\frac{\partial \phi_y^{(e)}}{\partial x} + \frac{\partial \phi_x^{(e)}}{\partial y}\right) + F_z\left(\frac{\partial \psi_y^{(e)}}{\partial x} + \frac{\partial \psi_x^{(e)}}{\partial y}\right) = \\
&= \gamma_{xy}^{0(e)} + zk_{xy}^{0(e)} + F_z \gamma_{xy}^{1(e)} = \mathbf{B}_x \mathbf{u}_y^{(e)} + \mathbf{B}_y \mathbf{u}_x^{(e)} + z\left(\mathbf{B}_x \boldsymbol{\phi}_y^{(e)} + \mathbf{B}_y \boldsymbol{\phi}_x^{(e)}\right) + F_z\left(\mathbf{B}_x \boldsymbol{\psi}_y^{(e)} + \mathbf{B}_y \boldsymbol{\psi}_x^{(e)}\right) \\
\gamma_{xz}^{(e)} &= \frac{\partial U_z^{(e)}}{\partial x} + \frac{\partial U_x^{(e)}}{\partial z} = \frac{\partial u_z^{(e)}}{\partial x} + \phi_x^{(e)} + \frac{\partial F_z}{\partial z}\psi_x^{(e)} = \gamma_{xz}^{0(e)} + \frac{\partial F_z}{\partial z}\gamma_{xz}^{1(e)} = \mathbf{B}_x \mathbf{u}_z^{(e)} + \overline{\mathbf{N}} \boldsymbol{\phi}_x^{(e)} + \frac{\partial F_z}{\partial z} \overline{\mathbf{N}} \boldsymbol{\psi}_x^{(e)} \\
\gamma_{yz}^{(e)} &= \frac{\partial U_z^{(e)}}{\partial y} + \frac{\partial U_y^{(e)}}{\partial z} = \frac{\partial u_z^{(e)}}{\partial y} + \phi_y^{(e)} + \frac{\partial F_z}{\partial z}\psi_y^{(e)} = \gamma_{yz}^{0(e)} + \frac{\partial F_z}{\partial z}\gamma_{yz}^{1(e)} = \mathbf{B}_y \mathbf{u}_z^{(e)} + \overline{\mathbf{N}} \boldsymbol{\phi}_y^{(e)} + \frac{\partial F_z}{\partial z} \overline{\mathbf{N}} \boldsymbol{\psi}_y^{(e)},
\end{aligned} \tag{26}$$

where $\varepsilon_{xx}^{(e)}, \varepsilon_{yy}^{(e)}, \gamma_{xy}^{(e)}$ are the membrane strains, and $\gamma_{xz}^{(e)}, \gamma_{yz}^{(e)}$ are the transverse shear strains. It should be observed that the normal strain along the thickness direction $\varepsilon_{zz}^{(e)}$ is omitted since the plane-strain assumption entails that $\varepsilon_{zz}^{(e)} = 0$. The generalized strains related to the plate middle surface can be easily defined from relations (26). The superscript "0" denotes those quantities that are included also in the well-known RM theory, whereas the terms related to the zig-zag effect are identified by the superscript "1". In particular, it should be noted that $\varepsilon_x^{0(e)}, \varepsilon_y^{0(e)}, \gamma_{xy}^{0(e)}$ are the well-known membrane strains, $k_x^{0(e)}, k_y^{0(e)}, k_{xy}^{0(e)}$ the bending and twisting curvatures, and $\gamma_{xz}^{0(e)}, \gamma_{yz}^{0(e)}$ the shear strains, which are also defined in the RM theory [36].

The constitutive equations are now used to characterize the stress components in the k-th layer of the laminate. The following definitions imply that the plane-stress assumption $\sigma_{zz}^{k(e)} = 0$ is assumed by hypothesis, whereas the other stress components are given by:

$$\begin{aligned}
\sigma_{xx}^{k(e)} &= \overline{Q}_{11}^{(k)} \varepsilon_{xx}^{(e)} + \overline{Q}_{12}^{(k)} \varepsilon_{yy}^{(e)} + \overline{Q}_{16}^{(k)} \gamma_{xy}^{(e)} \\
\sigma_{yy}^{k(e)} &= \overline{Q}_{12}^{(k)} \varepsilon_{xx}^{(e)} + \overline{Q}_{22}^{(k)} \varepsilon_{yy}^{(e)} + \overline{Q}_{26}^{(k)} \gamma_{xy}^{(e)} \\
\tau_{xy}^{k(e)} &= \overline{Q}_{16}^{(k)} \varepsilon_{xx}^{(e)} + \overline{Q}_{26}^{(k)} \varepsilon_{yy}^{(e)} + \overline{Q}_{66}^{(k)} \gamma_{xy}^{(e)} \\
\tau_{xz}^{k(e)} &= \overline{Q}_{44}^{(k)} \gamma_{xz}^{(e)} + \overline{Q}_{45}^{(k)} \gamma_{yz}^{(e)} \\
\tau_{yz}^{k(e)} &= \overline{Q}_{45}^{(k)} \gamma_{xz}^{(e)} + \overline{Q}_{55}^{(k)} \gamma_{yz}^{(e)}
\end{aligned} \tag{27}$$

in which $\overline{Q}_{ij}^{(k)}$ denotes the stiffnesses of the k-th orthotropic layer evaluated in the geometric reference system. These parameters have the same meaning in each finite element, since the mechanical properties of the structure do not vary in the plate middle surface. Their well-known definition, which depends on the parameters $Q_{ij}^{(k)}$ expressed as a function of the engineering constants of the k-layer defined below, can be found in the book by Reddy [36]:

$$Q_{11}^{(k)} = \frac{E_{11}^{(k)}}{1-\nu_{12}^{(k)}\nu_{21}^{(k)}},\ Q_{22}^{(k)} = \frac{E_{22}^{(k)}}{1-\nu_{12}^{(k)}\nu_{21}^{(k)}},\ Q_{12}^{(k)} = \frac{\nu_{12}^{(k)}E_{22}^{(k)}}{1-\nu_{12}^{(k)}\nu_{21}^{(k)}},\ Q_{66}^{(k)} = G_{12}^{(k)},\ Q_{44}^{(k)} = G_{13}^{(k)},\ Q_{55}^{(k)} = G_{23}^{(k)}. \tag{28}$$

Quantities in (28) for $k = 1,3$ depend on the thickness coordinate z, since the reinforcing fibers of the face-sheets are characterized by a non-uniform distribution in this direction. Each orthotropic layer can be also characterized by an arbitrary orientation $\theta^{(k)}$. The notation $\left(\theta^{(1)}/\text{core}/\theta^{(3)}\right)$ is used in the next Sections to specify the orientations of the reinforcing fibers in the face-sheets and the consequent lamination scheme. Conventionally, in the above notation the layer numbering always starts from the bottom surface of the plate.

At this point, the Hamilton's variational principle should be applied to obtain the governing equations for the dynamic problem under consideration [36,101]. If t_1, t_2 specify the boundary values of the considered time interval, the variational principle assumes the following aspect within the e-th discrete element:

$$\int_{t_1}^{t_2} \left(\delta \mathbf{K}^{(e)} - \delta \mathbf{\Phi}^{(e)}\right) dt = 0 \tag{29}$$

where $\delta\mathbf{K}^{(e)}$ is the variation of the kinetic energy, whereas $\delta\mathbf{\Phi}^{(e)}$ represents the variation of the elastic strain energy. The kinetic energy of the sandwich structure $\delta\mathbf{K}^{(e)}$ is given by:

$$\delta\mathbf{K}^{(e)} = -\sum_{k=1}^{3} \int_x \int_y \int_{z_k}^{z_{k+1}} \rho^{(k)} \left(\delta U_x^{(e)} \ddot{U}_x^{(e)} + \delta U_y^{(e)} \ddot{U}_y^{(e)} + \delta U_y^{(e)} \ddot{U}_y^{(e)}\right) dxdydz \tag{30}$$

where the double-dot notation specifies the second-order derivatives with respect to the time variable. On the other hand, the elastic strain energy $\delta\mathbf{\Phi}^{(e)}$ is defined as follows:

$$\delta\mathbf{\Phi}^{(e)} = \sum_{k=1}^{3} \int_x \int_y \int_{z_k}^{z_{k+1}} \left(\sigma_{xx}^{k(e)} \delta\varepsilon_{xx}^{(e)} + \sigma_{yy}^{k(e)} \delta\varepsilon_{yy}^{(e)} + \tau_{xy}^{k(e)} \delta\gamma_{xy}^{(e)} + \tau_{xz}^{k(e)} \delta\gamma_{xz}^{(e)} + \tau_{yz}^{k(e)} \delta\gamma_{yz}^{(e)}\right) dxdydz \tag{31}$$

The proper mathematical manipulations of the elastic strain energy provide the definitions of the stress resultants as the through-the-thickness integrals of the stress components. The stress resultants can be written in matrix form by means of the following relations, which provide their definitions as a function of the generalized strain components introduced before:

$$\begin{bmatrix} N_x^{0(e)} \\ N_y^{0(e)} \\ N_{xy}^{0(e)} \\ M_x^{0(e)} \\ M_y^{0(e)} \\ M_{xy}^{0(e)} \\ T_x^{0(e)} \\ T_y^{0(e)} \\ N_x^{1(e)} \\ N_y^{1(e)} \\ N_{xy}^{1(e)} \\ T_x^{1(e)} \\ T_y^{1(e)} \end{bmatrix} = \begin{bmatrix} A_{11} & A_{12} & A_{16} & B_{11} & B_{12} & B_{16} & 0 & 0 & F_{11} & F_{12} & F_{16} & 0 & 0 \\ A_{12} & A_{22} & A_{26} & B_{12} & B_{22} & B_{26} & 0 & 0 & F_{12} & F_{22} & F_{26} & 0 & 0 \\ A_{16} & A_{26} & A_{66} & B_{16} & B_{26} & B_{66} & 0 & 0 & F_{16} & F_{26} & F_{66} & 0 & 0 \\ B_{11} & B_{12} & B_{16} & D_{11} & D_{12} & D_{16} & 0 & 0 & G_{11} & G_{12} & G_{16} & 0 & 0 \\ B_{12} & B_{22} & B_{26} & D_{12} & D_{22} & D_{26} & 0 & 0 & G_{12} & G_{22} & G_{26} & 0 & 0 \\ B_{16} & B_{26} & B_{66} & D_{16} & D_{26} & D_{66} & 0 & 0 & G_{16} & G_{26} & G_{66} & 0 & 0 \\ 0 & 0 & 0 & 0 & 0 & 0 & \kappa A_{44} & \kappa A_{45} & 0 & 0 & 0 & \kappa H_{44} & \kappa H_{45} \\ 0 & 0 & 0 & 0 & 0 & 0 & \kappa A_{45} & \kappa A_{55} & 0 & 0 & 0 & \kappa H_{45} & \kappa H_{55} \\ F_{11} & F_{12} & F_{16} & G_{11} & G_{12} & G_{16} & 0 & 0 & L_{11} & L_{12} & L_{16} & 0 & 0 \\ F_{12} & F_{22} & F_{26} & G_{12} & G_{22} & G_{26} & 0 & 0 & L_{12} & L_{22} & L_{26} & 0 & 0 \\ F_{16} & F_{26} & F_{66} & G_{16} & G_{26} & G_{66} & 0 & 0 & L_{16} & L_{26} & L_{66} & 0 & 0 \\ 0 & 0 & 0 & 0 & 0 & 0 & \kappa H_{44} & \kappa H_{45} & 0 & 0 & 0 & \kappa P_{44} & \kappa P_{45} \\ 0 & 0 & 0 & 0 & 0 & 0 & \kappa H_{45} & \kappa H_{55} & 0 & 0 & 0 & \kappa P_{45} & \kappa P_{55} \end{bmatrix} \begin{bmatrix} \varepsilon_x^{0(e)} \\ \varepsilon_y^{0(e)} \\ \gamma_{xy}^{0(e)} \\ k_x^{0(e)} \\ k_y^{0(e)} \\ k_{xy}^{0(e)} \\ \gamma_{xz}^{0(e)} \\ \gamma_{yz}^{0(e)} \\ \varepsilon_x^{1(e)} \\ \varepsilon_y^{1(e)} \\ \gamma_{xy}^{1(e)} \\ \gamma_{xz}^{1(e)} \\ \gamma_{yz}^{1(e)} \end{bmatrix}. \quad (32)$$

The superscripts "0" and "1" have the same meaning discussed previously. In particular, $N_x^{0(e)}, N_y^{0(e)}, N_{xy}^{0(e)}$ are the membrane forces, $M_x^{0(e)}, M_y^{0(e)}, M_{xy}^{0(e)}$ the bending and twisting moments, and $T_x^{0(e)}, T_y^{0(e)}$ the shear forces, which are included also in the RM theory [11]. The other terms are related to the zig-zag effect. The elements of the constitutive operator in (32) are now discussed. Firstly, the following terms appear also in the RM model and have the same meaning [11]:

$$A_{ij} = \sum_{k=1}^{3} \int_{z_k}^{z_{k+1}} \overline{Q}_{ij}^{(k)} dz, \quad B_{ij} = \sum_{k=1}^{3} \int_{z_k}^{z_{k+1}} z\overline{Q}_{ij}^{(k)} dz, \quad D_{ij} = \sum_{k=1}^{3} \int_{z_k}^{z_{k+1}} z^2\overline{Q}_{ij}^{(k)} dz. \quad (33)$$

On the other hand, the stress resultants related to the zig-zag effect require the following definitions, which include the Murakami's function and its derivative with respect to the thickness coordinate:

$$\begin{gathered} F_{ij} = \sum_{k=1}^{3} \int_{z_k}^{z_{k+1}} F_z\overline{Q}_{ij}^{(k)} dz, \quad G_{ij} = \sum_{k=1}^{3} \int_{z_k}^{z_{k+1}} zF_z\overline{Q}_{ij}^{(k)} dz, \quad H_{ij} = \sum_{k=1}^{3} \int_{z_k}^{z_{k+1}} \frac{\partial F_z}{\partial z}\overline{Q}_{ij}^{(k)} dz, \\ L_{ij} = \sum_{k=1}^{3} \int_{z_k}^{z_{k+1}} F_z^2\overline{Q}_{ij}^{(k)} dz, \quad P_{ij} = \sum_{k=1}^{3} \int_{z_k}^{z_{k+1}} \left(\frac{\partial F_z}{\partial z}\right)^2\overline{Q}_{ij}^{(k)} dz. \end{gathered} \quad (34)$$

It should be specified that the integrals in (33)–(34) are computed numerically since the stiffnesses $\overline{Q}_{ij}^{(k)}$ can be arbitrary functions of z, due to the dependency on the thickness coordinate introduced by relation (11). Finally, it is important to specify that the shear forces need the shear correction factor κ. The value of 5/6 is used to this aim.

The system of dynamic equations for the problem under consideration represents the main result of the application of the Hamilton's principle. The fundamental equation for the generic element e is given by:

$$\mathbf{K}^{(e)}\overline{\mathbf{u}}^{(e)} + \mathbf{M}^{(e)}\ddot{\overline{\mathbf{u}}}^{(e)} = 0 \quad (35)$$

where $\mathbf{K}^{(e)}$ is the element stiffness matrix, $\mathbf{M}^{(e)}$ the element mass matrix, and $\ddot{\overline{\mathbf{u}}}^{(e)}$ the vector of the second-order time derivatives of the element degrees of freedom included in $\overline{\mathbf{u}}^{(e)}$. The fundamental operators $\mathbf{K}^{(e)}$ assumes the following definitions:

$$\mathbf{K}^{(e)} = \begin{bmatrix} \mathbf{K}_{11} & \cdots & \cdots & \mathbf{K}_{17} \\ \vdots & \ddots & & \vdots \\ \vdots & & \ddots & \vdots \\ \mathbf{K}_{71} & \cdots & \cdots & \mathbf{K}_{77} \end{bmatrix}. \tag{36}$$

On the other hand, the mass matrix $\mathbf{M}^{(e)}$ is given by:

$$\mathbf{M}^{(e)} = \begin{bmatrix} \mathbf{M}_{11} & \cdots & \cdots & \mathbf{M}_{17} \\ \vdots & \ddots & & \vdots \\ \vdots & & \ddots & \vdots \\ \mathbf{M}_{71} & \cdots & \cdots & \mathbf{M}_{77} \end{bmatrix}. \tag{37}$$

The matrices $\mathbf{K}_{ij}$ and $\mathbf{M}_{ij}$, for $i,j = 1,\ldots,7$ are defined in the Appendix A. In particular, the following inertia terms are required to compute the mass matrix:

$$\begin{aligned} I_0 &= \sum_{k=1}^{3} \int_{z_k}^{z_{k+1}} \rho^{(k)} dz, \quad I_1 = \sum_{k=1}^{3} \int_{z_k}^{z_{k+1}} z\rho^{(k)} dz, \quad I_2 = \sum_{k=1}^{3} \int_{z_k}^{z_{k+1}} z^2\rho^{(k)} dz, \\ I_3 &= \sum_{k=1}^{3} \int_{z_k}^{z_{k+1}} F_z\rho^{(k)} dz, \quad I_4 = \sum_{k=1}^{3} \int_{z_k}^{z_{k+1}} zF_z\rho^{(k)} dz, \quad I_5 = \sum_{k=1}^{3} \int_{z_k}^{z_{k+1}} F_z^2\rho^{(k)} dz. \end{aligned} \tag{38}$$

It can be easily observed that the terms I_0, I_1, I_2 are included also in the RM theory. On the other hand, the parameters I_3, I_4, I_5 are linked to the zig-zag effect and include the Murakami's function. It should be specified that these integrals are computed numerically, since the density $\rho^{(k)}$, for $k = 1,3$, is an arbitrary function of the thickness coordinate.

The fundamental Equation (35) is valid for each discrete subdomain at the element level. The well-known assembly procedure is required to obtain the corresponding global system of equations. This approach allows to enforce automatically the displacement continuity at the element interfaces. Therefore, a C^0 compatibility requirement is satisfied in the current approach. Once the global matrices are obtained, the fundamental system of equations assumes the following aspect:

$$\mathbf{K}\mathbf{u} + \mathbf{M}\ddot{\mathbf{u}} = 0 \tag{39}$$

in which $\mathbf{K}, \mathbf{M}$ are the global stiffness and mass matrices, respectively. The degrees of freedom of the whole structure are collected in the vector $\mathbf{u}$. The nodal displacements are listed following the scheme specified by the dashed line in Figure 4, where an example of a discrete plate domain is also depicted.

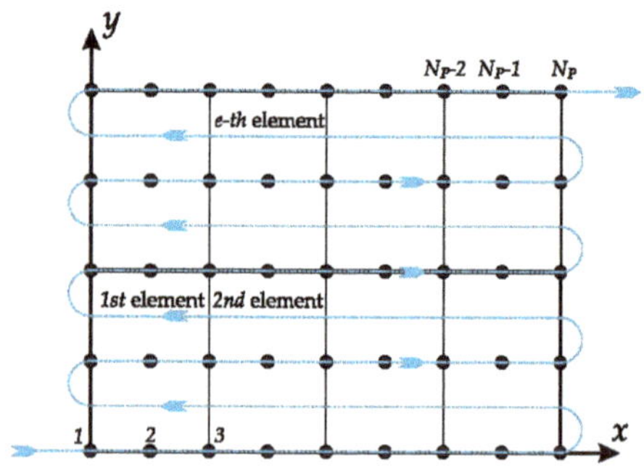

Figure 4. Discrete domain, identification of the elements and node numbering.

If N_P denotes the number of nodes, the model is characterized by $N_{dofs} = 7N_P$ as number of degrees of freedom. Consequently, the vector $\mathbf{u}$ can be written as follows:

$$\mathbf{u} = \Big[\underbrace{u_{x,1} \cdots u_{x,N_P}}_{1\rightarrow N_P} \quad \underbrace{u_{y,1} \cdots u_{y,N_P}}_{N_P+1\rightarrow 2N_P} \quad \underbrace{u_{z,1} \cdots u_{z,N_P}}_{2N_P+1\rightarrow 3N_P} \\ \underbrace{\phi_{x,1} \cdots \phi_{x,N_P}}_{3N_P+1\rightarrow 4N_P} \quad \underbrace{\phi_{y,1} \cdots \phi_{y,N_P}}_{4N_P+1\rightarrow 5N_P} \quad \underbrace{\psi_{x,1} \cdots \psi_{x,N_P}}_{5N_P+1\rightarrow 6N_P} \quad \underbrace{\psi_{y,1} \cdots \psi_{y,N_P}}_{6N_P+1\rightarrow 7N_P} \Big]^T. \tag{40}$$

The second-order time derivatives of these quantities are collected in the vector $\ddot{\mathbf{u}}$ following the same scheme. Finally, it should be recalled that the size of the fundamental operators $\mathbf{K}, \mathbf{M}$ is given by $N_{dofs} \times N_{dofs}$. The discrete system can be solved once the proper boundary conditions along the edges of the domain are enforced. In the present paper, since only fully clamped plates are considered, all the nodal displacements related to the boundary edges are all equal to zero.

3.1. Numerical Computation of the Fundamental Matrices

The Gauss-Legendre quadrature rule is employed to compute the fundamental matrices $\mathbf{K}, \mathbf{M}$. By definition, the integral of a generic function $G(x, y)$ defined in a two-dimensional domain can be evaluated as follows:

$$\int_x \int_y G(x,y)dxdy = \int_{-1}^{1} \int_{-1}^{1} G(\xi, \eta)\det\mathbf{J} d\xi d\eta \tag{41}$$

in which the determinant of the Jacobian matrix $\det\mathbf{J}$ is introduced. Therefore, the integral is computed in the parent space in which the reference system is given by natural coordinates ξ, η. From the numerical point of view, this integral can be converted into the following weighted linear sum:

$$\int_{-1}^{1} \int_{-1}^{1} G(\xi, \eta)\det\mathbf{J} d\xi d\eta \approx \sum_{I=1}^{M} \sum_{J=1}^{N} G\big(\xi_I, \eta_J\big) \det\mathbf{J}|_{\xi_I, \eta_J} W_I W_J \tag{42}$$

where W_I, W_J are the weighting coefficients, whereas ξ_I, η_J are the points in which the integral is computed. These nodes are the roots of Legendre polynomials [11]. The full integration is performed considering 9 evaluation points, whereas the reduced one is carried out in four points only. The position of these nodes is shown in Figure 3. It should be specified that the reduced integration is employed only to compute the elements of the stiffness matrix related to the shear forces in order to avoid the shear locking issue [11]. The analytical values of the weighting coefficients for each root of the Legendre polynomials can be found in the book by Reddy [11].

3.2. Evaluation of the Natural Frequencies

The free vibration analysis is based on a generalized eigenvalue problem from the analytical point of view. In particular, the following relation allows to compute the circular frequencies ω of the structures under consideration:

$$\left(\mathbf{K} - \omega^2 \mathbf{M}\right)\mathbf{d} = 0 \tag{43}$$

where the modal amplitudes are collected in the vector $\mathbf{d}$. Once relation (43) is solved, the natural frequencies f measured in Hz can be easily computed as $f = \omega / 2\pi$.

4. Numerical Applications

The numerical applications presented in this Section aim to evaluate the natural frequencies of several fully clamped sandwich plates. The geometric features are the same in each computation. In particular, a square domain defined by $L_x = L_y = 2.5$ m is considered. The thickness of the external layers is $h_s = 0.02$ m, whereas the soft-core is defined by $h_c = 0.06$ m. Each structure is subdivided into 100 finite elements as far as the discrete domain is concerned.

Firstly, the validity of the current approach based on the RMZ model is proved and compared with the results that could be obtained by means of the well-known RM theory. To this aim, a three-dimensional FE model is built through a commercial code. Secondly, the model is also validated with respect to the application of non-uniform distributions of the reinforcing fibers along the plate thickness. For this purpose, the results are compared with the ones available in the literature. Then, several parametric investigations are presented to discuss the effects of the damage, the through-the-thickness distributions of the reinforcing fibers, the lamination scheme and the in-plane orientation of the fibers, the mechanical properties of fibers and CNTs on the vibrational response.

4.1. Influence of the Murakami's Function and Validation of the RMZ

The first test aims to prove the need of the Murakami's function when the mechanical behavior of sandwich structures with an inner soft-core must be analyzed. In this application, the core is made by a virgin material ($D = 0$) and the fibers are uniformly placed along the thickness of the external face-sheets. The mechanical characterization is fully accomplished by setting $w_C = 0.05$ and $w_F = 0.80$. The natural frequencies are obtained by using the RM and the RMZ models, for three different lamination schemes. The same structures are investigated by means of a three-dimensional FE commercial code (twenty-node brick elements), denoted by 3D-FE in the following. The software Strand7 is employed for this purpose. The first ten natural frequencies for the sandwich plate with an inner soft-core under consideration are shown in Table 3, where the percentage differences (%diff) of the RMZ and RM solutions with respect to the 3D-FE results are also highlighted. The following aspects can be observed:

- The RMZ theory provides natural frequencies that are close to the results given by the reference solution (3D-FE). In fact, the maximum percentage difference is about 5% for higher modes. This difference is satisfactory having in mind the approximation introduced by a two-dimensional ESL theory;
- The computational cost is very different. In particular, the number of degrees of freedom in the 3D-FE model is ten times the one needed by the RMZ theory to obtain similar values;
- The RM model is not adequate to evaluate the natural frequencies of a sandwich soft-core structure, as it can be observed by the percentage differences with respect to the reference solution. The number of degrees of freedom in this circumstance is even lower if compared to the other models, but the computational saving cannot justify the poor approximation of the solution.

Table 3. First ten natural frequencies of a sandwich plate with an inner soft-core for three different lamination schemes: comparison with the 3D-FE solution.

Mode	3D-FE N_{dofs}=29673	RMZ N_{dofs}=3087	RM N_{dofs}=2205	%diff (RMZ)	%diff (RM)
		Lamination scheme: (0°/core/0°)			
1	142.126	137.126	169.127	3.52%	19.00%
2	195.625	189.516	224.819	3.12%	14.92%
3	289.303	280.546	324.708	3.03%	12.24%
4	303.824	288.429	408.390	5.07%	34.42%
5	346.073	329.618	453.375	4.75%	31.01%
6	414.415	401.529	463.476	3.11%	11.84%
7	421.211	402.429	533.774	4.46%	26.72%
8	498.657	469.706	635.351	5.81%	27.41%
9	527.513	504.420	650.839	4.38%	23.38%
10	534.573	505.028	678.379	5.53%	26.90%
		Lamination scheme: (30°/core/45°)			
1	129.776	125.889	147.134	2.99%	13.38%
2	211.140	203.739	244.133	3.51%	15.63%
3	265.283	253.703	327.302	4.37%	23.38%
4	299.738	288.099	354.309	3.88%	18.21%
5	371.903	354.527	462.224	4.67%	24.29%
6	398.033	381.561	476.960	4.14%	19.83%
7	426.714	404.619	562.840	5.18%	31.90%
8	483.191	459.287	616.601	4.95%	27.61%
9	506.505	484.458	616.724	4.35%	21.76%
10	541.210	512.228	704.809	5.36%	30.23%
		Lamination scheme: (−45°/core/45°)			
1	107.180	105.297	112.930	1.76%	5.36%
2	206.687	201.053	225.234	2.73%	8.97%
3	206.687	201.516	225.234	2.50%	8.97%
4	290.914	281.838	322.709	3.12%	10.93%
5	346.351	334.314	392.846	3.48%	13.42%
6	348.741	337.043	395.240	3.35%	13.33%
7	419.280	403.125	479.272	3.85%	14.31%
8	419.280	404.002	479.272	3.64%	14.31%
9	515.903	494.071	608.811	4.23%	18.01%
10	515.903	496.715	608.811	3.72%	18.01%

4.2. Validation of the Model with Respect to Non-Uniform Distributions of the Reinforcing Fibers

The current approach is validated also with respect to the application of non-uniform distribution of the reinforcing fibers in the thickness direction. In the paper by Lei et al. [71], a fully clamped square plate, characterized by the aspect ratio $L_x/h = 10$, is analyzed by means of the kp-Ritz method. The structure is made of an epoxy resin ($E_M = 2.1\,\text{GPa}$, $\nu_M = 0.34$, $\rho_M = 1150\,\text{kg/m}^3$) reinforced by aligned fibers of CNTs. This configuration can be modeled as a two-phase composite, in which the Carbon fibers are characterized by the following mechanical properties $E_{11}^F = 5.6466\,\text{TPa}$, $E_{22}^F = 7.0800\,\text{TPa}$, $G_{12}^F = 1.9445\,\text{TPa}$, $\nu_{12}^F = 0.175$, $\rho_F = 1400\,\text{kg/m}^3$. It should be noted that the approach presented in this paper can be used to evaluate the overall mechanical properties of this structure by neglecting the effect of the randomly oriented CNT particles scattered in the matrix. In other words, one gets $E_M^* = E_M$, $\nu_M^* = \nu_M$ and $\rho_M^* = \rho_M$. On the other hand, the aligned CNTs assume the same role of the straight fibers. Equation (10) is employed to obtain the value of w_F which provides the constant $\widetilde{V}_F = 0.11$ specified in the reference paper.

In order to validate the current methodology, the plate is made of two orthotropic layers of equal thickness (no soft-core is included), characterized by non-uniform distributions of the CNT fibers. "Case 1" is obtained by $f_1^{(1)}, f_2^{(2)}$ with $\alpha = 1$, whereas "Case 2" is given by $f_2^{(1)}, f_1^{(2)}$ with $\alpha = 1$. The frequencies are presented in Table 4 in dimensionless form as

$$\varpi = \omega \frac{L_x^2}{h} \sqrt{\frac{\rho_M}{E_M}} \tag{44}$$

in which ω denotes the circular frequencies. As specified in the previous sections, the Halpin-Tsai (HT) model is applied for the mechanical characterization of the structure. Nevertheless, the reference solutions are obtained by means of the rule of the mixture (MIX). The same approach has been used in the paper by Bacciocchi and Tarantino [39] for similar purposes. Therefore, only in the next application these two models (MIX and HT) are considered for the sake of comparison. Small differences are obtained depending on the homogenization procedure used in the computation.

Table 4. Comparison of the frequency parameter ϖ for a two phase fully clamped plate.

Mode	Lei et al. [71] (kp-Ritz)	Lei et al. [71] (Commercial FE)	RMZ (MIX)	RMZ (HT)
		Case 1		
1	16.667	16.707	16.671	17.383
2	22.138	22.253	22.098	23.381
3	32.237	32.378	32.211	33.630
4	32.424	32.857	32.242	34.397
5	35.674	35.809	35.652	37.437
6	37.367	37.447	37.403	40.615
		Case 2		
1	18.045	18.083	18.055	19.071
2	23.498	23.606	23.448	25.152
3	33.915	34.338	33.712	36.116
4	34.361	34.467	34.324	36.449
5	37.367	37.447	37.403	39.892
6	37.693	37.786	37.637	40.616

It should be emphasized that the current application does not aim to investigate the effect of the homogenization method but only the comparison with the reference solution of the frequency parameter. As it can be noted from the results shown in Table 4, a good agreement is observed with the reference solution, especially if the rule of the mixture is employed as it could be expected. In the same table, the FE results provided by a commercial code are also presented. Finally, it should be specified that the RMZ theory is considered as far as the theoretical model of the present solutions is concerned.

4.3. *Effect of Damage*

The same geometric and mechanical configurations are analyzed also in this paragraph. Nevertheless, the soft-core of the structures is affected by a decay of the mechanical properties and the effect of an increasing damage parameter $D \in [0,1]$ on the natural frequencies is discussed. For each lamination scheme, four different through-the-thickness distributions of the reinforcing fibers in the face-sheets are investigated, including the uniform one. "Scheme 1" denotes the uniform distribution of the fibers; "Scheme 2" is obtained by setting $f_2^{(1)} = f_2^{(3)}$ with $\alpha = 1$; "Scheme 3" is accomplished by using $f_2^{(1)}, f_1^{(3)}$ with $\alpha = 1$; finally, "Scheme 4" has the same functions of the previous one, assuming $\alpha = 2$. These configurations are graphically depicted in Figure 5. The volume fraction distribution of the fibers in the core is clearly equal to zero.

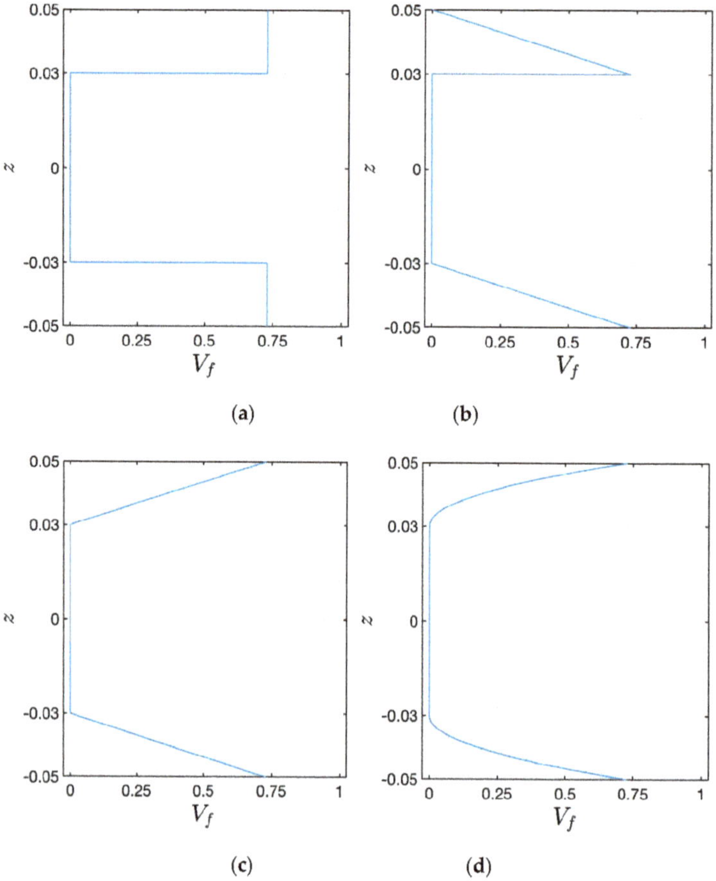

Figure 5. Through-the-thickness representation of the volume fraction distribution V_F for several schemes: (**a**) Scheme 1; (**b**) Scheme 2; (**c**) Scheme 3; (**d**) Scheme 4.

The results are presented in Figure 6, where it can be observed that the first natural frequency for each configuration depends noticeably on the parameter D. The following aspects should be noted, as well:

- The decrease of the frequency is clearly caused by the corresponding stiffness reduction of the structures. This expected tendency models accurately the physical behavior of structures with a lower value of stiffness. In fact, by increasing the value of D up to the unity (fully damaged core), the frequency would tend to zero;
- The same behavior is obtained for each volume fraction distribution, but the maximum value of the first frequency that can be reached depends on the through-the-thickness distributions of V_F (Figure 5);
- These aspects can be noted for each lamination scheme. Nevertheless, depending on the in-plane fiber orientation, the value of the first frequency could change. In addition, a peculiar choice of lamination scheme could reduce the influence of the through-the-thickness distributions of the volume fraction V_F, since the curves related to the various schemes are less detached;
- Finally, similar graphs could be obtained also for higher frequencies.

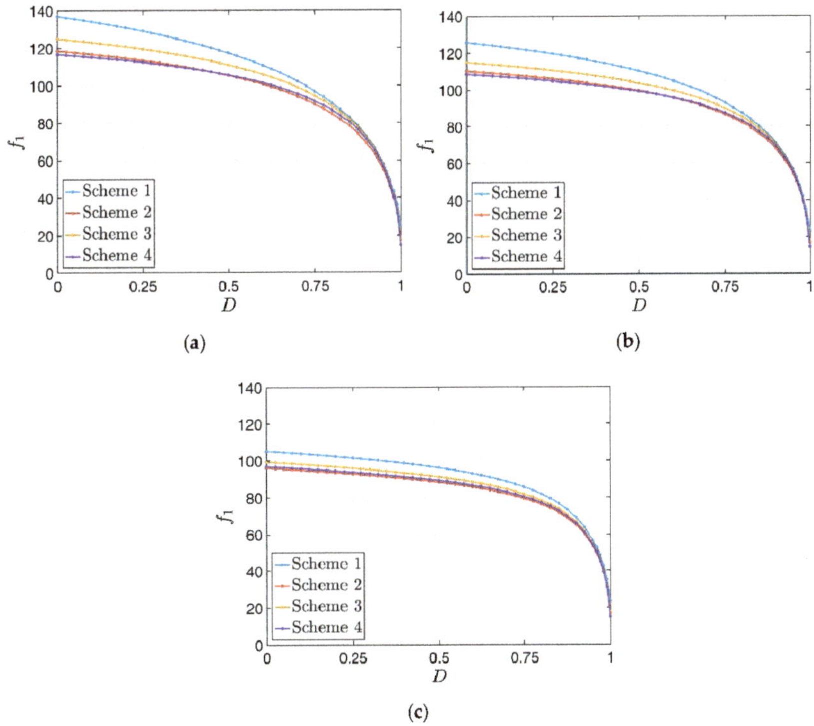

Figure 6. Variation of the first natural frequency f_1 [Hz] for sandwich plates with a damaged soft-core due to an increasing damage D, for three different lamination schemes: **(a)** (0°/core/0°); **(b)** (30° /core/45°); **(c)** (−45° /core/45°).

4.4. Influence of the Exponent of the Through-the-Thickness Distribution of the Fiber Volume Fraction

The current application deals with the effect of the exponent α that characterizes the through-the-thickness distribution of V_F. For this purpose, the Scheme 2 and Scheme 3 of the previous test are considered. Nevertheless, different configurations are obtained since $\alpha \in [0, \infty]$, as it can be seen from Figure 2.

The geometry of the sandwich plate is kept constant and three laminations schemes are considered: (0°/core/0°), (30°/core/45°) and (−45°/core/45°). A damaged core is modeled by setting $D = 0.5$. It should be recalled that the same values of the mass fraction of both CNTs and fibers (respectively $w_C = 0.05$ and $w_F = 0.80$) are employed. The proper choice of the exponent α allows to obtain also the following extreme cases. For $\alpha = 0$, the reinforcing fibers are uniformly distributed, and the Scheme 1 of the previous application is accomplished. On the other hand, if α tends to infinity, it is easy to verify that $V_F = 0$ and the face-sheets are made of an undamaged polymer matrix enforced only by CNTs. In this circumstance, the stiffness of the structure reaches its minimum value. In terms of natural frequencies, the results are included between these two boundary cases. The variation of the first three natural frequencies is depicted in Figure 7. The following features can be observed:

- Similar behaviors are obtained for the three lamination schemes under investigation. For lower values of α, the corresponding curves are detached and the natural frequencies that can be obtained assume different values depending on the fiber orientation. By increasing the exponent α, the effect of the fibers decreases since V_F draws near zero and the frequencies tends asymptotically to the same value;

- The initial choice of the through-the-thickness distribution of V_F (Scheme 2 or Scheme 3) affects the variation of the natural frequencies. In particular, this variation is faster for Scheme 2. In fact, the slopes of the related curves are steeper, whereas the frequency variation for Scheme 3 is a little bit more gradual;
- The biggest variation of frequencies is reached for lower values of α. The decrease of the value of natural frequencies for $\alpha > 20$ is negligible.

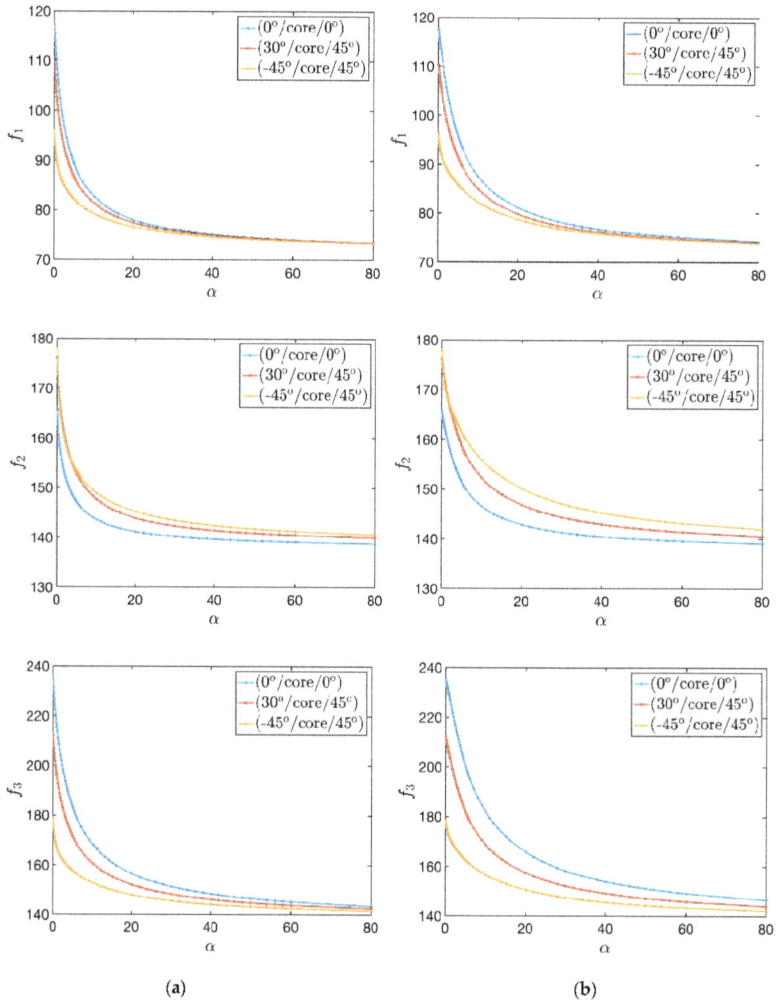

Figure 7. Variation of the first three natural frequencies f_1, f_2, f_3 [Hz] for sandwich plates with a damaged soft-core due to an increasing value of the exponent α of the through-the-thickness distribution of the fiber volume fraction, for two different schemes: (**a**) Scheme 2; (**b**) Scheme 3.

4.5. *Effect of the In-Plane Fiber Orientation*

A damaged sandwich plate with $D = 0.25$ is considered in the following application. The geometric features are kept constant with respect to the previous tests, whereas the engineering constants of the face-sheets can be computed assuming $w_C = 0.05$ and $w_F = 0.80$. A non-uniform distribution of the reinforcing fiber is defined according to the functions that describe Scheme 3, with $\alpha = 0.5$. The aim of this numerical application is to show the dependency of the natural frequencies on the in-plane fiber orientation. Therefore, several configurations are analyzed according to the value of an angular parameter θ. The variation of the first three natural frequencies is depicted in Figure 8 for various lamination schemes depending on θ. The graphs in Figure 8 prove the following results:

- As expected, the orientation of the fibers affects the dynamic response of the composite structures under consideration;
- If symmetric angle-ply or cross-ply laminates, as well as antisymmetric configurations, are considered, which are denoted by (θ/core/θ) and ($-\theta$/core/θ), the extreme values of frequencies can be obtained for $\theta = 0^\circ, 45^\circ, 90^\circ$. In addition, a symmetrical behavior is obtained after reaching the value of $\theta = 45^\circ$;
- This regular and symmetric behavior is lost if a laminate with a general stacking sequence, such as the last two lamination schemes, is analyzed.

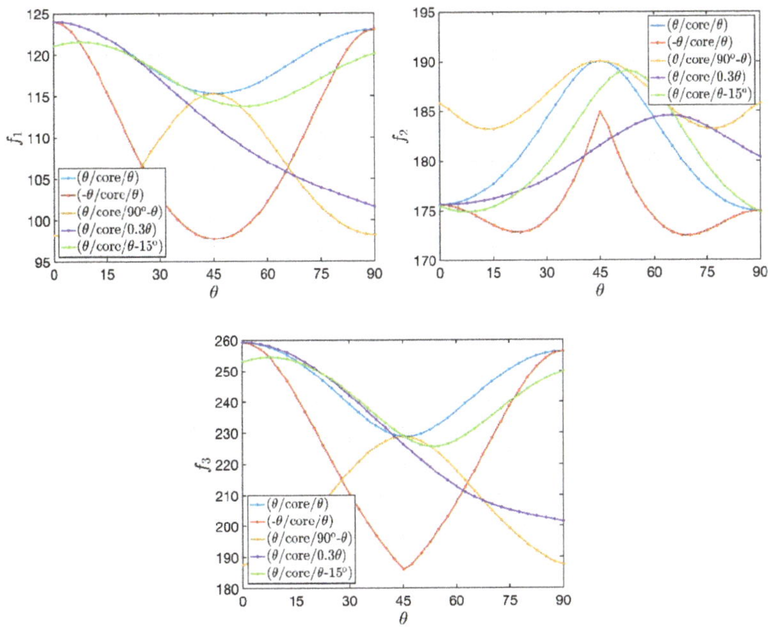

Figure 8. Variation of the first three natural frequencies f_1, f_2, f_3 [Hz] for a sandwich plates with a damaged soft-core for several lamination schemes depending on the angle parameter θ which defines the in-plane fiber orientation.

4.6. *Influence of the Material Properties*

Finally, this last application aims to discuss the effects of the CNT mass fraction w_C and the fiber mass fraction w_F on the structural response. The same geometric features are considered in this circumstance, assuming $D = 0.75$ as damage parameters. The through-the-thickness distribution of the reinforcing fibers V_F are defined by the functions used in Scheme 3 (Figure 5) by setting $\alpha = 4$. The lamination scheme is given by (30°/core/45°). Firstly, the effect of w_C is investigated, by keeping constant the value of $w_F = 0.80$, in the interval $w_C \in [0, 0.40]$.

It should be recalled that the polymer matrix is strengthened by only straight fibers if $w_C = 0$. Then, the opposite situation is taken into account. The variation of $w_F \in [0.40, 0.80]$ is studied for a fixed value of $w_C = 0.05$. The results are shown in Figure 9 for the first five natural frequencies. The following observations can be deduced:

- The influence of the CNT mass fraction w_C is greater than the corresponding variation of the fiber mass fraction w_F;
- For a small increase of w_C next to zero, the variation in terms of natural frequencies that can be obtained is relevant and the behavior is non-linear;
- On the other hand, bigger increases of w_F do not produce the same variation of natural frequencies. The behavior is linear in this case.

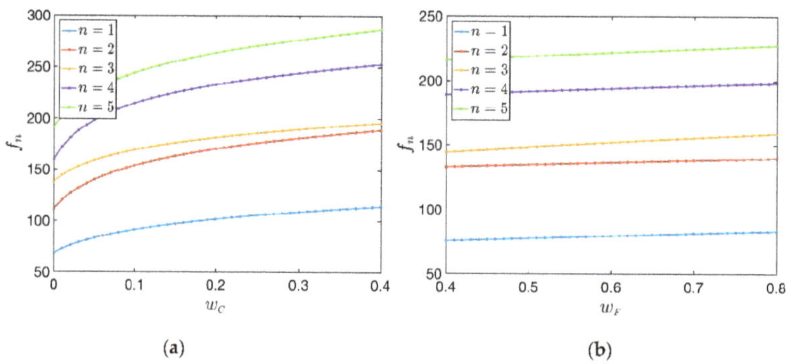

Figure 9. Variation of the first five natural frequencies f_n [Hz], for $n = 1, \ldots, 5$, of sandwich plates with a damaged soft-core for varying mechanical properties of the constituents: (**a**) Variation of CNT mass fraction; (**b**) Variation of fiber mass fraction.

4.7. Discussion on the Mode Shapes

Finally, a brief discussion on the dependency of the previous parameters on the mode shapes is presented. For this purpose, the same geometric features of the previous tests are considered. As far as the mechanical properties of the constituents are concerned, the plate is characterized by $w_C = 0.05$ and $w_F = 0.80$. Several configurations are analyzed in order to investigate the effect of the stacking sequence, of the damage and of the exponent of the through-the-thickness distribution of the fibers. The Scheme 3 depicted in Figure 5 is considered here, but similar results in terms of mode shapes could be obtained also with the other schemes. The cases under considerations are summarized in Table 5 for conciseness purposes.

Table 5. Mechanical configurations investigated to show the variation of the mode shapes due to the stacking sequence, the damage parameter and the exponent of the reinforcing fiber distribution.

Case	Stacking Sequence	Damage D	Exponent α
1	(0°/core/0°)	0.00	1
2	(0°/core/0°)	0.50	1
3	(0°/core/0°)	0.50	12
4	(30°/core/45°)	0.00	1
5	(30°/core/45°)	0.50	1
6	(30°/core/45°)	0.50	4
7	(30°/core/45°)	0.50	12

The contour plots of the first five mode shapes related to the corresponding natural frequencies are shown in Figure 10 for the seven cases introduced in Table 5. The examples analyzed in this paragraph allow to deduce the following considerations:

- In general, the increase of the damage parameter D does not cause any variation of the mode shapes;
- The mode shapes are highly affected by the orientation of the reinforcing fibers and by the stacking sequence. This aspect can be noted by comparing the same configurations in terms of D and α, but characterized by different lamination schemes (Case 1 and Case 4, for instance);
- As stated in the previous paragraphs, the increase of the exponent α reduces the influence of the reinforcing straight fibers. Therefore, the anisotropic behavior of a laminate with a general stacking sequence can be decreased. For example, the mode shapes of Case 7 tend to the ones related to Case 3 for $\alpha = 12$, even if they are characterized by different fiber orientations;
- The presence of a thicker isotropic core is predominant in the modal amplitudes and only noticeably variations of these mechanical parameters can define some changes in the mode shapes.

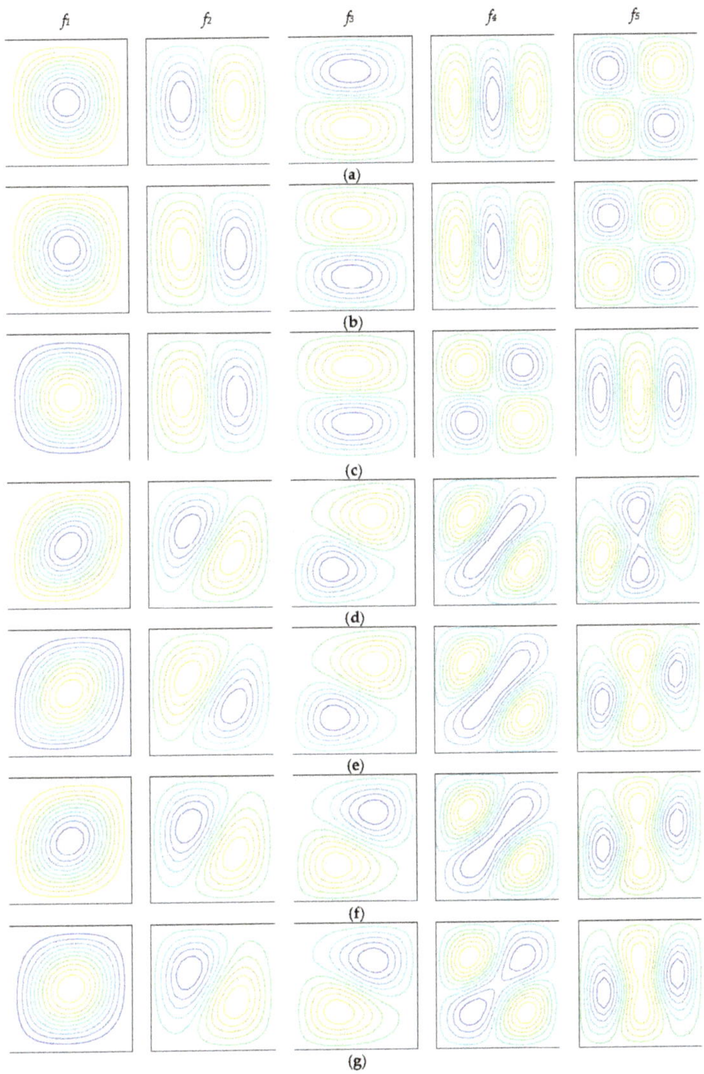

Figure 10. First five mode shapes of the mechanical configurations defined in Table 4: (**a**) Case 1; (**b**) Case 2; (**c**) Case 3; (**d**) Case 4; (**e**) Case 5; (**f**) Case 6; (**g**) Case 7.

5. Conclusions

A set of numerical investigations has been presented to describe the mechanical behavior of laminated sandwich plates with a damaged soft-core. The FE formulation has been developed by using a nine-node quadratic rectangular Lagrange element, whereas the theoretical model for laminated plates based on the Reissner-Mindlin model has been enriched by introducing the Murakami's function. As a consequence, a FE plate theory with 7 degrees of freedom per node has been presented. The external face-sheets are made of composite materials: The polymer matrix has been strengthened by randomly oriented CNTs and oriented straight fibers. Their mechanical characterization has been carried out by using a three-phase model, which include the Eshelby-Mori-Tanaka scheme and the Halpin-Tsai approach. Several parametric tests have been performed to analyze the effect of the damage, the through-the-thickness distribution of the reinforcing fibers, the orientation of the fibers, the stacking sequence, and the mechanical features. The main achievements of the paper are summarized below:

- The use of the Murakami's function is required to capture the effective mechanical behavior of sandwich structures with an inner soft-core. This aspect is very important especially if a FE commercial code is employed. In fact, it should be recalled that this function is not embedded in plate/shell formulations. Therefore, the results that can be obtained in these circumstances could be inaccurate, unless a 3D-FE modelling is pursued. Nevertheless, this approach is onerous in terms of computational time and resources;
- A non-uniform distribution of the fibers along the thickness of the face-sheets could be employed to model the effective distribution of the reinforcing phase that could occur during the manufacturing process or during the structural life. This research prove that the mechanical response is affected by this parameter;
- A progressive damage in the core causes a corresponding decrease of the natural frequencies, which becomes faster and faster for higher values of damages. The reinforcing layers could recover this situation. If a three-phase composite material is employed to this aim, the design of such layers could be carried out taking into account two parameters, which are the mass fractions of both CNTs and fibers. Nevertheless, a small increase of the CNT mass fraction can cause a quicker and more remarkable variation of the fundamental frequency with respect to the one that could be obtained by controlling the mass fraction of the straight fibers;
- The optimal structural response can be also obtained by choosing accurately the in-plane orientation of the straight fibers. The stacking sequence, in fact, affects the value of the natural frequencies, as well as of the mode shapes.

These comments should be taken into account during the analysis of the mechanical behavior of sandwich structures subjected to a progressive damage, as well as during the process manufacturing if an optimal design has to be pursued.

Author Contributions: Conceptualization, M.B., R.L., C.M. and A.M.T.; methodology, M.B., R.L., C.M. and A.M.T.; software, M.B., R.L., C.M. and A.M.T.; validation, M.B., R.L., C.M. and A.M.T.; formal analysis, M.B., R.L., C.M. and A.M.T.; investigation, M.B., R.L., C.M. and A.M.T.; resources, M.B., R.L., C.M. and A.M.T.; data curation, M.B., R.L., C.M. and A.M.T.; writing—original draft preparation, M.B., R.L., C.M. and A.M.T.; writing—review and editing, M.B., R.L., C.M. and A.M.T.; visualization, M.B., R.L., C.M. and A.M.T.; supervision, M.B., R.L., C.M. and A.M.T.; project administration, M.B., R.L., C.M. and A.M.T.

Funding: This research received no external funding.

Conflicts of Interest: The authors declare no conflict of interest.

Appendix A

The submatrices $\mathbf{K}_{ij}^{(e)}$, for $i, j = 1, \ldots, 7$, which appear in the element stiffness matrix $\mathbf{K}^{(e)}$ defined in (36), can be computed using the following definitions. The submatrices are presented below row-by-row

$$
\begin{gathered}
\mathbf{K}_{11} = \int_x \int_y \left(\mathbf{B}_x^T\left(A_{11}\mathbf{B}_x + A_{16}\mathbf{B}_y\right) + \mathbf{B}_y^T\left(A_{16}\mathbf{B}_x + A_{66}\mathbf{B}_y\right)\right)dxdy \\
\mathbf{K}_{12} = \int_x \int_y \left(\mathbf{B}_x^T\left(A_{12}\mathbf{B}_y + A_{16}\mathbf{B}_x\right) + \mathbf{B}_y^T\left(A_{26}\mathbf{B}_y + A_{66}\mathbf{B}_x\right)\right)dxdy \\
\mathbf{K}_{13} = 0 \\
\mathbf{K}_{14} = \int_x \int_y \left(\mathbf{B}_x^T\left(B_{11}\mathbf{B}_x + B_{16}\mathbf{B}_y\right) + \mathbf{B}_y^T\left(B_{16}\mathbf{B}_x + B_{66}\mathbf{B}_y\right)\right)dxdy \\
\mathbf{K}_{15} = \int_x \int_y \left(\mathbf{B}_x^T\left(B_{12}\mathbf{B}_y + B_{16}\mathbf{B}_x\right) + \mathbf{B}_y^T\left(B_{26}\mathbf{B}_y + B_{66}\mathbf{B}_x\right)\right)dxdy \\
\mathbf{K}_{16} = \int_x \int_y \left(\mathbf{B}_x^T\left(F_{11}\mathbf{B}_x + F_{16}\mathbf{B}_y\right) + \mathbf{B}_y^T\left(F_{16}\mathbf{B}_x + F_{66}\mathbf{B}_y\right)\right)dxdy \\
\mathbf{K}_{17} = \int_x \int_y \left(\mathbf{B}_x^T\left(F_{12}\mathbf{B}_y + F_{16}\mathbf{B}_x\right) + \mathbf{B}_y^T\left(F_{26}\mathbf{B}_y + F_{66}\mathbf{B}_x\right)\right)dxdy
\end{gathered}
\tag{A1}
$$

$$
\begin{gathered}
\mathbf{K}_{21} = \mathbf{K}_{12}^T \\
\mathbf{K}_{22} = \int_x \int_y \left(\mathbf{B}_y^T\left(A_{22}\mathbf{B}_y + A_{26}\mathbf{B}_x\right) + \mathbf{B}_x^T\left(A_{26}\mathbf{B}_y + A_{66}\mathbf{B}_x\right)\right)dxdy \\
\mathbf{K}_{23} = 0 \\
\mathbf{K}_{24} = \int_x \int_y \left(\mathbf{B}_y^T\left(B_{12}\mathbf{B}_x + B_{26}\mathbf{B}_y\right) + \mathbf{B}_x^T\left(B_{16}\mathbf{B}_x + B_{66}\mathbf{B}_y\right)\right)dxdy \\
\mathbf{K}_{25} = \int_x \int_y \left(\mathbf{B}_y^T\left(B_{22}\mathbf{B}_y + B_{26}\mathbf{B}_x\right) + \mathbf{B}_x^T\left(B_{26}\mathbf{B}_y + B_{66}\mathbf{B}_x\right)\right)dxdy \\
\mathbf{K}_{26} = \int_x \int_y \left(\mathbf{B}_y^T\left(F_{12}\mathbf{B}_x + F_{26}\mathbf{B}_y\right) + \mathbf{B}_x^T\left(F_{16}\mathbf{B}_x + F_{66}\mathbf{B}_y\right)\right)dxdy \\
\mathbf{K}_{27} = \int_x \int_y \left(\mathbf{B}_y^T\left(F_{22}\mathbf{B}_y + F_{26}\mathbf{B}_x\right) + \mathbf{B}_x^T\left(F_{26}\mathbf{B}_y + F_{66}\mathbf{B}_x\right)\right)dxdy
\end{gathered}
\tag{A2}
$$

$$
\begin{gathered}
\mathbf{K}_{31} = \mathbf{K}_{13}^T, \quad \mathbf{K}_{32} = \mathbf{K}_{23}^T \\
\mathbf{K}_{33} = \int_x \int_y \left(\mathbf{B}_x^T\left(\kappa A_{44}\mathbf{B}_x + \kappa A_{45}\mathbf{B}_y\right) + \mathbf{B}_y^T\left(\kappa A_{45}\mathbf{B}_x + \kappa A_{55}\mathbf{B}_y\right)\right)dxdy \\
\mathbf{K}_{34} = \int_x \int_y \left(\mathbf{B}_x^T\left(\kappa A_{44}\overline{\mathbf{N}}\right) + \mathbf{B}_y^T\left(\kappa A_{45}\overline{\mathbf{N}}\right)\right)dxdy \\
\mathbf{K}_{35} = \int_x \int_y \left(\mathbf{B}_x^T\left(\kappa A_{45}\overline{\mathbf{N}}\right) + \mathbf{B}_y^T\left(\kappa A_{55}\overline{\mathbf{N}}\right)\right)dxdy \\
\mathbf{K}_{36} = \int_x \int_y \left(\mathbf{B}_x^T\left(\kappa H_{44}\overline{\mathbf{N}}\right) + \mathbf{B}_y^T\left(\kappa H_{45}\overline{\mathbf{N}}\right)\right)dxdy \\
\mathbf{K}_{37} = \int_x \int_y \left(\mathbf{B}_x^T\left(\kappa H_{45}\overline{\mathbf{N}}\right) + \mathbf{B}_y^T\left(\kappa H_{55}\overline{\mathbf{N}}\right)\right)dxdy
\end{gathered}
\tag{A3}
$$

$$
\begin{gathered}
\mathbf{K}_{41} = \mathbf{K}_{14}^T, \quad \mathbf{K}_{42} = \mathbf{K}_{24}^T, \quad \mathbf{K}_{43} = \mathbf{K}_{34}^T \\
\mathbf{K}_{44} = \int_x \int_y \left(\mathbf{B}_x^T\left(D_{11}\mathbf{B}_x^e + D_{16}\mathbf{B}_y\right) + \mathbf{B}_y^T\left(D_{16}\mathbf{B}_x + D_{66}\mathbf{B}_y\right)\right)dxdy + \int_x \int_y \overline{\mathbf{N}}^T \kappa A_{44}\overline{\mathbf{N}}dxdy \\
\mathbf{K}_{45} = \int_x \int_y \left(\mathbf{B}_x^T\left(D_{12}\mathbf{B}_y + D_{16}\mathbf{B}_x\right) + \mathbf{B}_y^T\left(D_{26}\mathbf{B}_y + D_{66}\mathbf{B}_x\right)\right)dxdy + \int_x \int_y \overline{\mathbf{N}}^T \kappa A_{45}\overline{\mathbf{N}}dxdy \\
\mathbf{K}_{46} = \int_x \int_y \left(\mathbf{B}_x^T\left(G_{11}\mathbf{B}_x + G_{16}\mathbf{B}_y\right) + \mathbf{B}_y^T\left(G_{16}\mathbf{B}_x + G_{66}\mathbf{B}_y\right)\right)dxdy + \int_x \int_y \overline{\mathbf{N}}^T \kappa H_{44}\overline{\mathbf{N}}dxdy \\
\mathbf{K}_{47} = \int_x \int_y \left(\mathbf{B}_x^T\left(G_{12}\mathbf{B}_y + G_{16}\mathbf{B}_x\right) + \mathbf{B}_y^T\left(G_{26}\mathbf{B}_y + G_{66}\mathbf{B}_x\right)\right)dxdy + \int_x \int_y \overline{\mathbf{N}}^T \kappa H_{45}\overline{\mathbf{N}}dxdy
\end{gathered}
\tag{A4}
$$

$$\begin{aligned}
&\mathbf{K}_{51} = \mathbf{K}_{15}^T, \quad \mathbf{K}_{52} = \mathbf{K}_{25}^T, \quad \mathbf{K}_{53} = \mathbf{K}_{35}^T, \quad \mathbf{K}_{54} = \mathbf{K}_{45}^T \\
&\mathbf{K}_{55} = \int_x\int_y \left(\mathbf{B}_y^T\left(D_{22}\mathbf{B}_y + D_{26}\mathbf{B}_x\right) + \mathbf{B}_x^T\left(D_{26}\mathbf{B}_y + D_{66}\mathbf{B}_x\right)\right)dxdy + \int_x\int_y \overline{\mathbf{N}}^T \kappa A_{55}\overline{\mathbf{N}}dxdy \\
&\mathbf{K}_{56} = \int_x\int_y \left(\mathbf{B}_y^T\left(G_{12}\mathbf{B}_x + G_{26}\mathbf{B}_y\right) + \mathbf{B}_x^T\left(G_{16}\mathbf{B}_x + G_{66}\mathbf{B}_y\right)\right)dxdy + \int_x\int_y \overline{\mathbf{N}}^T \kappa H_{45}\overline{\mathbf{N}}dxdy \\
&\mathbf{K}_{57} = \int_x\int_y \left(\mathbf{B}_y^T\left(G_{22}\mathbf{B}_y + G_{26}\mathbf{B}_x\right) + \mathbf{B}_x^T\left(G_{26}\mathbf{B}_y + G_{66}\mathbf{B}_x\right)\right)dxdy + \int_x\int_y \overline{\mathbf{N}}^T \kappa H_{55}\overline{\mathbf{N}}dxdy
\end{aligned} \tag{A5}$$

$$\begin{aligned}
&\mathbf{K}_{61} = \mathbf{K}_{16}^T, \quad \mathbf{K}_{62} = \mathbf{K}_{26}^T, \quad \mathbf{K}_{63} = \mathbf{K}_{36}^T, \quad \mathbf{K}_{64} = \mathbf{K}_{46}^T, \quad \mathbf{K}_{65} = \mathbf{K}_{56}^T \\
&\mathbf{K}_{66} = \int_x\int_y \left(\mathbf{B}_x^T\left(L_{11}\mathbf{B}_x + L_{16}\mathbf{B}_y\right) + \mathbf{B}_y^T\left(L_{16}\mathbf{B}_x + L_{66}\mathbf{B}_y\right)\right)dxdy + \int_x\int_y \overline{\mathbf{N}}^T \kappa P_{44}\overline{\mathbf{N}}dxdy \\
&\mathbf{K}_{67} = \int_x\int_y \left(\mathbf{B}_x^T\left(L_{12}\mathbf{B}_y + L_{16}\mathbf{B}_x\right) + \mathbf{B}_y^T\left(L_{26}\mathbf{B}_y + L_{66}\mathbf{B}_x\right)\right)dxdy + \int_x\int_y \overline{\mathbf{N}}^T \kappa P_{45}\overline{\mathbf{N}}dxdy
\end{aligned} \tag{A6}$$

$$\begin{aligned}
&\mathbf{K}_{71} = \mathbf{K}_{17}^T, \quad \mathbf{K}_{72} = \mathbf{K}_{27}^T, \quad \mathbf{K}_{73} = \mathbf{K}_{37}^T, \quad \mathbf{K}_{74} = \mathbf{K}_{47}^T, \quad \mathbf{K}_{75} = \mathbf{K}_{57}^T, \quad \mathbf{K}_{76} = \mathbf{K}_{67}^T \\
&\mathbf{K}_{77} = \int_x\int_y \left(\mathbf{B}_y^T\left(L_{22}\mathbf{B}_y + L_{26}\mathbf{B}_x\right) + \mathbf{B}_x^T\left(L_{26}\mathbf{B}_y + L_{66}\mathbf{B}_x\right)\right)dxdy + \int_x\int_y \overline{\mathbf{N}}^T \kappa P_{55}\overline{\mathbf{N}}dxdy
\end{aligned} \tag{A7}$$

Analogously, the submatrices $\mathbf{M}_{ij}^{(e)}$, for $i, j = 1, \ldots, 7$, which appear in the element mass matrix $\mathbf{M}^{(e)}$ introduced in (37), are defined below. The submatrices are presented row-by-row as in the previous case

$$\mathbf{M}_{11} = \int_x\int_y \overline{\mathbf{N}}^T I_0\overline{\mathbf{N}}dxdy, \quad \mathbf{M}_{14} = \int_x\int_y \overline{\mathbf{N}}^T I_1\overline{\mathbf{N}}dxdy, \quad \mathbf{M}_{16} = \int_x\int_y \overline{\mathbf{N}}^T I_3\overline{\mathbf{N}}dxdy \tag{A8}$$

$$\mathbf{M}_{22} = \int_x\int_y \overline{\mathbf{N}}^T I_0\overline{\mathbf{N}}dxdy, \quad \mathbf{M}_{25} = \int_x\int_y \overline{\mathbf{N}}^T I_1\overline{\mathbf{N}}dxdy, \quad \mathbf{M}_{27} = \int_x\int_y \overline{\mathbf{N}}^T I_3\overline{\mathbf{N}}dxdy \tag{A9}$$

$$\mathbf{M}_{33} = \int_x\int_y \overline{\mathbf{N}}^T I_0\overline{\mathbf{N}}dxdy \tag{A10}$$

$$\mathbf{M}_{41} = \mathbf{M}_{14}^T, \quad \mathbf{M}_{44} = \int_x\int_y \overline{\mathbf{N}}^T I_2\overline{\mathbf{N}}dxdy, \quad \mathbf{M}_{46} = \int_x\int_y \overline{\mathbf{N}}^T I_4\overline{\mathbf{N}}dxdy \tag{A11}$$

$$\mathbf{M}_{52} = \mathbf{M}_{25}^T, \quad \mathbf{M}_{55} = \int_x\int_y \overline{\mathbf{N}}^T I_2\overline{\mathbf{N}}dxdy, \quad \mathbf{M}_{57} = \int_x\int_y \overline{\mathbf{N}}^T I_4\overline{\mathbf{N}}dxdy \tag{A12}$$

$$\mathbf{M}_{61} = \mathbf{M}_{16}^T, \quad \mathbf{M}_{64} = \mathbf{M}_{46}^T, \quad \mathbf{M}_{66} = \int_x\int_y \overline{\mathbf{N}}^T I_5\overline{\mathbf{N}}dxdy \tag{A13}$$

$$\mathbf{M}_{72} = \mathbf{M}_{27}^T, \quad \mathbf{M}_{75} = \mathbf{M}_{57}^T, \quad \mathbf{M}_{77} = \int_x\int_y \overline{\mathbf{N}}^T I_5\overline{\mathbf{N}}dxdy \tag{A14}$$

The null submatrices ($\mathbf{M}_{ij}^{(e)} = 0$) are omitted in definitions (A8)–(A14). The full integration scheme is always used to compute the elements of the mass matrix.

It can be easily noted from definitions (A1)–(A14) that the stiffness and mass matrices are both symmetrical operators, thus $\mathbf{K}_{ij}^{(e)} = \mathbf{K}_{ji}^{(e)}$, $\mathbf{M}_{ij}^{(e)} = \mathbf{M}_{ji}^{(e)}$. It should be recalled that the same definitions $\mathbf{K}_{ij}^{(e)}$, $\mathbf{M}_{ij}^{(e)}$ are valid also for the well-known RM theory, for $i, j = 1, \ldots, 5$.

References

1. Duncan, W.J.; Collar, A.R. A method for the solution of oscillations problems by matrices. *Phil. Mag.* **1934**, *17*, 865–909. [CrossRef]
2. Duncan, W.J.; Collar, A.R. Matrices applied to the motions of damped systems. *Phil. Mag.* **1935**, *19*, 197–219. [CrossRef]
3. Hrennikoff, A. Solution of Problems of Elasticity by the Frame–Work Method. *ASME J. Appl. Mech.* **1941**, *8*, A619–A715.
4. Courant, R. Variational methods for the solution of problems of equilibrium and vibration. *B. Am. Math. Soc.* **1943**, *49*, 1–23. [CrossRef]
5. Clough, R.W. The finite element method in plane stress analysis. In Proceedings of the 2nd ASCE conference in electronics computation, Pittsburgh, PA, USA, 8–9 September 1960.
6. Melosh, R.J. Basis for derivation of matrices for the direct stiffness method. *AIAA J.* **1963**, *1*, 1631–1637. [CrossRef]
7. Oden, J.T. *Finite Elements of Nonlinear Continua*; McGraw-Hill: New York, NY, USA, 1972.
8. Oden, J.T.; Reddy, J.N. *An Introduction to the Mathematical Theory of Finite Elements*; John Wiley: New York, NY, USA, 1976.
9. Hinton, E. *Numerical Methods and Software for Dynamic Analysis of Plates and Shells*; Pineridge Press: Swansea, UK, 1988.
10. Zienkiewicz, O.C. *The Finite Element Method*; McGraw–Hill: New York, NY, USA, 1991.
11. Reddy, J.N. *An Introduction to the Finite Element Method*; McGraw–Hill: New York, NY, USA, 1993.
12. Hughes, T.J.R. *The Finite Element Method-Linear Static and Dynamic Finite Element Analysis*; Dover Publications: New York, NY, USA, 2000.
13. Ferreira, A.J.M. *MATLAB Codes for Finite Element Analysis*; Springer: New York, NY, USA, 2008.
14. Martínez-Pañeda, E. On the finite element implementation of functionally graded materials. *Materials* **2019**, *12*, 287. [CrossRef]
15. Nguyen, H.N.; Nguyen, T.Y.; Tran, K.V.; Tran, T.T.; Nguyen, T.T.; Phan, V.D.; Do, T.V. A finite element model for dynamic analysis of triple-layer composite plates with layers connected by shear connectors subjected to moving load. *Materials* **2019**, *12*, 598. [CrossRef] [PubMed]
16. Leonetti, L.; Fantuzzi, N.; Trovalusci, P.; Tornabene, F. Scale effects in orthotropic composite assemblies as micropolar continua: A comparison between weak-and strong-form finite element solutions. *Materials* **2019**, *12*, 758. [CrossRef]
17. Liu, P.; Bui, T.Q.; Zhu, D.; Yu, T.T.; Wang, J.W.; Yin, S.H.; Hirose, S. Buckling failure analysis of cracked functionally graded plates by a stabilized discrete shear gap extended 3-node triangular plate element. *Compos. Part B Eng.* **2015**, *77*, 179–193. [CrossRef]
18. Hosseini, S.S.; Bayesteh, H.; Mohammadi, S. Thermo-mechanical XFEM crack propagation analysis of functionally graded materials. *Mat. Sci. Eng. A* **2013**, *561*, 285–302. [CrossRef]
19. Yin, S.; Yu, T.; Bui, T.Q.; Liu, P.; Hirose, S. Buckling and vibration extended isogeometric analysis of imperfect graded Reissner-Mindlin plates with internal defects using NURBS and level sets. *Comput. Struct.* **2016**, *177*, 23–38. [CrossRef]
20. Singh, S.K.; Singh, I.V.; Mishra, B.K.; Bhardwaj, G.; Singh, S.K. Analysis of cracked plate using higher-order shear deformation theory: Asymptotic crack-tip fields and XIGA implementation. *Comput. Method. Appl. M.* **2018**, *336*, 594–639. [CrossRef]
21. Lemaitre, J.; Chaboche, J.L. *Mechanics of Solid Materials*; Cambridge University Press: New York, NY, USA, 1990.
22. Reddy, J.N.; Miravete, A. *Practical Analysis of Composite Laminates*; CRC Press: Boca Raton, FL, USA, 1995.
23. Tarantino, A.M. Equilibrium paths of a hyperelastic body under progressive damage. *J. Elast.* **2014**, *114*, 225–250. [CrossRef]
24. Lanzoni, L.; Tarantino, A.M. Damaged hyperelastic membranes. *Int. J. Nonlinear Mech.* **2014**, *60*, 9–22. [CrossRef]
25. Lanzoni, L.; Tarantino, A.M. Equilibrium configurations and stability of a damaged body under uniaxial tractions. *Z. Angew. Math. Phys.* **2015**, *66*, 171–190. [CrossRef]
26. Lanzoni, L.; Tarantino, A.M. A simple nonlinear model to simulate the localized necking and neck propagation. *Int. J. Nonlinear Mech.* **2016**, *84*, 94–104. [CrossRef]

27. Savino, V.; Lanzoni, L.; Tarantino, A.M.; Viviani, M. Simple and effective models to predict the compressive and tensile strength of HPFRC as the steel fiber content and type changes. *Compos. Part B Eng.* **2018**, *137*, 153–162. [CrossRef]
28. Falope, F.O.; Lanzoni, L.; Tarantino, A.M. Modified hinged beam test on steel fabric reinforced cementitious matrix (SFRCM). *Compos. Part B Eng.* **2018**, *146*, 232–243. [CrossRef]
29. Dezi, L.; Menditto, G.; Tarantino, A.M. Homogeneous structures subjected to successive structural system changes. *J. Eng. Mech. ASCE* **1990**, *116*, 1723–1732. [CrossRef]
30. Dezi, L.; Tarantino, A.M. Time dependent analysis of concrete structures with variable structural system. *ACI Mater. J.* **1991**, *88*, 320–324.
31. Dezi, L.; Menditto, G.; Tarantino, A.M. Viscoelastic heterogeneous structures with variable structural system. *J. Eng. Mech. ASCE* **1992**, *119*, 238–250. [CrossRef]
32. Dezi, L.; Tarantino, A.M. Creep in continuous composite beams. Part II: Parametric study. *J. Eng. Mech. ASCE* **1993**, *119*, 2112–2133.
33. Tarantino, A.M. Homogeneous equilibrium configurations of a hyperelastic compressible cube under equitriaxial dead-load tractions. *J. Elast.* **2008**, *92*, 227–254. [CrossRef]
34. Vinson, J.R. *The Behavior of Shells Composed of Isotropic and Composite Materials*; Springer: New York, NY, USA, 1993.
35. Jones, R.M. *Mechanics of Composite Materials*, 2nd ed.; Taylor & Francis: Philadelphia, PA, USA, 1999.
36. Reddy, J.N. *Mechanics of Laminated Composite Plates and Shells-Theory and Analysis*, 2nd ed.; CRC Press: Boca Raton, FL, USA, 2004.
37. Barbero, E.J. *Introduction to Composite Materials Design*; CRC Press: Boca Raton, FL, USA, 2011.
38. Tornabene, F.; Bacciocchi, M.; Fantuzzi, N.; Reddy, J.N. Multiscale approach for three-phase cnt/polymer/fiber laminated nanocomposite structures. *Polym. Compos.* **2019**, *40*, E102–E126. [CrossRef]
39. Bacciocchi, M.; Tarantino, A.M. Time-dependent behavior of viscoelastic three-phase composite plates reinforced by carbon nanotubes. *Compos. Struct.* **2019**, *216*, 20–31. [CrossRef]
40. Popov, V.N.; Van Doren, V.E. Elastic properties of single-walled carbon nanotubes. *Phys. Rev. B* **2000**, *61*, 3078–3084. [CrossRef]
41. Qian, D.; Wagner, G.J.; Liu, W.K.; Yu, M.F.; Ruoff, R.S. Mechanics of carbon nanotubes. *Appl. Mech. Rev.* **2002**, *55*, 495–533. [CrossRef]
42. Fidelus, J.D.; Wiesel, E.; Gojny, F.H.; Schulte, K.; Wagner, H.D. Thermo–mechanical properties of randomly oriented carbon/epoxy nanocomposites. *Compos. Part A Appl. S.* **2005**, *36*, 1555–1561. [CrossRef]
43. Ray, M.C.; Batra, R.C. Effective Properties of Carbon Nanotube and Piezoelectric Fiber Reinforced Hybrid Smart Composites. *J. App. Mech. T. ASME* **2009**, *76*, 034503. [CrossRef]
44. Song, Y.S.; Youn, J.R. Modeling of effective elastic properties for polymer based carbon nanotube composites. *Polymer* **2006**, *47*, 1741–1748. [CrossRef]
45. Coiai, S.; Passaglia, E.; Pucci, A.; Ruggeri, G. Nanocomposites based on thermoplastic polymers and functional nanofiller for sensor applications. *Materials* **2015**, *8*, 3377–3427. [CrossRef]
46. Bhattacharya, M. Polymer Nanocomposites-a comparison between carbon nanotubes, graphene, and clay as nanofillers. *Materials* **2016**, *9*, 262. [CrossRef] [PubMed]
47. Acierno, S.; Barretta, R.; Luciano, R.; Marotti de Sciarra, F.; Russo, P. Experimental evaluations and modeling of the tensile behavior of polypropylene/single-walled carbon nanotubes fibers. *Compos. Struct.* **2017**, *174*, 12–18. [CrossRef]
48. Wang, G.; Wang, Y.; Luo, Y.; Luo, S. Carbon nanomaterials based smart fabrics with selectable characteristics for in–line monitoring of high-performance composites. *Materials* **2018**, *11*, 1677. [CrossRef] [PubMed]
49. Arena, M.; Viscardi, M.; Barra, G.; Vertuccio, L.; Guadagno, L. Multifunctional performance of a nano–modified fiber reinforced composite aeronautical panel. *Materials* **2019**, *12*, 869. [CrossRef] [PubMed]
50. Odegard, G.M.; Gates, T.S.; Wise, K.E.; Park, C.; Siochi, E.J. Constitutive modeling of nanotube–reinforced polymer composites. *Compos. Sci. Technol.* **2003**, *63*, 1671–1687. [CrossRef]
51. Shi, D.L.; Huang, Y.Y.; Hwang, K.C.; Gao, H. The effect of nanotube waviness and agglomeration on the elastic property of carbon nanotube–reinforced composites. *J. Eng. Mater. T. ASME* **2004**, *126*, 250–257. [CrossRef]
52. Eshelby, J.D. The determination of the elastic field of an ellipsoidal inclusion, and related problems. *P. Roy. Soc. Lond. A Mat.* **1957**, *241*, 376–396.
53. Mori, T.; Tanaka, K. Average stress in matrix and average elastic energy of materials with misfitting inclusions. *Acta Metall.* **1973**, *21*, 571–574. [CrossRef]

54. Safaei, B.; Moradi-Dastjerdi, R.; Qin, Z.; Behdinan, K.; Chu, F. Determination of thermoelastic stress wave propagation in nanocomposite sandwich plates reinforced by clusters of carbon nanotubes. *J. Sandw. Struct. Mater.* **2019**. [CrossRef]
55. Halpin, J.C. *Effects of Environmental Factors on Composite Materials*; Technical Report AFML-TR-67-423; Air Force Materials Lab Wright-Patterson AFB: Greene, OH, USA, 1969.
56. Tsai, S.W. *Structural Behavior of Composite Materials*; Philco Corporation: Newport Beach, CA, USA, 1964.
57. Tsai, S.W. *Strength Characteristics of Composite Materials*; Philco Corporation: Newport Beach, CA, USA, 1965.
58. Hill, R. Theory of mechanical properties of fibre-strengthened materials: I. Elastic behavior. *J. Mech. Phys. Solids* **1964**, *12*, 199–212. [CrossRef]
59. Hill, R. Theory of mechanical properties of fibre–strengthened materials: II. Inelastic behavior. *J. Mech. Phys. Solids* **1964**, *12*, 213–218. [CrossRef]
60. Thostenson, E.T.; Li, W.Z.; Wang, D.Z.; Ren, Z.F.; Chou, T.W. Carbon nanotube/carbon fiber hybrid multiscale composites. *J. Appl. Phys.* **2002**, *91*, 6034–6037. [CrossRef]
61. Bekyarova, E.; Thostenson, E.T.; Yu, A.; Kim, H.; Gao, J.; Tang, J.; Hahn, H.T.; Chou, T.W.; Itkis, M.E.; Haddon, R.C. Multiscale carbon nanotube-carbon fiber reinforcement for advanced epoxy composites. *Langmuir* **2007**, *23*, 3970–3974. [CrossRef] [PubMed]
62. Kim, M.; Park, Y.B.; Okoli, O.I.; Zhang, C. Processing, characterization, and modeling of carbon nanotube-reinforced multiscale composites. *Compos. Sci. Technol.* **2009**, *69*, 335–342. [CrossRef]
63. Rafiee, M.; He, X.Q.; Mareishi, S.; Liew, K.M. Modeling and stress analysis of smart CNTs/fiber/polymer multiscale composite plates. *Int. J. Appl. Mech.* **2014**, *6*, 1450025. [CrossRef]
64. Chamis, C.C.; Sendeckyj, G.P. Critique on theories predicting thermoelastic properties of fibrous composites. *J. Compos. Mater.* **1968**, *2*, 332–358. [CrossRef]
65. Hill, R. Theory of mechanical properties of fibre–strengthened materials: III. Self–consistent model. *J. Mech. Phys. Solids* **1964**, *13*, 189–198. [CrossRef]
66. Chou, T.W. A self-consistent approach to the elastic stiffness of short–fiber composites. *J. Compos. Mater.* **1980**, *14*, 178–188. [CrossRef]
67. Hashin, Z. The elastic moduli of heterogeneous materials. *J. Appl. Mech. T. ASME* **1962**, *29*, 143–150. [CrossRef]
68. Hashin, Z.; Rosen, B.W. The elastic moduli of fiber-reinforced materials. *J. Appl. Mech. T. ASME* **1964**, *31*, 223–232. [CrossRef]
69. Ekvall, J.C. *Elastic Properties of Orthotropic Monofilament Laminates*; Lockheed Aircraft Corporation: Burbank, CA, USA, 1961; 61-AV-56.
70. Chen, C.H.; Cheng, S. Mechanical properties of fiber reinforced composites. *J. Compos. Mater.* **1967**, *1*, 30–41. [CrossRef]
71. Lei, Z.X.; Liew, K.M.; Yu, J.L. Free vibration analysis of functionally graded carbon nanotube-reinforced composite plates using the element-free kp-Ritz method in thermal environment. *Compos Struct.* **2013**, *106*, 128–138. [CrossRef]
72. Gusella, F.; Cluni, F.; Gusella, V. Homogenization of dynamic behaviour of heterogeneous beams with random Young's modulus. *Eur. J. Mech. A Solid.* **2019**, *73*, 260–267. [CrossRef]
73. Gusella, F.; Cluni, F.; Gusella, V. Homogenization of the heterogeneous beam dynamics: The influence of the random Young's modulus mixing law. *Compos. Part B Eng* **2019**, *167*, 608–614. [CrossRef]
74. Reddy, J.N.; Chin, C.D. Thermomechanical analysis of functionally graded cylinders and plates. *J. Therm. Stresses* **1998**, *21*, 593–626. [CrossRef]
75. Reddy, J.N. Analysis of functionally graded plates. *Int. J. Numer. Meth. Engng.* **2000**, *47*, 663–684. [CrossRef]
76. Vel, S.S.; Batra, R.C. Three-dimensional exact solution for the vibration of functionally graded rectangular plates. *J. Sound Vib.* **2004**, *272*, 703–730. [CrossRef]
77. Batra, R.C.; Jin, J. Natural frequencies of a functionally graded rectangular plate. *J. Sound Vib.* **2005**, *282*, 509–516. [CrossRef]
78. Kim, J.; Reddy, J.N. A general third-order theory of functionally graded plates with modified couple stress effect and the von Kármán nonlinearity: Theory and finite element analysis. *Acta Mech.* **2015**, *226*, 2973–2998. [CrossRef]
79. Kim, J.; Reddy, J.N. Modeling of functionally graded smart plates with gradient elasticity effects. *Mech. Adv. Mater. Struct.* **2017**, *24*, 437–447. [CrossRef]

80. Alexandrov, S.; Wang, Y.C.; Lang, L. A theory of elastic/plastic plane strain pure bending of FGM sheets at large strain. *Materials* **2019**, *12*, 456. [CrossRef] [PubMed]
81. Tornabene, F. Free vibration analysis of functionally graded conical, cylindrical shell and annular plate structures with a four-parameter power-law distribution. *Comput. Method. Appl. Mech. Eng.* **2009**, *198*, 2911–2935. [CrossRef]
82. Tornabene, F.; Viola, E. Free vibration analysis of functionally graded panels and shells of revolution. *Meccanica* **2009**, *44*, 255–281. [CrossRef]
83. Sofiyev, A.H.; Kuruoglu, N. Dynamic instability of three-layered cylindrical shells containing an FGM interlayer. *Thin Wall. Struct.* **2015**, *93*, 10–21. [CrossRef]
84. Alibeigloo, A. Thermo elasticity solution of sandwich circular plate with functionally graded core using generalized differential quadrature method. *Compos. Struct.* **2016**, *136*, 229–240. [CrossRef]
85. Civalek, Ö.; Baltacıoglu, A.K. Free vibration analysis of laminated and FGM composite annular sector plates. *Compos. Part B Eng.* **2019**, *157*, 182–194. [CrossRef]
86. Nguyen, H.N.; Tan, T.C.; Luat, D.T.; Phan, V.D.; Thom, D.V.; Minh, P.V. Research on the buckling behavior of functionally graded plates with stiffeners based on the third-order shear deformation theory. *Materials* **2019**, *12*, 1262. [CrossRef] [PubMed]
87. Lanc, D.; Vo, T.P.; Turkalj, G.; Lee, J. Buckling analysis of thin-walled functionally graded sandwich box beams. *Thin Wall. Struct.* **2015**, *86*, 148–156. [CrossRef]
88. Lanc, D.; Turkalj, G.; Vo, T.; Brnic, J. Nonlinear buckling behaviours of thin-walled functionally graded open section beams. *Compos. Struct.* **2016**, *152*, 829–839. [CrossRef]
89. Kim, J.; Zur, K.K.; Reddy, J.N. Bending, free vibration, and buckling of modified couples stress-based functionally graded porous micro-plates. *Compos. Struct.* **2019**, *209*, 879–888. [CrossRef]
90. Barretta, R.; Feo, L.; Luciano, R.; Marotti de Sciarra, F.; Penna, R. Functionally graded Timoshenko nanobeams: A novel nonlocal gradient formulation. *Compos. Part B Eng.* **2016**, *100*, 208–219. [CrossRef]
91. Apuzzo, A.; Barretta, R.; Faghidian, S.A.; Luciano, R.; Marotti de Sciarra, F. Nonlocal strain gradient exact solutions for functionally graded inflected nano-beams. *Compos. Part B Eng.* **2019**, *164*, 667–674. [CrossRef]
92. Carrera, E. C^0 reissner-mindlin multilayered plate elements including Zig-Zag and interlaminar stress continuity. *Int. J. Numer. Meth. Eng.* **1996**, *39*, 1797–1820. [CrossRef]
93. Carrera, E. Developments, ideas and evaluations based upon the Reissner's mixed theorem in the modeling of multilayered plates and shells. *Appl. Mech. Rev.* **2001**, *54*, 301–329. [CrossRef]
94. Carrera, E. Historical review of Zig-Zag theories for multilayered plates and shells. *Appl. Mech. Rev.* **2003**, *56*, 287–308. [CrossRef]
95. Carrera, E. Theories and finite elements for layered plates and shells: A unified compact formulation with numerical assessment and benchmarking. *Arch. Comput. Meth. Eng.* **2003**, *10*, 215–296. [CrossRef]
96. Carrera, E. On the use of the Murakami's Zig-Zag function in the modeling of layered plates and shells. *Comput. Struct.* **2004**, *82*, 541–554. [CrossRef]
97. Maturi, D.A.; Ferreira, A.J.M.; Zenkour, A.M.; Mashat, D.S. Analysis of laminated shells by murakami's Zig-Zag theory and radial basis functions collocation. *J. Appl. Math.* **2013**, *2013*, 14. [CrossRef]
98. Brischetto, S.; Carrera, E.; Demasi, L. Improved bending analysis of sandwich plates using a Zig-Zag function. *Compos. Struct.* **2009**, *89*, 408–415. [CrossRef]
99. Hu, H.; Belouettar, S.; Daya, E.M.; Potier-Ferry, M. Evaluation of kinematic formulations for viscoelastically damped sandwich beam modeling. *J. Sandw. Struct. Mater.* **2006**, *8*, 477–495. [CrossRef]
100. Murakami, H. Laminated composite plate theory with improved in-plane responses. *J. Appl. Mech.* **1986**, *53*, 661–666. [CrossRef]
101. Lenci, S.; Tarantino, A.M. Chaotic dynamics of an elastic beam resting on a Winkler-type soil. *Chaos Soliton. Fract.* **1996**, *7*, 1601–1614. [CrossRef]

materials

Article

Rate-Dependent Cohesive Zone Model for Fracture Simulation of Soda-Lime Glass Plate

Dong Li [1,*] and Demin Wei [1,2]

1 School of Civil Engineering and Transportation, South China University of Technology, Guangzhou 510640, China; dmwei@scut.edu.cn
2 School of Civil Engineering, Guangzhou College of South China University of Technology, Guangzhou 510800, China
* Correspondence: ctdlee@mail.scut.edu.cn; Tel.: +86-155-2142-3684

Received: 29 December 2019; Accepted: 1 February 2020; Published: 6 February 2020

Abstract: In this paper, rate-dependent cohesive zone model was established to numerical simulate the fracture process of soda-lime glass under impact loading. Soda-lime glass is widely used in architecture and automobile industry due to its transparency. To improve the accuracy of fracture simulation of soda-lime glass under impact loading, strain rate effect was taken into consideration and a rate-dependent cohesive zone model was established. Tensile-shear mixed mode fracture was also taken account. The rate-dependent cohesive zone model was implemented in the commercial finite element code ABAQUS/Explicit with the user subroutine VUMAT. The fracture behavior of a monolithic glass plate impacted by a hemispherical impactor was simulated. The simulation results demonstrated that the rate-dependent cohesive zone model is more suitable to describe the impact failure characteristics of a monolithic glass plate, compared to cohesive zone model without consideration of strain rate. Moreover, the effect of the strain rate sensitivity coefficient C, the mesh size of glass plate and the impact velocity on the fracture characteristics were studied.

Keywords: soda-lime glass; cohesive zone model; rate-dependent; impact loading

1. Introduction

Because of its good transparency, soda-lime glass is often used in window panes and curtain wall systems in modern buildings. However, soda lime glass is threatened by extreme loads such as impact loading and blast loading. Under these extreme conditions, due to brittleness, soda-lime glass may suddenly crack, fragment, or even shatter. For example, statistics shows that windborne debris as a major contributor to glass damage when encountering typhoon weather [1]. The study of soda-lime glass materials under impact loading is vital. For this reason, it is necessary to investigate the cracking mechanism of glass window panes.

In recent years, a large number of experimental [2–11], analytic [12–16], and numerical [17–44] studies have been devoted to the fracture of soda lime glass. Nie and Chen [2–5] conducted SHPB experiments to study the effect of the shear stress, mechanical confinement, and temperature on the strength of glass under dynamic loading. Daryadel et al. [6] investigate the influence of the surface flaw on the strength. Johnson et al. [7] and Zhang et al. [9–11] reveals that the compressive strength of soda-lime glass is rate-dependent. Further research shows that not only the compressive strength but also the tensile strength could increase with the increase of strain rate [8,11]. Ji and Dharani [15] developed analytic approaches to simulate the damage probability in laminated glass subjected to low velocity small missile impacts. Dharani et al. [14] proposed a two-parameter Weibull distribution to characterize the cumulative probability of inner glass ply breakage. Zhou et al. [13] proposed an empirical formula to calculate the average fragment size. Overend et al. [12] delivered a general crack growth model based on established statistical failure theory and linear elastic fracture mechanics.

Nurhuda et al. [16] put forward a model to estimation of strengths in large annealed glass panels which contain widely spaced flaws. A large amount of numerical method is proposed to simulate the cracking process of soda-lime glass. For example, André et al. [39] and Braun et al. [40] adopt the discrete element method (DEM) to conduct the simulation. However, finite element method (FEM) is the most commonly used method. The discontinuity modeling is the most important part when FEM is adopted to the fracture simulation of the soda-lime glass. Element delete method (EDM) is the most common way to model the discontinuity. For instance, Bois et al. [41] simulate an impact of a sphere into a glass plate using an element delete method. Liu et al. [22] adopted EDM to investigate the energy absorption process of windshield under different impact speeds and angles. Extended finite element method (XFEM) [23–26] is also applied to the fracture simulation of soda-lime glass.

Cohesive zone model (CZM) has been found widely used in the cracking process simulation [27–38,44]. Barenblatt [31] and Dugdale [32] were the first to proposed the CZM. After that, application of CZM in the simulation of cracking of glass is more and more widespread. Repetto et al. [28] simulated the dynamic fracture and fragmentation of glass rod under impact loading with CZM. A mixed-mode formula is proposed to combined the tensile and shear cracking mode [29]. Intrinsic and extrinsic [30,34–37] cohesive zone models were used to simulate the cracking process of laminated glass.

So far, however, there has been little discussion about the rate-dependent cohesive zone model applied to the failure process of soda-lime glass. However, as the experimental results [8–11] showed that the strength of soda-lime glass could increase with the increase of strain rate. As a result, to ensure the accuracy, a rate-dependent CZM in fracture simulation of soda-lime glass is important and necessary.

The aim of this study is to improve the accuracy of fracture simulation of glass by employing a rate-dependent CZM. The overall structure of the study takes the form of five sections, including this introductory chapter. Section 2 begins by laying out the theoretical methodology of the rate-dependent cohesive zone model. In Section 3, simulation results are presented, including a successful validation against experimental results of a drop-weight test with monolithic glass plate. Besides, the effects of the strain rate sensitivity coefficient C, mesh size and impact velocity are investigated in Sections 4.1–4.3, respectively. Finally, the conclusions are drawn in Section 5.

2. Rate-Dependent Cohesive Zone Model

2.1. Traction–separation Law

To study the cracking process of soda-lime glass plate under impact loading, the rate-dependent cohesive zone model is employed to simulate the discontinuity in the finite element model. Cohesive zone model assumes that there is a small cohesive zone behind the crack tip, as shown in Figure 1 In the cohesive zone model, there is a mathematical relationship between the separation δ and traction t, as shown in Equation (1). This mathematical relationship is called traction–separation law (TSL).

$$t = f(\delta) \tag{1}$$

A rate-dependent traction–separation law is employed in this research, and its normal law is showed in the Figure 2. The tensile stiffness of the cohesive zone will be degraded when the normal separation δ_{n} exceeds to a critical value, δ_{n}^{0} under quasi-static or $\delta_{\mathrm{n,dyna}}^{0}$ under dynamic for example. As the separation continues to increase to the maximum normal separation ($\delta_{\mathrm{n}}^{\mathrm{f}}$ under quasi-static and $\delta_{\mathrm{n,dyna}}^{\mathrm{f}}$ under dynamic), the tensile stiffness decreases to 0 completely, which indicates the formation of macro cracks. The energy required to form a new crack is called fracture energy. The dissipated energy (G) due to failure is equal to the integration of $f(\delta_{\mathrm{n}})$ and δ_{n}. Therefore, the normal fracture energy release rate in Figure 2 is equal to the area enclosed by the traction–separation curve and δ_{n} axis. The normal penalty stiffness K_{n} is the slope of the curve in the undamaged stage.

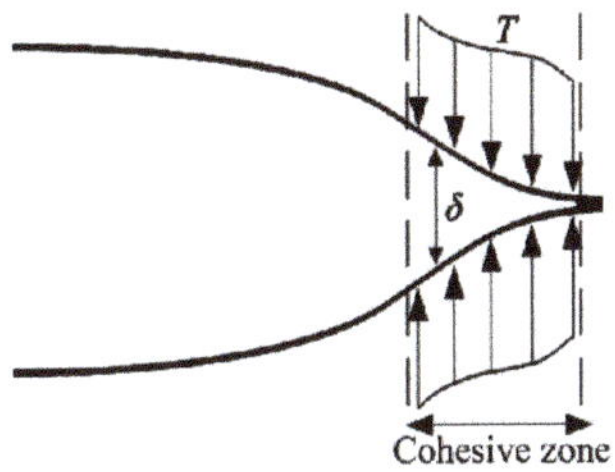

Figure 1. Schematic diagram of cohesive model.

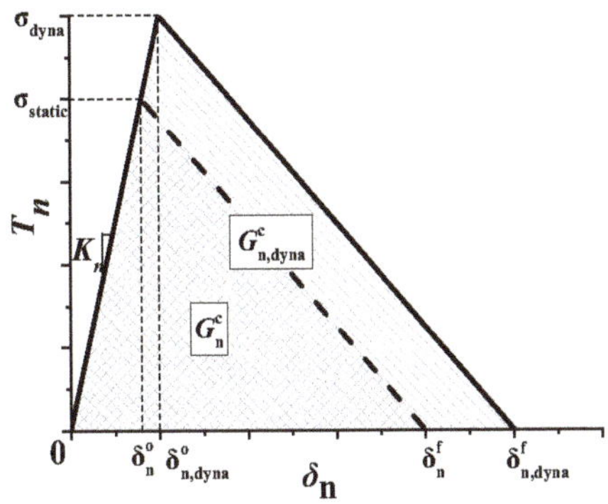

Figure 2. Rate-dependent traction–separation law in normal direction.

Similar to the normal traction–separation law, the shear traction–separation law is defined by a linear relationship between shear traction (τ_s) and the shear separation (δ_s), as shown in Figure 3. The shear stiffness of the cohesive zone will be degraded when the shear separation δ_s exceeds to a critical value, δ_s^0 under quasi-static or $\delta_{s,dyna}^0$ under dynamic for example. As the separation continues to increase to the maximum shear separation δ_s^f (under quasi-static and $\delta_{s,dyna}^f$ under dynamic), the shear stiffness decreases to 0 completely, which indicates the formation of macro cracks. The dissipated energy (G) due to failure is equal to the integration of $f(\delta_s)$ and δ_s. Therefore, the shear fracture energy release rate in Figure 2 is equal to the area enclosed by the traction–separation curve and δ_s axis. The shear penalty stiffness K_s is the slope of the curve in the undamaged stage.

Under dynamic loading, the soda-lime glass is rate dependent [8,10,11]. In JH-2 constitutive model, Holmquist et al. [7] suggested the DIF (dynamic increment factor) should be formulated in Equation (2).

$$DIF = 1.0 + C\ln(\dot{\varepsilon}) \tag{2}$$

where the C is the strain rate sensitivity coefficient.

Therefore, the dynamic strength could be expressed by the Equations (3) and (4).

$$\sigma_{dyna} = DIF\sigma_{static} \tag{3}$$

$$\tau_{dyna} = DIF\tau_{static} \tag{4}$$

The maximum normal separation δ_n^f and the maximum shear separation δ_s^f under quasi-static could be calculated by the critical energy divided by stress. By multiplying DIF, we could obtain the dynamic maximum separation $\delta_{n,dyna}^f$ and $\delta_{s,dyna}^f$, respectively, as shown in Equations (5) and (6).

$$\delta_{n,dyna}^f = \frac{2G_n^c}{\sigma_{static}}DIF \tag{5}$$

$$\delta^{f}_{s,dyna} = \frac{2G^{c}_{s}}{\tau_{static}} DIF \tag{6}$$

The dynamic critical energy could be calculated by Equations (7) and (8), according to its definition.

$$G^{c}_{n,dyna} = 2\sigma_{dyna}\delta^{f}_{n,dyna} \tag{7}$$

$$G^{c}_{s,dyna} = 2\tau_{dyna}\delta^{f}_{s,dyna} \tag{8}$$

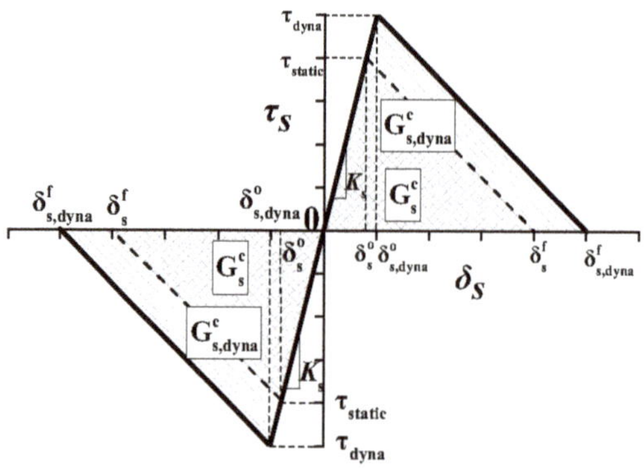

Figure 3. Rate-dependent TSL for shear direction.

2.2. Mixed-Mode Failure

The mode mix of the deformation fields in the cohesive zone quantify the relative proportions of normal and shear deformation. Under impact loading, mixed-mode cracking is very common. In mixed-mode failure, the maximum separation until failure δ^{f}_{m} is expressed by Equation (9).

$$\delta^{f}_{m} = \frac{2G^{c}_{m}}{\sigma^{0}_{m}} \tag{9}$$

where the G^{c}_{m} is mixed-mode critical failure energy, calculated by the Benzeggagh–Kenane fracture criterion [45]. The G^{c}_{m} could be calculated by Equation (10). In Equation (9), the σ^{0}_{m} is the mixed-mode separation corresponding the damage initiation, which could be calculated by Equation (11) [29].

$$G^{c}_{m} = G^{c}_{n,dyna} + (G^{c}_{s,dyna} - G^{c}_{n,dyna})\left(\frac{\beta^{2}}{1+\beta^{2}}\right)^{\eta} \tag{10}$$

$$\sigma^{0}_{m} = K_{n}\delta^{0}_{s,dyna}\delta^{0}_{n,dyna}\sqrt{\frac{1+\beta^{2}}{\left(\delta^{0}_{s,dyna}\right)^{2} + \left(\beta^{2}\delta^{0}_{n,dyna}\right)^{2}}} \tag{11}$$

where $\beta = \frac{\delta_{s}}{\delta_{n}}$. $\delta^{0}_{n,dyna}$ and $\delta^{0}_{s,dyna}$ are initial damage separations under single failure mode, which could be calculated by Equations (12) and (13).

$$\delta^{0}_{n,dyna} = \frac{\sigma_{dyna}}{K_{n}} \tag{12}$$

$$\delta^{0}_{s,dyna} = \frac{\tau_{dyna}}{K_{s}} \tag{13}$$

where K_{n} and K_{s} are the penalty stiffness of normal and shear mode, respectively.

Under mixed-mode failure, the damage is defined as

$$D = \frac{\delta_m^f \left(\delta_m^{max} - \delta_m^0\right)}{\delta_m^{max}\left(\delta_m^f - \delta_m^0\right)} \tag{14}$$

where the δ_m^{max} refers to the maximum value of the effective separation δ_m attained during the loading history. δ_m could be expressed by

$$\delta_m = \sqrt{\langle\delta_n\rangle^2 + (\delta_s)^2} \tag{15}$$

where δ_n and δ_s are the immediate normal and shear separation, respectively, and $<x>$ is the Macauley operator, which could be defined as

$$\langle x\rangle = \begin{cases} 0, x < 0 \\ x, x > 0 \end{cases} \tag{16}$$

3. Simulation Results and Experimental Validation

3.1. Set Up of Experimental Test

Pauw [46] conducted drop-weight test with a monolithic glass specimen, which is a circular soda-lime glass plate with a radius of 50 mm and a thickness of 4 mm. The impactor is a steel cylinder with a 10 mm radius ending in a spherical tip [46]. If the boundary of the fixed plate is too rigid, it is easy to lead to stress concentration, which causes the plate to failure prematurely. Therefore, soft cushions are placed around the upper and lower parts of the plate. The cushions are annular, with a thickness of 3 mm, an inner radius of 46 mm and an outer radius of 54 mm. The schematic diagram of the experimental device is shown in Figure 4. Due to the symmetry, only one-quarter model is showed, and the model is symmetrical about the plane of *x0y* and *y0z*.

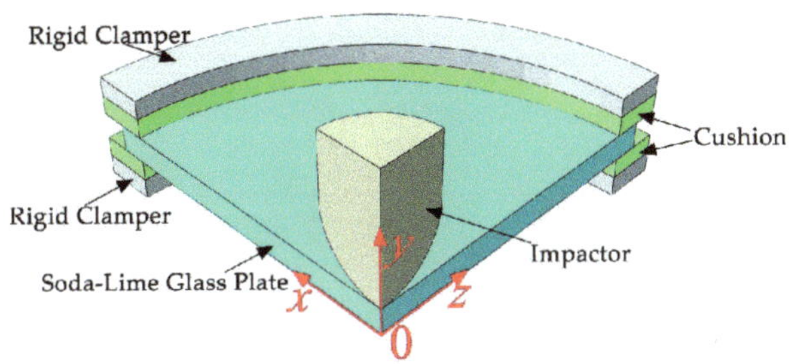

Figure 4. Schematic diagram of basic test device.

During the test, the acceleration time history curve of the punch is recorded by the sensor. The drop height of weight is 200 mm, and the velocity of contact plate is 1.98 m/s. The punch mass is 6.84 kg. The theoretical impact kinetic energy can be calculated as 13407.7 mJ by the equation $E_k = \frac{1}{2}mv^2$.

3.2. Finite Element Model

Eight solid node elements are employed to discrete the soda-lime glass plate. The element size of glass plate is 0.7 mm × 0.7 mm. The mesh model is showed in the Figure 5. Due to the symmetry, only one-quarter model is showed. Because the location of cracks cannot be predicted, zero thickness cohesive elements are inserted between every two solid elements.

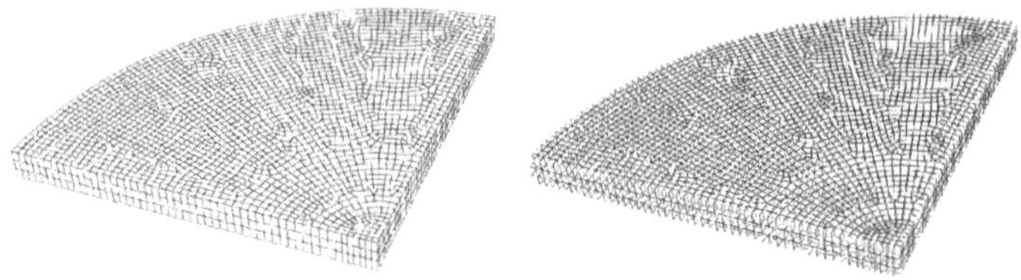

(a) (b)

Figure 5. One-quarter mesh model of the plate (**a**) solid element mesh; (**b**) cohesive element mesh.

The frictionless hard contact is defined between the impactor and plate, and between the plate and the soft cushion, and between the soft cushion and the rigid clamper. The rigid clamper is fully fitted in translational and rotational degrees of freedom of x-, y-, and z-axis. The boundary conditions of the glass plate and the soft cushion are free. The initial velocity of impactor is 1.98 m/s, which is consistent with the experimental test. Symmetric boundary conditions are set on all symmetric surfaces.

Due to their stiffness is a lot larger than other parts, rigid bodies are employed to model impactor and the clamper. The material of soft cushion is polypropylene and a linear elastic constitutive model is adopted in simulation. Wedge element is used to mesh the impactor. If the size of wedge element is too large, there would be a loss of mass, which could result in the loss of kinetic energy. Table 1 depicts the errors corresponding to different mesh size of impactor. As could be seen in Table 1, the error will become larger when the mesh size of impactor increase. In order to ensure the accuracy, the wedge element size should be controlled lower than 2 mm. In this study, the element size of impactor near the impact area is set to 0.7 mm × 0.7 mm × 0.7 mm, whereas the other area is 2 mm × 2 mm × 2 mm.

Table 1. Relationship between the mesh size of impactor and the kinetic energy

Mesh Size of Impactor (mm)	Kinetic Energy (mJ)	Error (%)
1	13382.3	0.19
2	13277.1	0.97
3	12922.2	3.62
4	12864.9	4.05

3.3. *Result Analysis*

3.3.1. Elastic Result

To verify the accuracy of the FE model, including boundary conditions and so on, a simulation with elastic model is conducted. The constants of elastic model are shown in Table 2. Figure 6 shows the acceleration time curves of impactor. Compared to the experimental test results, the simulation result is a bit higher but keep the similar pattern. The reflection of stress wave on the boundary leads to the fluctuation of the curve, which could be found both in the experimental result and the simulation result.

Table 2. Constants of the linear elastic model for glass and polypropylene

Material	Density (kg/m^3)	Young's Modulus (GPa)	Poisson's Ratio
Soda-lime glass	2530	73	0.3
Polypropylene	1000	2	0.3

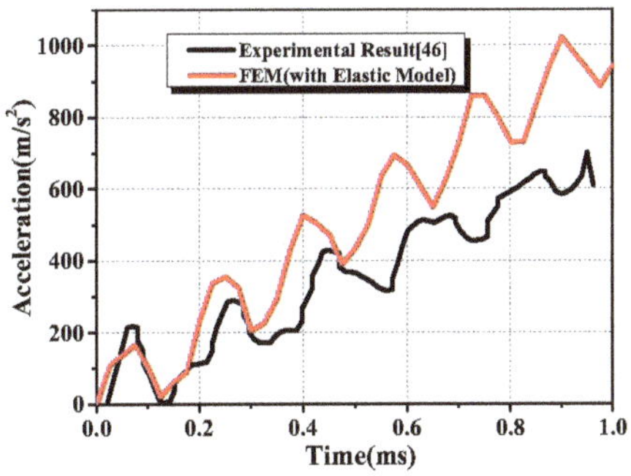

Figure 6. Acceleration-time curves of impactor.

3.3.2. Rate-Dependent CZM Results

Table 3 depicts the constants of rate-dependent cohesive zone model adopted in this study. In this section, numerical simulations are carried out with the rate-dependent CZM. In the simulation, cohesive element will be removed if damage is equal to 1. The rate-dependent cohesive zone model was implemented in the commercial finite element code ABAQUS/Explicit with the user subroutine VUMAT.

Table 3. Constants of rate-dependent cohesive zone model.

Properties	Reference Parameters
σ_{static}	60 MPa [25,37]
τ_{static}	250 MPa [25,37]
G_n^c	0.01 mJ/mm^2 [25]
G_s^c	0.05 mJ/mm^2 [25]
K_{n}	1.8 × 10^6 MPa/mm [37]
K_{s}	6.25 × 10^6 MPa/mm [37]
C	0.03

Figure 7 shows the simulation result conducted with rate-dependent cohesive zone model. The fracture mode of the simulation result agrees well with the experimental result. In the simulation result, the impact area of the plate is found obvious smash. Circle crack is found near the impact area, which thoroughly penetrated the thickness of the plate. Therefore, lots of fragments are formed. In the experimental test results, circle of cracks is also could be found, which results in a large area of vacancy. This phenomenon has also been well simulated in numerical test. In addition, in the experimental test, radial cracks appeared in the plate, which penetrated the whole plate and extended to the edge of the plate. The numerical results have the same crack pattern. To sum up, it can be concluded that the rate-dependent cohesive zone model can accurately simulate the fracture mode of the impact failure of the soda-lime glass plate.

After treated by the image processing method, we could obtain the front view of Figure 7, as shown in Figure 8. The numerical test results are symmetrical. However, due to the irregular microcracks on the surface of the glass plate in the laboratory test, the results of the laboratory test are not symmetrical. In Figure 8a, because of the symmetry, the cracks are regular. The fragments in Figure 8a in the middle are approximately triangular, and the maximum size of the fragment is 25.0 mm. The maximum size of the fragment obtained in the experimental test is about 25.2 mm. Therefore, we could conclude that the rate-dependent cohesive zone model could accurately calculate the size of the fragments under impact loading.

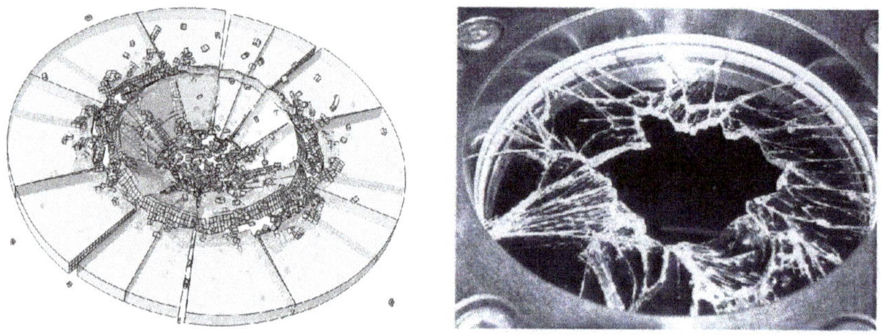

(a) (b)

Figure 7. Comparison of the numerical test result and laboratory test result. (**a**) Numerical result; (**b**) experimental result [46].

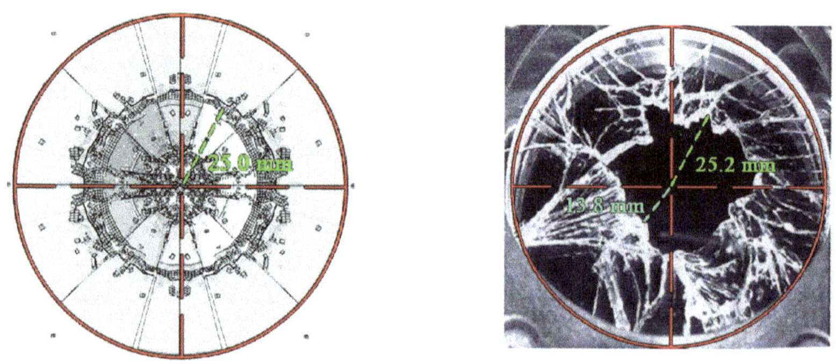

(a) (b)

Figure 8. Comparison of the size of fragments. (**a**) Numerical result; (**b**) experimental result [46].

Figure 9 depicts the velocity-time curves of impactor. Between 0.0 ms and 0.2 ms, the curve of experimental test, simulation result with rate-dependent CZM and simulation result with CZM agree well. However, after 0.2 ms, the velocity obtained by CZM is obvious lower than the experimental results. The simulation result conducted with rate-dependent CZM is able to predict the velocity well as shown in Figure 9.

Figure 10 shows the fracture process of the glass plate. As could be seen in the Figure 10a, the radial cracks are the first to appear. After that, inner circle cracks could be found in the impact area, as shown in Figure 10c. Outer circle cracks are found at around 2.0 ms. At this moment, all the cracks are formed basically. We could infer that the formation of radial cracks is mainly because of the discontinuity of angular velocity. The formation of inner circle cracks is because of the punch shear, whereas the formation of outer circle cracks is because of the bending.

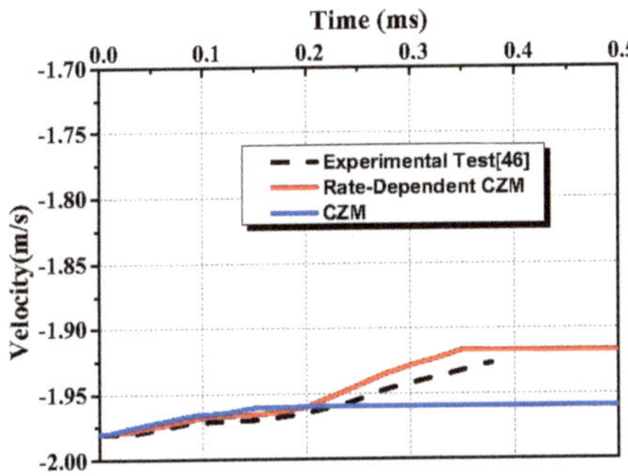

Figure 9. Velocity–time curves of impactor.

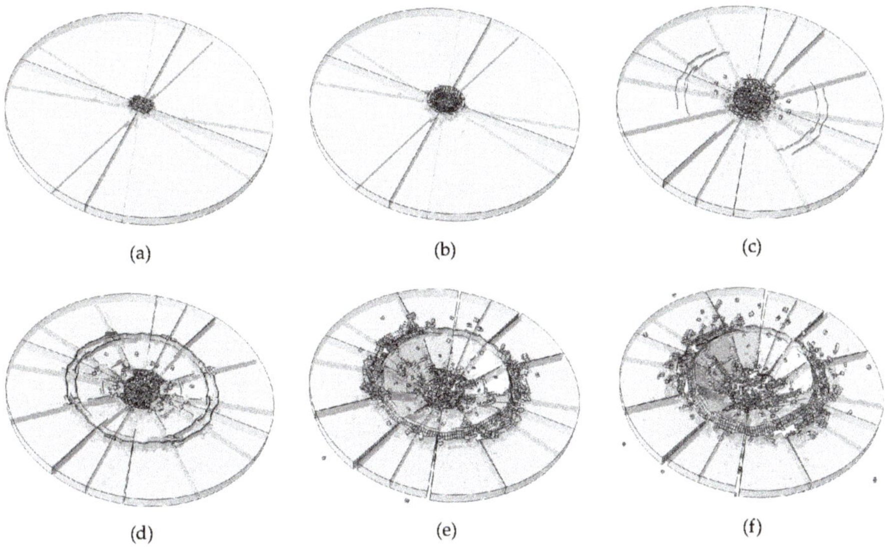

Figure 10. Failure process of the plate (**a**) 0.5 ms; (**b**) 1.0 ms; (**c**) 2.0 ms; (**d**) 2.5 ms; (**e**) 4.0 ms; (**f**) 5.0 ms.

4. Parametric Studies

4.1. Strain Rate Sensitivity Coefficient C

In Equation (2), we employed a strain rate sensitivity coefficient C to consider the effect of strain rate on the fracture strength. Five numerical tests are carried out in this section, in order to study the effect of C on the cracking pattern. The numerical results are shown in Figure 11. In Figure 11a, the coefficient C is 0, which indicates that the strength of glass is rate independent. It could be found in Figure 11a that the fracture gathers near the impact area, which is inconsistent with the actual situation shown in Figure 7b. In Figure 11b, most of the energy dissipation is gather near the impact point as well even though the strain rate effect is put into consideration. When the coefficient C is 0.02, we could see that the outer circle crack is obvious. The crack pattern when C = 0.03 is the most consistent with the experimental result. The strain rate effect forms a shielding zone in the region with high strain rate, which makes the energy propagate outward. However, when C = 0.04, the strength of soda-lime glass becomes too high making the fracture area too small compared with Figure 11d. We could also find that the number of small fragments when C = 0.04 is obviously less than that in Figure 11d. This can also be attributed to the strength of the glass is too high when C = 0.04.

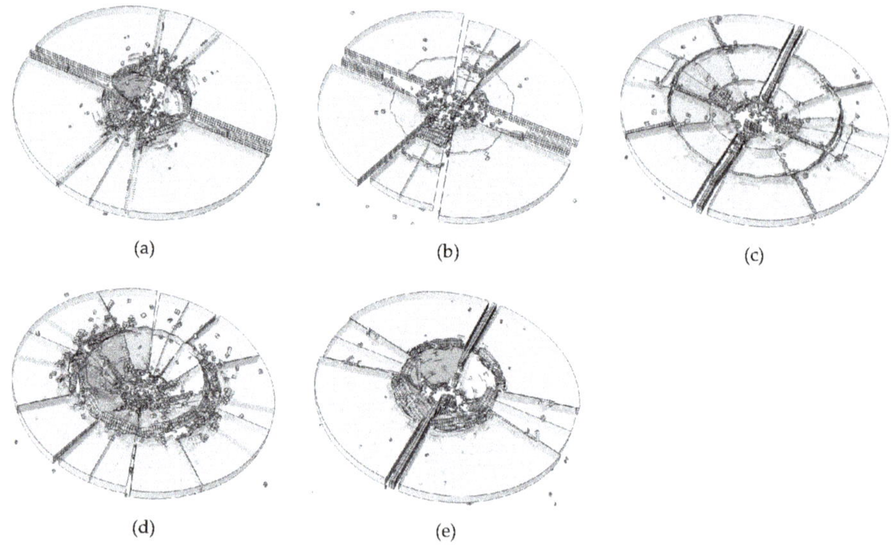

Figure 11. Effect of C on the results (a) C = 0.00; (b) C = 0.01; (c) C = 0.02; (d) C = 0.03; (e) C = 0.04.

4.2. Mesh Size

Two different mesh is conducted in this section, including a fine mesh with element size of 0.7 mm and a coarse mesh with element size of 2 mm. The results are shown in Figure 12. The crack pattern of glass plate is similar. Radial cracks and circle crack can be found in both of numerical result. Fewer small fragments are found in the coarse mesh result. However, the large fragments of both the two results show agreement. We could conclude that the rate-dependent cohesive zone model has little dependence on the mesh size.

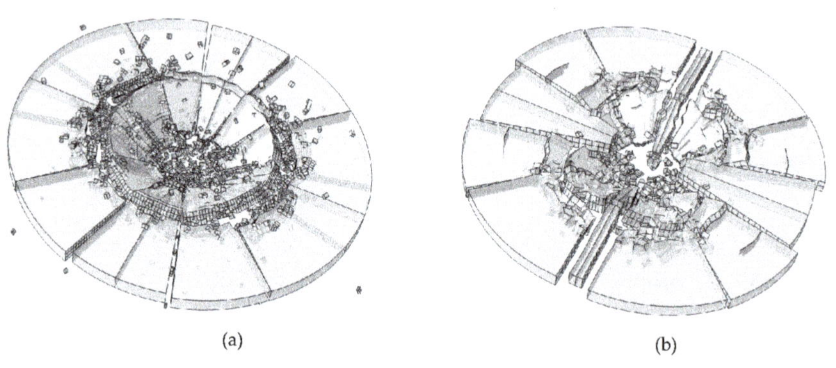

Figure 12. Effect of mesh size on the results. (a) Fine mesh; (b) coarse mesh.

4.3. Impact Velocity

Figure 13 depicts the numerical results conducted with 5 different impact velocity. When the velocity is 0.50 m/s and 1.00 m/s, no outer circle cracks could be found in the result, as the Figure 13a,b shown. As the velocity increase, the outer circle cracks appear when velocity is 1.98 m/s, 3.00 m/s and 4.00 m/s. Thus, we can conclude that more kinetic energy is needed to form a circular crack compared with a radial crack.

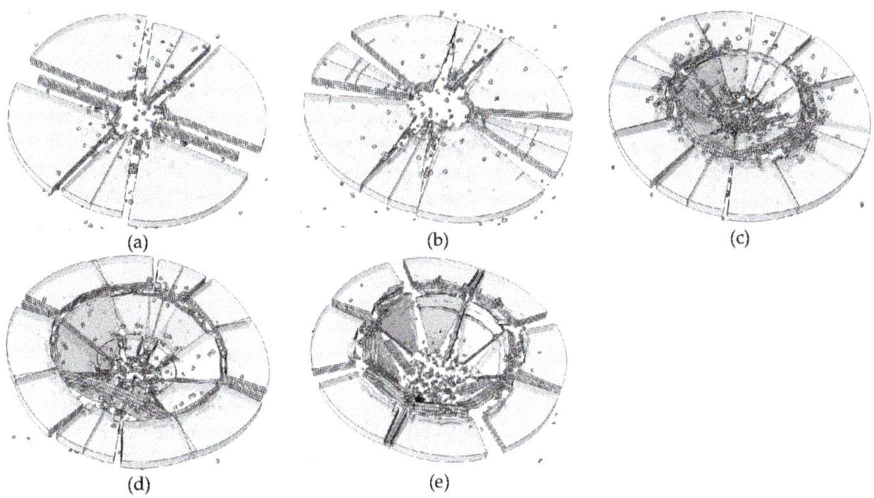

Figure 13. Effect of initial velocity on the results (**a**) v0 = 0.50 m/s; (**b**) v0 = 1.00 m/s; (**c**) v0 = 1.98 m/s; (**d**) v0 = 3.00 m/s; (**e**) v0 = 4.00 m/s.

5. Conclusions

In this paper, we have established a rate-dependent cohesive zone model for the fracture simulation of soda-lime glass plate. A monolithic glass plate is simulated in this paper conducted with the rate-dependent CZM and the simulation has been validated by the experimental results. Finally, parametric studies are carried out. The main conclusions found of this research are as follows:

(1) Soda-lime glass is a rate dependent material. It is necessary to consider the rate dependence of glass when simulating fracture under impact loading. The numerical simulation results show that the rate-dependent CZM considering the strain rate is more accurate.
(2) The rate-dependent CZM is applied to the finite element simulation by VUMAT. In the rate-dependent CZM, the maximum traction and the maximum separation are rate dependent, which results in that the cohesive element dissipate more energy when it is damaged. The strain rate effect forms a shielding zone in the region with high strain rate, which makes the energy propagate outward.
(3) The rate-dependent cohesive zone model has little dependence on the mesh size. Furthermore, more kinetic energy is needed to form a circular crack compared with a radial crack.

Author Contributions: Conceptualization, D.L.; methodology, D.L.; software, D.L.; validation, D.L., D.W; formal analysis, D.W.; investigation, D.W.; resources, D.L.; data curation, D.L.; writing—original draft preparation, D.L.; writing—review and editing, D.L.; visualization, D.L.; supervision, D.W.; project administration, D.W.; funding acquisition, D.W. All authors have read and agreed to the published version of the manuscript.

Funding: The authors would like to gratefully acknowledge research grants from the National Natural Science Foundation of China (Project no. 11672110).

Conflicts of Interest: The authors declare that there is no conflict of interest regarding the publication of this paper.

References

1. Minor, J.E. Lessons Learned from Failures of the Building Envelope in Windstorms. *J. Archit. Eng.* **2005**, *11*, 10–13. [CrossRef]
2. Nie, X.; Chen, W.W.; Sun, X.; Templeton, D.W. Dynamic Failure of Borosilicate Glass Under Compression/Shear Loading Experiments. *J. Am. Ceram. Soc.* **2007**, *90*, 2556–2562. [CrossRef]

3. Nie, X.; Chen, W. High-Rate Progressive Failure of Borosilicate Glass under Mechanical Confinement at High Temperatures. *Exp. Mech.* **2013**, *53*, 67–75. [CrossRef]
4. Nie, X.; Chen, W.W. Dynamic Equibiaxial Flexural Strength of Borosilicate Glass at High Temperatures. *Exp. Mech.* **2012**, *52*, 135–143. [CrossRef]
5. Sun, X.; Liu, W.; Chen, W.; Templeton, D. Modeling and characterization of dynamic failure of borosilicate glass under compression/shear loading. *Int. J. Impact. Eng.* **2009**, *36*, 226–234. [CrossRef]
6. Daryadel, S.S.; Mantena, P.R.; Kim, K.; Stoddard, D.; Rajendran, A.M. Dynamic response of glass under low-velocity impact and high strain-rate SHPB compression loading. *J. Non Cryst. Solids* **2016**, *432*, 432–439. [CrossRef]
7. Holmquist, T.J.; Johnson, G.R.; Lopatin, C.M.; Grandy, D.E. High strain rate properties and constitutive modeling of glass. In Proceedings of the 15th International Symposium on Ballistics, Jerusalem, Israel, 21–24 May 1995; pp. 237–244.
8. Peroni, M.; Solomos, G.; Pizzinato, V.; Larcher, M. Experimental Investigation of High Strain-Rate Behaviour of Glass. *Appl. Mech. Mater.* **2011**, *68*, 63–68. [CrossRef]
9. Zhang, X.; Zou, Y.; Hao, H.; Li, X.; Ma, G.; Liu, K. Laboratory Test on Dynamic Material Properties of Annealed Float Glass. *Int. J. Prot. Struct.* **2012**, *3*, 407–430. [CrossRef]
10. Zhang, X.; Hao, H.; Ma, G. Dynamic material model of annealed soda-lime glass. *Int. J. Impact Eng.* **2015**, *77*, 108–119. [CrossRef]
11. Zhang, X.; Hao, H.; Ma, G. Laboratory test and numerical simulation of laminated glass window vulnerability to debris impact. *Int. J. Impact Eng.* **2013**, *55*, 49–62. [CrossRef]
12. Overend, M.; Parke, G.A.; Buhagiar, D. Predicting Failure in Glass-A General Crack Growth Model. *J. Struct. Eng.* **2007**, *133*, 1146–1155. [CrossRef]
13. Zhou, F.; Molinari, J.; Ramesh, K.T. Characteristic fragment size distributions in dynamic fragmentation. *Appl. Phys. Lett.* **2006**, *88*, 261918. [CrossRef]
14. Dharani, L.R.; Ji, F.; Behr, R.A.; Minor, J.E.; Kremer, P.A. Breakage Prediction of Laminated Glass Using the "Sacrificial Ply" Design Concept. *J. Archit. Eng.* **2004**, *10*, 126–135. [CrossRef]
15. Ji, F.S.; Dharani, L.R.; Behr, R.A. Damage probability in laminated glass subjected to low velocity small missile impacts. *J. Mater. Sci.* **1998**, *33*, 4775–4782. [CrossRef]
16. Nurhuda, I.; Lam, N.T.K.; Gad, E.F.; Calderone, I. Estimation of strengths in large annealed glass panels. *Int. J. Solids Struct.* **2010**, *47*, 2591–2599. [CrossRef]
17. Pyttel, T.; Liebertz, H.; Cai, J. Failure criterion for laminated glass under impact loading and its application in finite element simulation. *Int. J. Impact Eng.* **2011**, *38*, 252–263. [CrossRef]
18. Xu, X.Q.; Liu, B.H.; Wang, Y.; Li, Y.B.; Xu, J. A numerical model on PVB laminated windshield subjected to headform low-speed impact. *J. Phys. Conf. Ser.* **2013**, *451*, 12016. [CrossRef]
19. Zhang, X.; Hao, H.; Ma, G. Parametric study of laminated glass window response to blast loads. *Eng. Struct.* **2013**, *56*, 1707–1717. [CrossRef]
20. Peng, Y.; Yang, J.; Deck, C.; Willinger, R. Finite element modeling of crash test behavior for windshield laminated glass. *Int. J. Impact Eng.* **2013**, *57*, 27–35. [CrossRef]
21. Hidallana-Gamage, H.D.; Thambiratnam, D.P.; Perera, N.J. Numerical modelling and analysis of the blast performance of laminated glass panels and the influence of material parameters. *Eng. Fail. Anal.* **2014**, *45*, 65–84. [CrossRef]
22. Liu, B.; Xu, T.; Xu, X.; Wang, Y.; Sun, Y.; Li, Y. Energy absorption mechanism of polyvinyl butyral laminated windshield subjected to head impact: Experiment and numerical simulations. *Int. J. Impact Eng.* **2016**, *90*, 26–36. [CrossRef]
23. Jaśkowiec, J. Numerical Modeling Mechanical Delamination in Laminated Glass by XFEM. *Procedia Eng.* **2015**, *108*, 293–300. [CrossRef]
24. Xu, J.; Li, Y.; Chen, X.; Yan, Y.; Ge, D.; Zhu, M.; Liu, B. Characteristics of windshield cracking upon low-speed impact: Numerical simulation based on the extended finite element method. *Comp. Mater. Sci.* **2010**, *48*, 582–588. [CrossRef]
25. Xu, X.; Xu, J.; Chen, J.; Li, P.; Liu, B.; Li, Y. Investigation of dynamic multi-cracking behavior in PVB laminated glass plates. *Int. J. Impact Eng.* **2017**, *100*, 62–74. [CrossRef]
26. Ghaffari, D.; Rash Ahmadi, S.; Shabani, F. XFEM simulation of a quenched cracked glass plate with moving convective boundaries. *C. R. Mécanique* **2016**, *344*, 78–94. [CrossRef]

27. Xu, X.P.; Needleman, A. Void nucleation by inclusion debonding in a crystal matrix. *Model. Simul. Mater. Sci. Eng.* **1993**, *1*, 111–132. [CrossRef]
28. Repetto, E.A.; Radovitzky, R.; Ortiz, M. Finite element simulation of dynamic fracture and fragmentation of glass rods. *Comput. Methods Appl Mech. Eng.* **2000**, *183*, 3–14. [CrossRef]
29. Camanho, P.P.; Davila, C.G.; de Moura, M.F. Numerical Simulation of Mixed-Mode Progressive Delamination in Composite Materials. *J. Compos. Mater.* **2003**, *37*, 1415–1438. [CrossRef]
30. Gao, W.; Xiang, J.; Chen, S.; Yin, S.; Zang, M.; Zheng, X. Intrinsic cohesive modeling of impact fracture behavior of laminated glass. *Mater. Design* **2017**, *127*, 321–335. [CrossRef]
31. Barenblatt, G.I. The Mathematical Theory of Equilibrium Cracks in Brittle Fracture. *Adv. Appl. Mech.* **1962**, *7*, 32. [CrossRef]
32. Dugdale, D.S. Yielding of steel sheets containing slits. *J. Mech. Phys. Solids* **1960**, *8*, 100–104. [CrossRef]
33. Celes, W.; Paulino, G.H.; Espinha, R. A compact adjacency-based topological data structure for finite element mesh representation. *Int. J. Numer. Meth. Eng.* **2005**, *64*, 1529–1556. [CrossRef]
34. Zhang, Z.J.; Paulino, G.H.; Celes, W. Extrinsic cohesive modelling of dynamic fracture and microbranching instability in brittle materials. *Int. J. Numer. Meth. Eng.* **2007**, *72*, 893–923. [CrossRef]
35. Chen, S.; Zang, M.; Xu, W. A three-dimensional computational framework for impact fracture analysis of automotive laminated glass. *Comput. Methods Appl. Mech. Eng.* **2015**, *294*, 72–99. [CrossRef]
36. Chen, S.; Zang, M.; Wang, D.; Zheng, Z.; Zhao, C. Finite element modelling of impact damage in polyvinyl butyral laminated glass. *Compos. Struct.* **2016**, *138*, 1–11. [CrossRef]
37. Lin, D.; Wang, D.; Chen, S.; Zang, M. Numerical simulations of impact fracture behavior of an automotive windshield glazing: An intrinsic cohesive approach. *Compos. Struct.* **2018**, *186*, 79–93. [CrossRef]
38. Musto, M.; Alfano, G. A fractional rate-dependent cohesive-zone model. *Int. J. Numer. Meth. Eng.* **2015**, *103*, 313–341. [CrossRef]
39. André, D.; Jebahi, M.; Iordanoff, I.; Charles, J.; Néauport, J. Using the discrete element method to simulate brittle fracture in the indentation of a silica glass with a blunt indenter. *Comput. Methods Appl. Mech. Eng.* **2013**, *265*, 136–147. [CrossRef]
40. Braun, M.; Fernández-Sáez, J. A new 2D discrete model applied to dynamic crack propagation in brittle materials. *Int. J. Solids Struct.* **2014**, *51*, 3787–3797. [CrossRef]
41. Du Bois, P.A.; Kolling, S.; Fassnacht, W. Modelling of safety glass for crash simulation. *Comp. Mater. Sci.* **2003**, *28*, 675–683. [CrossRef]
42. Timmel, M.; Kolling, S.; Osterrieder, P.; Du Bois, P.A. A finite element model for impact simulation with laminated glass. *Int. J. Impact Eng.* **2007**, *34*, 1465–1478. [CrossRef]
43. Yao, J.; Yang, J.; Otte, D. Investigation of head injuries by reconstructions of real-world vehicle-versus-adult-pedestrian accidents. *Safety Sci.* **2008**, *46*, 1103–1114. [CrossRef]
44. Xu, X.P.; Needleman, A. Numerical simulations of fast crack growth in brittle solids. *J. Mech. Phys. Solids* **1994**, *42*, 1397–1434. [CrossRef]
45. Benzeggagh, M.L.; Kenane, M. Measurement of mixed-mode delamination fracture toughness of unidirectional glass/epoxy composites with mixed-mode bending apparatus. *Compos. Sci. Technol.* **1996**, *56*, 439–449. [CrossRef]
46. De Pauw, S. Experimental and Numerical Study of Impact on Window Glass Fitted with Safety Window Film. Ph.D. Thesis, Ghent University, Ghent, Belgium, 2010.

materials

MDPI

Article

Material-Oriented Shape Functions for FGM Plate Finite Element Formulation

Wojciech Gilewski * and Jan Pełczyński

Faculty of Civil Engineering, Warsaw University of Technology, 00-637 Warsaw, Poland; j.pelczynski@il.pw.edu.pl

* Correspondence: w.gilewski@il.pw.edu.pl; Tel.: +48-22-234-5753

Received: 16 December 2019; Accepted: 6 February 2020; Published: 10 February 2020

Abstract: A four-noded finite element of a moderately thick plate made of functionally graded material (FGM) is presented. The base element is rectangular and can be extended to any shape using a transformation based on NURBS functions. The proposed 2D shape functions are consistent with the physical interpretation and describe the states of element displacement caused by unit displacements of nodes. These functions depend on the FGM's material parameters and are called material-oriented. The shape function matrix is based on a superposition displacement field of two plate strips with 1D exact shape functions. A characteristic feature of the proposed formulation is full coupling of the membrane and bending states in the plate. The analytical form of the stiffness matrix and the nodal load vector was obtained, which leads to the numerical efficiency of the formulation. The element has been incorporated into Abaqus software with the use of Maple program. The finite element shows good convergence properties for different FGM models in the transverse direction to the middle plane of the plate. During derivation of the 2D plate element the formally exact 1D finite element for transverse nonhomogeneous FGM plate strip was developed.

Keywords: finite element; FGM; plate; material-oriented shape functions; NURBS

1. Introduction

Functionally graded materials (FGMs) belong to a class of innovative materials with varying properties over a changing dimension [1,2]. The materials both occur in nature and can be obtained in an artificial way. A good example of an FGM found in nature is bamboo wood [1,2]. Due to the natural fiber distribution in the stem cross-section, it has high bending strength under natural loads. The beginnings of conscious creation of artificial FGMs should be dated to the beginning of the 1980s, with applications for the construction of thermal shields with unprecedented parameters. FGM eliminates the sharp interfaces existing in composite materials, replacing them with a gradient interface to produce a smooth transition from one material to the next.

There are different kinds of fabrication processes for producing functionally graded materials. Thin FGM sections are produced by physical or chemical vapor depositions, plasma spraying, self-propagating high temperature synthesis, etc. Volume FGM members are produced using powder metallurgy technique, centrifugal casting method, solid freedom fabrication technology, etc. Further details can be found in the literature [1,2].

Applications of functionally graded materials are quite wide, from aerospace, energy and automobile, through mechanical and civil engineering, to medicine, sport, sensors and optoelectronic fields. As the fabrication process is improved, the overall process cost is reduced, hence expanding the applications of FGM.

Plate structures are an important area of FGM applications. It is possible to build effective two-dimensional theories in which material variation in the transverse direction is usually assumed.

Shear deformable as well as thin plate theories are to be applied. An overview of important papers in this field and the derivation of the equations of the first and higher order theories can be found in [3]. The scope of obtaining effective solutions by means of analytical methods is very limited (see [3]). It is necessary to use computational methods, among which the finite element method [4] dominates.

In the standard FEM model, the functional variability of material coefficients is represented in the elasticity matrix, which appears in the formulas for the finite element stiffness matrix. Shape functions are usually polynomial, with varying numbers of nodes and various techniques for improving formulation properties. A synthetic description of the most important models, together with citations of representative papers, is presented below.

Formulation within three-dimensional theory was applied to static analysis of thick or moderately thick rectangular plates [5] (eight-node brick element), to circular plates on elastic foundation [6] as well as for vibration analysis [7] (8- and 20-node isoparametric elements). Gradation of the finite elements' elastic properties can be also used in the 3D formulation.

Various two-dimensional theories with the influence of shear deformation are the most common approach. Usually the first-order formulations are applied.

Rectangular finite elements with linear shape functions were used for static analysis, piezothermoelastic dynamics [8], active deflection/vibration control [9] and free vibrations [10]. Eight-node rectangular elements were applied for nonlinear free flexural vibrations [11] and piezo-electric fiber reinforcement [12]. The enhanced strain formulation to avoid the locking effects in eight-node elements was proposed in [13] for structural stability and dynamics problems. The elements can be successfully implemented in the Abaqus package [14].

Triangular plate finite elements with the discrete shear gap improvement were analyzed in [15,16] for static and free vibration problems.

Axi-symmetric plate finite elements within buckling, vibration and thermal analysis were proposed in [17].

Nine-node quadrilateral finite element was proposed for higher order shear deformation theory for static and free vibration analysis in [18] and was extended for large amplitude problems in [19]. Modified discrete Kirchhoff four-node quadrilateral element was used for the third-order theory for skew plates in [20].

An example of the application of C-1 continuity finite element formulated for classical thin FGM plate theory and applied for natural frequency analysis can be found in [21].

Isogeometric finite element formulation with NURBS-based shape functions was presented in [22] for static, vibrations, buckling and flutter as well as in [23] for thermal buckling.

The finite element formulation can be also applied for FGM shell structures and FGM beams—see [24–28] as representative papers.

A variety of finite element formulations applied for FGM plates seems to be a sign that the development of an element suitable for the analysis is still an open problem. This was an inspiration for the proposal of the present paper. The common problems with FE formulations based on standard polynomial shape functions are shear locking (connected to the parasitic shear) and zero energy modes. Analogous problems occur in FGM modeling using FEM. In addition, standard polynomial approximation assumes decoupling of approximated displacement fields. This is not consistent with the physical interpretation of the shape function in FGM boards, which should describe the displacement fields caused by unit displacement of nodes and should be coupled. In the matrices of shape functions proposed in the work, we observe full displacement of displacement fields.

The proposed elements are free from zero energy modes. Earlier studies of homogeneous plates with similar physical shape functions [29] did not show shear locking even for very thin plates. This phenomenon was also not observed in the FGM model proposed in this paper. The new finite elements are based on the physical interpretation of the shape functions, according to which they describe the distribution of finite element displacements caused by unit displacement of nodes. Naturally, such distributions should depend on the physical properties of the element. This approach was proposed for the statics of homogeneous plate bending in [30] and generalized in [29,31,32] for other types of structures. The proposed base element is rectangular with fully coupled displacement fields (membrane and bending conditions), which are based on analytical solutions regarding plate strips constructed of material with any FGM properties. The developed finite element was incorporated into the Abaqus system with UEL element user subroutine (using Maple program procedures) and its convergence was tested. The basic four-node rectangular finite element can be extended to any shape via parametric transformation based on NURBS functions [33]. During the development of the rectangular element, a formally accurate finite element for plate strip was created.

2. FGM plate Finite Elements

The subject under consideration is a rectangular four-noded finite element of dimensions $2a \times 2b$ and thickness h in nondimensional coordinate system $\xi = \frac{x-x_e}{a}, \eta = \frac{y-y_e}{b}$. The element is presented in Figure 1.

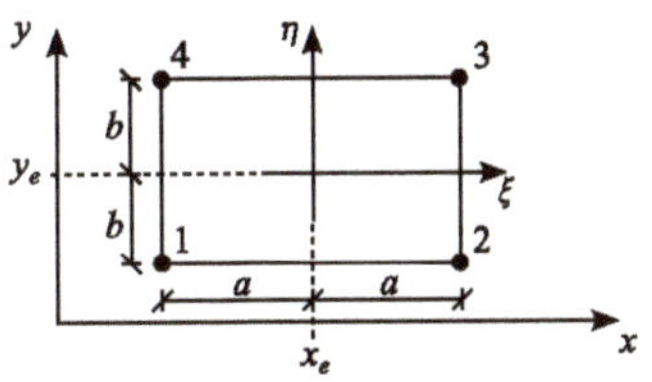

Figure 1. Four-noded rectangular plate element.

Generalized displacements of membrane and bending states as well as natural nodal displacements are defined as

$$\mathbf{u} = \left\{u, v, w, \varphi_x, \varphi_y\right\} \tag{1}$$

$$\mathbf{q} = \left\{\mathbf{q}_1, \mathbf{q}_2, \mathbf{q}_3, \mathbf{q}_4\right\}, \tag{2}$$

$$\mathbf{q}_i = \left\{u_i, v_i, \varphi_{xi}, \varphi_{yi}\right\},\ i = 1, 2, 3, 4 \tag{3}$$

Shape function matrix ($\mathbf{u} = \mathbf{N}\mathbf{q}$) is proposed in the form

$$\mathbf{N} = [\mathbf{N}_1 \mathbf{N}_2 \mathbf{N}_3 \mathbf{N}_4], \tag{4}$$

$$\mathbf{N}_{\mathrm{i}} = (\xi, \eta) == \begin{bmatrix} \hat{N}^i_{11}\hat{\hat{N}}^i_{11} & \hat{N}^i_{12}\hat{\hat{N}}^i_{22} & \hat{N}^i_{12}\hat{\hat{N}}^i_{22} & \hat{N}^i_{13}\hat{\hat{N}}^i_{22} & \hat{N}^i_{12}\hat{\hat{N}}^i_{23} \\ \hat{N}^i_{21}\hat{\hat{N}}^i_{12} & \hat{N}^i_{11}\hat{\hat{N}}^i_{11} & \hat{N}^i_{22}\hat{\hat{N}}^i_{12} & \hat{N}^i_{23}\hat{\hat{N}}^i_{12} & \hat{N}^i_{22}\hat{\hat{N}}^i_{13} \\ \hat{N}^i_{21}\hat{\hat{N}}^i_{22} & \hat{N}^i_{12}\hat{\hat{N}}^i_{21} & \hat{N}^i_{22}\hat{\hat{N}}^i_{22} & \hat{N}^i_{23}\hat{\hat{N}}^i_{22} & \hat{N}^i_{22}\hat{\hat{N}}^i_{23} \\ \hat{N}^i_{31}\hat{\hat{N}}^i_{22} & \hat{N}^i_{22}\hat{\hat{N}}^i_{21} & \hat{N}^i_{32}\hat{\hat{N}}^i_{22} & \hat{N}^i_{33}\hat{\hat{N}}^i_{22} & \hat{N}^i_{32}\hat{\hat{N}}^i_{23} \\ \hat{N}^i_{21}\hat{\hat{N}}^i_{32} & \hat{N}^i_{22}\hat{\hat{N}}^i_{31} & \hat{N}^i_{22}\hat{\hat{N}}^i_{32} & \hat{N}^i_{23}\hat{\hat{N}}^i_{32} & \hat{N}^i_{22}\hat{\hat{N}}^i_{33} \end{bmatrix} \tag{5}$$

The matrix (Equation (5)) is consistent with physical interpretation of the shape function, according to which it describes the distribution of finite element displacements caused by unit displacement of nodes. This approach is an extension of the concept proposed for the statics of homogeneous plates [29] and for other types of structures [30,31]. Fully coupled displacement fields for membrane and bending states can be observed. The proposed shape matrix is an overlay of one-variable distributions of displacements of two crossed plate strips (Figure 2) with imposed boundary displacements.

$$f = f(\xi),\ \hat{f} = \hat{f}(\eta) \tag{6}$$

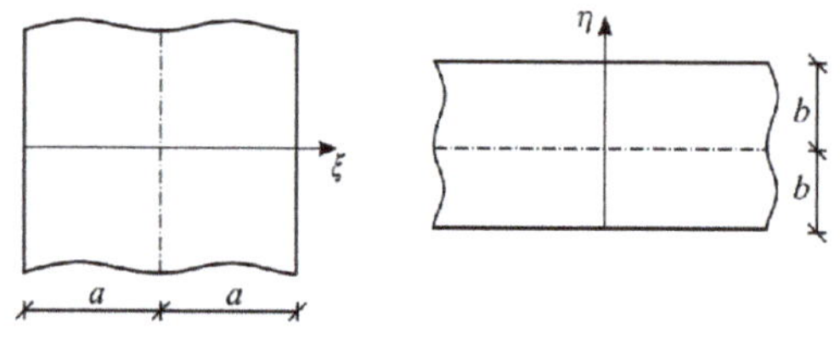

Figure 2. Crossed plate strips.

One-variable displacement fields are related to the FGM plate strips (Figure 2). If the first-order plate theory with shear deformation is used [3], the following six-level set of differential equations with physical boundary conditions is to be considered (for the x direction as the example):

$$\begin{bmatrix} A_0\frac{1}{a^2}\frac{d^2}{d\xi^2} & 0 & -A_1\frac{1}{a^2}\frac{d^2}{d\xi^2} \\ 0 & -B_0\frac{1}{a^2}\frac{d^2}{d\xi^2} & B_0\frac{1}{a}\frac{d}{d\xi} \\ -A_1\frac{1}{a^2}\frac{d^2}{d\xi^2} & B_0\frac{1}{a}\frac{d}{d\xi} & A_2\frac{1}{a^2}\frac{d^2}{d\xi^2} - B_0 \end{bmatrix} \begin{bmatrix} u \\ w \\ \varphi \end{bmatrix} = \begin{bmatrix} -p \\ -q \\ -m \end{bmatrix} \tag{7}$$

where

$$A_i = \frac{E_o}{1-v^2}\int_{\frac{-h}{2}}^{\frac{h}{2}}\left[P(z)\cdot z^i\right]dz,\ B_0 = \frac{5}{6}\frac{1-\nu}{2}A_0,\ i = 0,1,2, \tag{8}$$

and p, q and m are load components. Coefficients A_i and B_0 depend on the adopted FGM distribution represented for isotropic transverse nonhomogeneous material by the Young's modulus E_0, Poisson's ratio ν and function $P(z)$. In the above system of equations, one can observe a coupling of the membrane and bending states in the FGM panel. A similar system of equations for a moderately thick beam made of FGM was considered in the paper [26].

Solution of the homogeneous equations can be used to obtain exact shape functions for the strip, as follows:

$$u(\xi) = C_5 + C_6\xi + \frac{3A_1}{A_0 a}C_4\xi^2, \tag{9}$$

$$w(\xi) = C_1 + C_2\xi + C_3\xi^2 + C_4\xi^3, \tag{10}$$

$$\phi(\xi) = \frac{1}{a}\left[C_2 + 2C_3\xi + 3C_4(\xi^2 + \frac{2A_2}{B_0 a^2} - \frac{2A_1^2}{A_0 B_0 a^2})\right] \tag{11}$$

Two 1D displacement vectors $\hat{\mathbf{u}} = (\xi)$ and $\hat{\mathbf{u}} = (\eta)$ for plate strips (Figure 2) are defined as

$$\hat{\mathbf{u}} = (\xi) = \{u(\xi), w(\xi), \varphi_x(\xi)\},\ \hat{\mathbf{u}} = (\eta) = \{v(\eta), w(\eta), \varphi_y(\eta)\} \tag{12}$$

Formally exact shape function matrices can be developed for each plate strip:

$$\hat{\mathbf{u}} = (\xi) = \hat{\mathbf{N}}(\xi)\cdot\hat{\mathbf{q}}_e,\ \hat{\mathbf{u}} = (\eta) = \hat{\hat{\mathbf{N}}}(\eta)\cdot\hat{\hat{\mathbf{q}}}_e \tag{13}$$

with natural boundary parameters at each end:

$$\hat{\mathbf{q}}_e = \{u_1, w_1, \varphi_{x1}, u_2, w_2, \varphi_{x2}\},\ \hat{\hat{\mathbf{q}}}_e = \{v_1, w_1, \varphi_{y1}, v_2, w_2, \varphi_{y2}\} \tag{14}$$

The 1D shape function matrices can be expressed in the form

$$\hat{\mathbf{N}} = \begin{bmatrix} \hat{\mathbf{N}}_1(\xi) & \hat{\mathbf{N}}_2(\xi) \end{bmatrix},\ \hat{\hat{\mathbf{N}}} = \begin{bmatrix} \hat{\hat{\mathbf{N}}}_1(\eta) & \hat{\hat{\mathbf{N}}}_2(\eta) \end{bmatrix} \tag{15}$$

$$\hat{\mathbf{N}}^i(\xi) = \begin{bmatrix} \hat{N}^i_{11}(\xi) & \hat{N}^i_{12}(\xi) & \hat{N}^i_{13}(\xi) \\ 0 & \hat{N}^i_{22}(\xi) & \hat{N}^i_{23}(\xi) \\ 0 & \hat{N}^i_{32}(\xi) & \hat{N}^i_{33}(\xi) \end{bmatrix}, \hat{\hat{\mathbf{N}}}^i(\eta) = \begin{bmatrix} \hat{\hat{N}}^i_{11}(\eta) & \hat{\hat{N}}^i_{12}(\eta) & \hat{\hat{N}}^i_{13}(\eta) \\ 0 & \hat{\hat{N}}^i_{22}(\eta) & \hat{\hat{N}}^i_{23}(\eta) \\ 0 & \hat{\hat{N}}^i_{32}(\eta) & \hat{\hat{N}}^i_{33}(\eta) \end{bmatrix} \tag{16}$$

where

$$\begin{aligned} \hat{N}^i_{11}(\xi) &= \tfrac{1}{2}(1+\xi_i\xi), \\ \hat{N}^i_{12}(\xi) &= \tfrac{3}{4}\frac{\xi_i A_1 B_0 a\left(1-\xi^2\right)}{3A_0A_2+A_0B_0a^2-3A_1^2}, \\ \hat{N}^i_{13}(\xi) &= \tfrac{3}{4}\frac{A_1 B_0 a^2(\xi^2-1)}{3A_0A_2+A_0B_0a^2-3A_1^2}, \\ \hat{N}^i_{22}(\xi) &= \tfrac{1}{2}+\tfrac{1}{2}\left(1+\tfrac{1}{2}\frac{A_0B_0a^2\left(1-\xi^2\right)}{3A_0A_2+A_0B_0a^2-3A_1^2}\right)\xi_i\xi, \\ \hat{N}^i_{23}(\xi) &= \tfrac{1}{4}\left(\frac{A_0B_0a^3\xi}{3A_0A_2+A_0B_0a^2-3A_1^2}+a\xi_i\right)(\xi^2-1), \\ \hat{N}^i_{32}(\xi) &= \tfrac{3}{4}\frac{\xi_i A_0 B_0 a(1-\xi^2)}{3A_0A_2+A_0B_0a^2-3A_1^2} \\ \hat{N}^i_{33}(\xi) &= \tfrac{1}{2}\left(\xi_i\xi+\tfrac{3}{2}\frac{A_0B_0a^2(\xi^2-1)}{3A_0A_2+A_0B_0a^2-3A_1^2}+1\right). \end{aligned} \tag{17}$$

The shape functions for the second direction are analogous. They can be obtained from Equation (17) by exchanging variables $\xi \to \eta, a \to b$. The functions (Equation (17)) strongly depend on the parameters of FGM, so the formulation is named "material-oriented".

Following the above formulation, the formally exact stiffness matrix as well as the load vector for two line-noded, 6 d.o.f. (horizontal displacements, vertical displacements and rotations) plate strip finite element can be expressed in the form (for the ξ direction as an example)

$$\mathbf{K}_e = \begin{bmatrix} k_{11} & 0 & k_{13} & -k_{11} & 0 & -k_{13} \\ 0 & k_{22} & ak_{22} & 0 & -k_{22} & ak_{22} \\ k_{13} & ak_{22} & k_{33} & -k_{13} & -ak_{22} & k_{36} \\ -k_{11} & 0 & -k_{13} & k_{11} & 0 & k_{13} \\ 0 & -k_{22} & -ak_{22} & 0 & k_{22} & -ak_{22} \\ -k_{13} & ak_{22} & k_{36} & k_{13} & -ak_{22} & k_{33} \end{bmatrix}, \tag{18}$$

$$\begin{aligned} k_{11} &= \tfrac{A_0}{2a}, \\ k_{13} &= -\tfrac{A_1}{2a}, \\ k_{22} &= \frac{3B_0\left(A_0A_2-A_1^2\right)}{2a\left[-3\left(A_1^2-A_0A_2\right)+a^2A_0B_0\right]}, \\ k_{33} &= \frac{-3A_1^2\left(A_2+a^2B_0\right)+A_0A_2\left(3A_2+4a^2B_0\right)}{2a\left[-3\left(A_1^2-A_0A_2\right)+a^2A_0B_0\right]}, \\ k_{36} &= \frac{3A_1^2\left(A_2-a^2B_0\right)-A_0A_2\left(3A_2-2a^2B_0\right)}{2a\left[-3\left(A_1^2-A_0A_2\right)+a^2A_0B_0\right]}, \end{aligned} \tag{19}$$

$$\mathbf{Q}_e = pa\begin{bmatrix} 1 \\ -\frac{aA_1B_0}{-3A_1^2+A_0+(3A_2+a^2B_0)} \\ -\frac{a^2A_1B_0}{-3A_1^2+A_0+(3A_2+a^2B_0)} \\ 1 \\ -\frac{aA_1B_0}{-3A_1^2+A_0+(3A_2+a^2B_0)} \\ -\frac{a^2A_1B_0}{-3A_1^2+A_0+(3A_2+a^2B_0)} \end{bmatrix} + qa\begin{bmatrix} 0 \\ 1 \\ \frac{1}{3}a \\ 0 \\ 1 \\ -\frac{1}{3}a \end{bmatrix} + ma\begin{bmatrix} 0 \\ -\frac{aA_0B_0}{-3A_1^2+A_0+(3A_2+a^2B_0)} \\ -\frac{-3A_1^2+3A_1A_2}{-3A_1^2+A_0+(3A_2+a^2B_0)} \\ 0 \\ -\frac{aA_0B_0}{-3A_1^2+A_0+(3A_2+a^2B_0)} \\ -\frac{-3A_1^2+3A_1A_2}{-3A_1^2+A_0+(3A_2+a^2B_0)} \end{bmatrix} \tag{20}$$

The results to be obtained within the above formulation are formally exact and no convergence study is necessary.

The strain matrix for the rectangular finite element can be expressed as

$$\mathbf{B} = [\mathbf{B}_1\mathbf{B}_2\mathbf{B}_3\mathbf{B}_4], \ \mathbf{B}_i = \mathbf{D}\mathbf{N}_i, \tag{21}$$

where

$$\mathbf{D} = \begin{bmatrix} \mathbf{D}_1 & 0 \\ 0 & \mathbf{D}_2 \end{bmatrix}, \mathbf{D}_1 = \begin{bmatrix} \frac{1}{a}\frac{\partial}{\partial\xi} & 0 \\ 0 & \frac{1}{b}\frac{\partial}{\partial\eta} \\ \frac{1}{b}\frac{\partial}{\partial\eta} & \frac{1}{a}\frac{\partial}{\partial\xi} \end{bmatrix}, \mathbf{D}_2 = \begin{bmatrix} 0 & -\frac{1}{a}\frac{\partial}{\partial\xi} & 0 \\ 0 & 0 & -\frac{1}{b}\frac{\partial}{\partial\eta} \\ 0 & -\frac{1}{b}\frac{\partial}{\partial\eta} & -\frac{1}{a}\frac{\partial}{\partial\xi} \\ \frac{1}{a}\frac{\partial}{\partial\xi} & -1 & 0 \\ \frac{1}{b}\frac{\partial}{\partial\eta} & 0 & -1 \end{bmatrix} \tag{22}$$

The stiffness matrix and load vector can be received in the standard FEM procedure as follows:

$$\mathbf{K}_e = ab\int_{-1}^{1}\int_{-1}^{1}\mathbf{B}^T\mathbf{E}\mathbf{B}d\xi d\eta, \tag{23}$$

$$\mathbf{Q}_e = ab\int_{-1}^{1}\int_{-1}^{1}\mathbf{N}^T\mathbf{p}d\xi d\eta, \tag{24}$$

where elasticity matrix is expressed as

$$\mathbf{E} = \begin{bmatrix} \mathbf{E}_1 & \mathbf{E}_2 & 0 \\ \mathbf{E}_2^T & \mathbf{E}_3 & 0 \\ 0 & 0 & \mathbf{E}_4 \end{bmatrix} \tag{25}$$

$$\mathbf{E}_1 = \begin{bmatrix} A_0 & vA_0 & 0 \\ vA_0 & A_0 & 0 \\ 0 & 0 & \frac{1-v}{2}A_0 \end{bmatrix}, \mathbf{E}_2 = \begin{bmatrix} A_1 & vA_1 & 0 \\ vA_1 & A_1 & 0 \\ 0 & 0 & 0 \end{bmatrix}, \tag{26}$$

$$\mathbf{E}_3 = \begin{bmatrix} A_2 & vA_2 & 0 \\ vA_2 & A_2 & 0 \\ 0 & 0 & \frac{1-v}{2}A_2 \end{bmatrix}, \mathbf{E}_4 = \begin{bmatrix} B_0 & 0 \\ 0 & B_0 \end{bmatrix}, \tag{27}$$

and

$$\mathbf{p} = \{p_x, p_y, p_z, m_x, m_y\} \tag{28}$$

is the load vector for the panel.

The analytical form of stiffness matrix can be obtained with the use of *Maple* software—the expressions are relatively long and exceed the limitations of the present paper.

The analytical form of the load vector can be expressed as

$$\mathbf{Q}_e = \begin{bmatrix} \mathbf{Q}_1^e \\ \mathbf{Q}_2^e \\ \mathbf{Q}_3^e \\ \mathbf{Q}_4^e \end{bmatrix}, \tag{29}$$

$$
\mathbf{Q}_i^e = p_x ab \begin{bmatrix} 1 \\ 0 \\ \frac{aA_1B_0}{3A_0A_2+A_0B_0a^2-3A_1^2}\xi_i \\ -\frac{a^2A_1B_0}{3A_0A_2+A_0B_0a^2-3A_1^2} \\ -\frac{1}{3}\frac{abA_1B_0}{3A_0A_2+A_0B_0a^2-3A_1^2}\xi_i\eta_i \end{bmatrix} + p_y ab \begin{bmatrix} 0 \\ 1 \\ \frac{bA_1B_0}{3A_0A_2+A_0B_0a^2-3A_1^2}\eta_i \\ -\frac{1}{3}\frac{abA_1B_0}{3A_0A_2+A_0B_0a^2-3A_1^2}\xi_i\eta_i \\ -\frac{a^2A_1B_0}{3A_0A_2+A_0B_0a^2-3A_1^2} \end{bmatrix} + p_z ab \begin{bmatrix} 0 \\ 0 \\ 1 \\ -\frac{a\xi_i}{3} \\ -\frac{b\eta_i}{3} \end{bmatrix} +
$$
$$
+ m_x ab \begin{bmatrix} 0 \\ 0 \\ \frac{aA_0B_0}{3A_0A_2+A_0B_0a^2-3A_1^2}\xi_i \\ -\frac{a^2A_0B_0}{3A_0A_2+A_0B_0a^2-3A_1^2} \\ -\frac{1}{3}\frac{abA_0B_0}{3A_0A_2+A_0B_0a^2-3A_1^2}\xi_i\eta_i \end{bmatrix} + m_y ab \begin{bmatrix} 0 \\ 0 \\ \frac{bA_1B_0}{3A_0A_2+A_0B_0a^2-3A_1^2}\eta_i \\ -\frac{1}{3}\frac{abA_1B_0}{3A_0A_2+A_0B_0a^2-3A_1^2}\xi_i\eta_i \\ -\frac{a^2A_1B_0}{3A_0A_2+A_0B_0a^2-3A_1^2} \end{bmatrix}. \tag{30}
$$

The rectangular finite element with material-oriented shape functions satisfy rigid body motion, constant strain and ellipticity conditions [4,30].

The proposed finite element was introduced to the Abaqus system using the UEL element user subroutine (see [34,35] for details). The UEL procedure is written in FORTRAN. The stiffness matrix is entered into the code from Maple Software for Mathematica using the command Fortran (K, optimize, resultname = "AMATRX") [36] using Equations (18–20) or (21–30).

For the purpose of the present paper, the analysis is limited to the isotropic plates with functional transverse gradation of Young's modulus (E, E_0) and mass density (ρ, ρ_0):

$$E(z) = E_0 \cdot P(z), v = const., \ \rho(z) = \rho_0 \cdot P(z). \tag{31}$$

Three FGM distributions (Equations (32–34)) were taken into consideration:

$$P(z) = 1 + 4\frac{z^2}{h^2}\gamma, \tag{32}$$

$$P(z) = (\eta - 1)\left(\frac{z}{h} + \frac{1}{2}\right)^n + 1, \ \eta = \frac{P_t}{P_b}, \tag{33}$$

$$P(z) = \frac{\delta/\pi}{\delta^2 + \left(\frac{z}{h} - \xi_0\right)^2} + 1, \tag{34}$$

where $\gamma, P_t, P_b, n, \delta, \xi_0$ are constant parameters to model various FGM distributions. Here, "t" represents the top of the plate, "b" represents the bottom of the plate, $E_0 = E_b, \rho_0 = \rho_b$.

Examples of functions *P(z)* along thickness of the plate are presented in Figure 3.

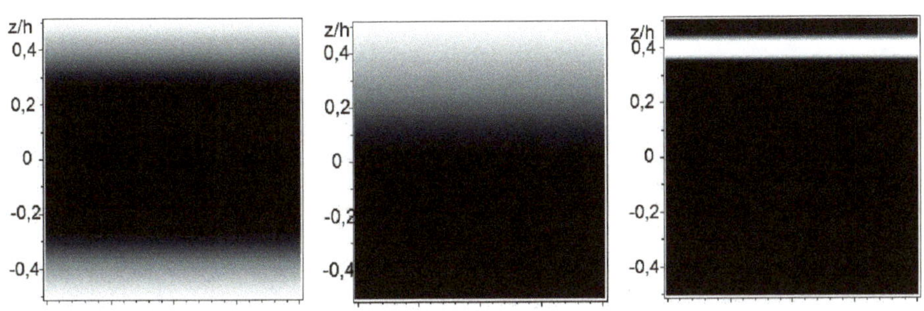

$\gamma = 15$ (32) $\quad n = 4, \ P_t = 2, \ P_b = 1$ (33) $\quad \delta = 0.02, \ \xi_0 = 0.4$ (34)

Figure 3. Selected transverse nonhomogeneous materials of the plate.

Light color represents material with a higher intensity of material properties.

The functions described by Equations (32) and (33) are normally used to describe FGMs. The function described by Equation (34) is original and can be used to describe plates with reinforced mats glued in.

3. Numerical Examples

The proposed finite element was adapted to the Abaqus software. To verify the finite element, several examples were made. Some of the examples made for convergence check are presented below. Square plates are considered (simply supported and clamped—Figure 4) with the following data: dimensions $L \times L$, thickness $h = L/10$, Young's modulus E_0, Poisson's ratio $v = 0.25$, concentrated force P.

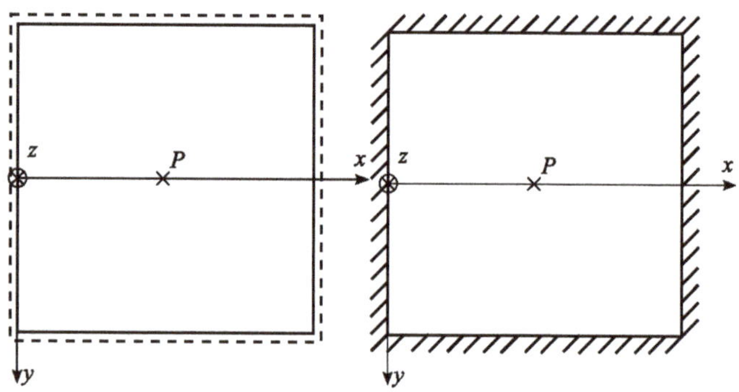

Figure 4. Plates taken for convergence analysis: simply supported and clamped with concentrated load in the middle.

In Figures 5 and 6 convergence results are shown for the maximum vertical deflection in the middle of the plates.

Model 1a is a transverse homogeneous plate (for $\gamma = 0$ according to Equation (32)). Model 1b is a transverse inhomogeneous plate according to Equation (32) for $\gamma = 15$. Model 2 is a transverse inhomogeneous plate according to Equation (33) for $n = 4, P_t = P_b$. Model 3 is a transverse inhomogeneous plate according to Equation (34) for $\delta = 0.02, \xi_0 = 0.4$.

One can observe that each example is characterized by a good rate of convergence for all models.

The influence of the FGM model applied in comparison to the homogeneous plate is visible. The considerable differences result from the adopted calculation parameters in the functions described by Equations (32)–(34). These parameters describe strong material variation—hence the differences compared to a homogeneous material. The rectangular element shows correct convergence.

Validation of the proposed model in the scope of FGM was carried out by comparison with the analytical solution available in the literature—the benchmark presented in the [37] was selected. A comparison regarding the maximum displacement of a simply supported plate subjected to uniform loading for the material model (Equation (33)) is shown in Table 1.

The numerical solution for homogeneous plates (k = 0 and k = infinity) is consistent with the analytical solution [37] in the range of five significant digits. For other parameters of n, the proposed finite element gives slightly less than analytical results at a level of 1%–3% error.

It is worth noting that the solution in the field of cylindrical bending using Equations (18) and (20) is formally accurate for any FGM.

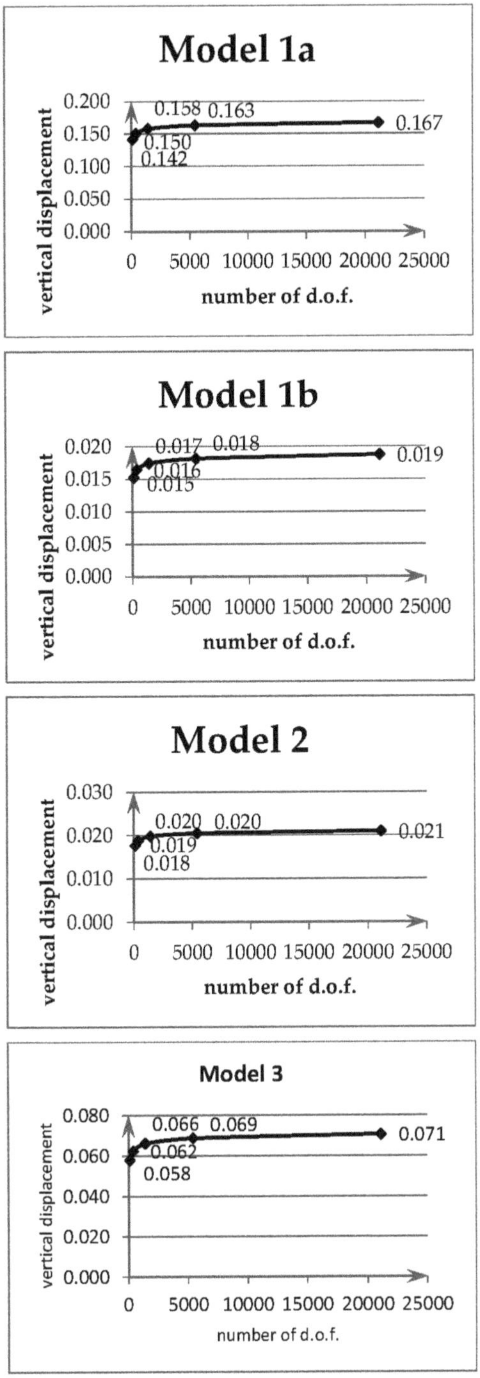

Figure 5. Convergence graphs for simply supported plate (multiplier $\frac{PL^2}{E_0h^3}$).

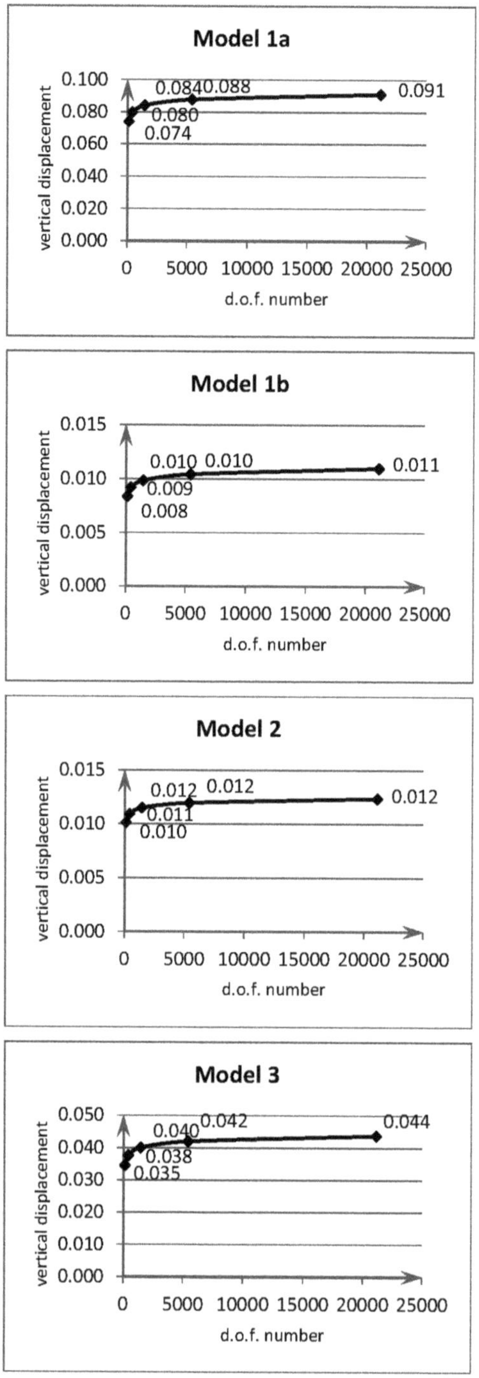

Figure 6. Convergence graphs for clamped plate (multiplier $\frac{PL^2}{E_0h^3}$).

Table 1. Comparison of the FEM results with analytical solutions [37] for a simply supported square plate ($L \times L$, $h/L = 1/10$) with uniformly distributed load q. Material model (Equation (33)), $E_t = 380GPa, E_b = 70GPa$.

	$\tilde{w} = 10\frac{E_t h^3}{qL^4} w(\frac{A}{2}, \frac{B}{2})$						
n	**0**	**1**	**2**	**3**	**5**	**10**	**Infinity**
Analytical solution [37]	0.4665	0.9421	1.2228	1.3530	1.4647	1.6054	2.5328
Present study	0.4665	0.9280	1.1903	1.3124	1.4202	1.5692	2.5328

4. Free-Form Plate Finite Element—the Use of NURBS Functions

An extension of the proposed finite element for unrestricted shape (Figure 7—examples with four and six nodes) can be done by coordinate system transformation with the use of NURBS functions [33]:

$$x = \sum_{i=1}^{N} F_i(\xi, \eta) x_i, y = \sum_{i=1}^{N} F_i(\xi, \eta) y_i, \tag{35}$$

where $F_i(\xi, \eta)$, $i = 1, \ldots, N$, are NURBS functions.

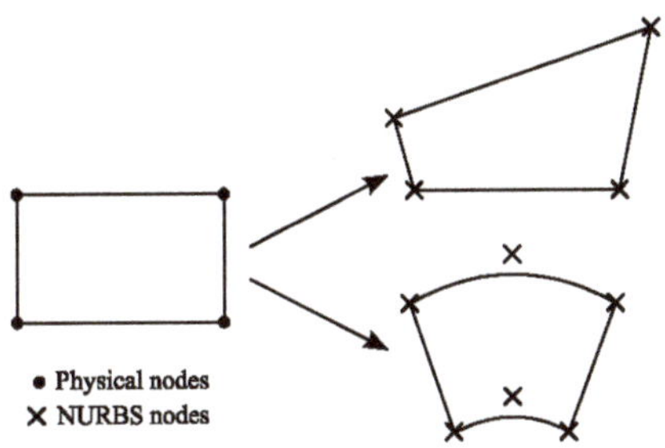

Figure 7. Example of free-form element shapes.

The formulation within description of geometry is close to superparametric [4]. Only four nodes are used in every element for representation of displacement fields (according to Equation (5)), as well as N nodes for description of a free-form element geometry.

The Jacobi and inverse Jacobi matrices can be obtained:

$$\mathbf{J} = \begin{bmatrix} \frac{\partial x}{\partial \xi} & \frac{\partial y}{\partial \xi} \\ \frac{\partial x}{\partial \eta} & \frac{\partial y}{\partial \eta} \end{bmatrix} = \begin{bmatrix} \sum_{i=1}^{N} \frac{\partial F_i}{\partial \xi} x_i & \sum_{i=1}^{N} \frac{\partial F_i}{\partial \xi} y_i \\ \sum_{i=1}^{N} \frac{\partial F_i}{\partial \eta} x_i & \sum_{i=1}^{N} \frac{\partial F_i}{\partial \eta} y_i \end{bmatrix}, \tag{36}$$

$$\mathbf{J}^{-1} = \begin{bmatrix} \frac{\partial \xi}{\partial x} & \frac{\partial \eta}{\partial x} \\ \frac{\partial \xi}{\partial y} & \frac{\partial \eta}{\partial y} \end{bmatrix} = \frac{1}{J} \begin{bmatrix} \frac{\partial y}{\partial \eta} & -\frac{\partial y}{\partial \xi} \\ -\frac{\partial x}{\partial \eta} & \frac{\partial x}{\partial \xi} \end{bmatrix} = \begin{bmatrix} J_{11}^{-1} & J_{12}^{-1} \\ J_{21}^{-1} & J_{22}^{-1} \end{bmatrix}. \tag{37}$$

where J is the Jacobian of this transformation

$$J = \frac{\partial x}{\partial \xi}\frac{\partial y}{\partial \eta} - \frac{\partial x}{\partial \eta}\frac{\partial y}{\partial \xi} \tag{38}$$

Now, one can calculate derivatives necessary for development of the strain matrix:

$$\frac{\partial(\ldots)}{\partial x} = \frac{\partial(\ldots)}{\partial \xi} J_{11}^{-1} + \frac{\partial(\ldots)}{\partial \eta} J_{12}^{-1}, \quad \frac{\partial(\ldots)}{\partial y} = \frac{\partial(\ldots)}{\partial \xi} J_{21}^{-1} + \frac{\partial(\ldots)}{\partial \eta} J_{22}^{-1}, \tag{39}$$

Parameters a and b, necessary for the strain matrix, are average values of the element dimensions. The strain matrix can be expressed in the form

$$\mathbf{B} = \begin{bmatrix} \mathbf{B}_1 & \mathbf{B}_2 & \mathbf{B}_3 & \mathbf{B}_4 \end{bmatrix}, \ \mathbf{B}_i = \mathbf{D}\mathbf{N}_i \tag{40}$$

The above formulae provide a way to calculate the stiffness matrix and the load vector of an element:

$$\mathbf{K}^e = \iint_{F^e} \mathbf{B}^T \mathbf{E} \mathbf{B} dF^e = \int_{-1}^{1} \int_{-1}^{1} \mathbf{B}^T(\xi,\eta) \mathbf{E} \mathbf{B}(\xi,\eta) J \, d\xi d\eta, \tag{41}$$

$$\mathbf{Q}^e = \iint_{F^e} \mathbf{N}^T \mathbf{P} dF^e = \int_{-1}^{1} \int_{-1}^{1} \mathbf{N}^T \mathbf{P} J \, d\xi d\eta, \tag{42}$$

where $\mathbf{E}$ is the elasticity matrix and $\mathbf{P}$ is the external load vector.

Following the above procedure, the limitation of the rectangular shape of the proposed finite element can be avoided.

5. Conclusions

The paper proposes a four-noded finite element of a moderately thick plate made of FGM. The base element is rectangular and can be extended to any shape using a transformation based on NURBS functions.

The shape functions in the considered elements are consistent with the physical interpretation and describe the states of element displacement caused by unit displacements of nodes. These functions depend on the FGM's material parameters and are called material-oriented. A characteristic feature of the proposed formulation is full coupling of the membrane and bending states in the plate and the dependence of the shape functions on the FGM model.

The finite element meets the conditions of rigid body motion and constant strain as well as the elliptical condition. No spurious zero-energy modes were identified.

The analytical form of the stiffness matrix and the nodal load vector was obtained, which leads to the numerical efficiency of the formulation. The element has been incorporated into Abaqus software with the use of Maple program.

The finite element with material-oriented shape functions shows good convergence properties for different FGM models in the transverse direction to the middle plane of the plate. The element is free from locking effects. Examples have been completed to compare the FEM solution with the analytical solution. Basic examples were used to demonstrate the convergence and correctness of the proposed model. The results are promising as a base for modeling more complex plates built of FGMs with various distribution of the properties over the thickness.

During derivation of the 2D plate element, the formally exact 1D finite element for a transverse nonhomogeneous FGM plate strip was developed.

Author Contributions: Conceptualization, W.G.; Methodology, W.G.; Software, J.P.; Validation, J.P.; Formal Analysis, W.G. and J.P.; Investigation, W.G. and J.P.; Resources, W.G. and J.P.; Writing-Original Draft Preparation, W.G. and J.P.; Writing-Review & Editing, W.G.; Visualization, W.G. and J.P.; Supervision, W.G. All authors have read and agreed to the published version of the manuscript.

Funding: This research received no external funding.

Conflicts of Interest: The authors declare no conflict of interest.

References

1. Miyamoto, Y.; Kaysser, W.A.; Rabin, B.H.; Kawasaki, A.; Ford, R.G. *Functionally Graded Materials: Design, Processing and Applications*; Kluwer Academy Publishers: Dordrecht/Boston/London, UK, 1999.
2. Mahamood, R.M.; Akinalbi, E.T.; Shukla, M.; Pityana, S. Functionally graded material: An overview. In Proceedings of the World Congress on Engineering, London, UK, 4–6 July 2012.
3. Reddy, J.N. Analysis of functionally graded plates. *Int. J. Numer. Methods Eng.* **2000**, *47*, 663–684. [CrossRef]
4. Taylor, R.L.; Zienkiewicz, O.C. *The Finite Element Method. Volume 2: Solid Mechanics*; Butterworth-Heinemann: Oxford, UK, 2000.
5. Asemi, K.; Selehi, M.; Akhlaghi, M. Three dimensional static analysis of two dimensional functionally graded plates. *Int. J. Recent Adv. Mech. Eng.* **2013**, *2*, 21–32.
6. Shariat, M.; Mohammadjani, R. Three-dimensional compatible finite element stress analysis of spinning two-directional FGM annular plates and disks with load and elastic foundation non-uniformities. *Lat. Am. J. Solids Struct.* **2013**, *10*, 859–890. [CrossRef]
7. Burlayenko, V.N.; Sadowski, T. Free vibrations and static analysis of functionally graded sandwich plates with three-dimensional finite elements. *Meccanica* **2019**, *1–19*. [CrossRef]
8. Liew, K.M.; He, X.Q.; Ng, T.Y.; Sivashanker, S. Active control of FGM plates subjected to a temperature gradient: Modelling via finite element method based on FSDT. *Int. J. Numer. Methods Eng.* **2001**, *52*, 1253–1271. [CrossRef]
9. He, X.Q.; Ng, T.Y.; Sivashanker, S.; Liew, K.M. Active control of FGM plates with integrated piezoelectric sensors and actuators. *Int. J. Solids Struct.* **2001**, *38*, 1641–1655. [CrossRef]
10. Kurtaran, H. Shape effect on free vibration of functionally graded plates. *Int. J. Eng. Appl. Sci.* **2014**, *6*, 52–67. [CrossRef]
11. Sundararajan, N.; Prakash, T.; Ganapathi, M. Nonlinear free flexural vibrations of functionally graded rectangular and skew plates under thermal environments. *Finite Elem. Anal. Des.* **2005**, *42*, 152–168. [CrossRef]
12. Ray, M.C.; Sachade, H.M. Finite element analysis of smart functionally graded plates. *Int. J. Solids Struct.* **2006**, *43*, 5468–5484. [CrossRef]
13. Park, W.T. Structural stability and dynamics of FGM plates using an improved 8-ANS finite element. *Adv. Mater. Sci. Eng.* **2016**, *2016*, 2821473. [CrossRef]
14. Martinez-Paneda, E. On the finite element implementation of functionally graded materials. *Materials* **2019**, *12*, 287. [CrossRef] [PubMed]
15. Nguyen-Xuan, H.; Tran, L.V.; Thai, C.H.; Nguyen-Thoi, T. Analysis of functionally graded plates by an efficient finite element method with node-based strain smoothing. *Thin-Walled Struct.* **2012**, *54*, 1–18. [CrossRef]
16. Natarajan, S.; Ferreira, A.J.M.; Bordas, S.; Carrera, E.; Cinefra, M.; Zenkour, A.M. Analysis of functionally graded material plates using triangular elements with cell-based smooth discrete shear gap method. *Math. Problems Eng.* **2014**, *2014*, 247932. [CrossRef]
17. Prakash, T.; Ganapathi, M. Asymmetric flexural vibration and thermoelastic stability of FGM circular plates using finite element method. *Compos. Part B* **2006**, *37*, 642–649. [CrossRef]
18. Talha, M.; Singh, B.N. Static response and free vibration analysis of FGM plates using higher order shear deformation theory. *Appl. Math. Model.* **2010**, *34*, 3991–4011. [CrossRef]
19. Talha, M.; Singh, B.N. Large amplitude free flexural vibration analysis of shear deformable FGM plates using nonlinear finite element method. *Finite Elem. Anal. Des.* **2011**, *47*, 394–401. [CrossRef]
20. Kulkarni, S.D.; Trivedi, C.J.; Ishi, R.G. Static free vibration analysis of functionally graded skew plates using a four node quadrilateral element. In *Advances in Structural Engineering*; Matsagar, V., Ed.; Springer India: New Delhi, India, 2015; pp. 15–24.
21. Ramu, I.; Mohanty, S.C. Modal analysis of functionally graded material plates using finite element method. *Proced. Mater. Sci.* **2014**, *6*, 460–467. [CrossRef]
22. Valizadeh, N.; Natarajan, S.; Gonzalez-Estrada, O.A.; Rabczuk, T.; Bui, T.Q.; Bordas, S.P.A. NURBS-based finite element analysis of functionally graded plates: Static bending, vibrations, buckling and flutter. *Compos. Struct.* **2013**, *99*, 309–326. [CrossRef]

23. Tran, L.V.; Thai, C.H.; Nguyen-Xuan, H. An isogeometric finite element formulation for thermal buckling analysis of functionally graded plates. *Finite Elem. Anal. Des.* **2013**, *73*, 65–76. [CrossRef]
24. Koteswara Rao, D.; Blessington, P.J.; Tarapada, R. Finite element modeling and analysis of functionally graded (FG) composite shell structures. *Proced. Eng.* **2012**, *38*, 3192–3199.
25. Daszkiewicz, K.; Chróścielewski, J.; Witkowski, W. Geometrically nonlinear analysis of functionally graded shells based on 2-D cosserat constitutive model. *Eng. Trans.* **2014**, *62*, 109–130.
26. Chakraborty, A.; Gopalakrishnan, S.; Reddy, J.N. A new beam finite element for the analysis of functionally graded materials. *Int. J. Mech. Sci.* **2003**, *45*, 519–539. [CrossRef]
27. Filippi, M.; Carrera, E.; Zenkour, A.M. Static analyses of FGM beams by various theories and finite elements. *Compos. Part B* **2015**, *72*, 1–9. [CrossRef]
28. Kahya, V.; Turan, M. Finite element model for vibration and buckling of functionally graded beams based on the first-order shear deformation theory. *Compos. Part B* **2017**, *109*, 108–115. [CrossRef]
29. Gilewski, W.; Gomuliński, A. Physical shape functions in finite element analysis of moderately thick plates. *Int. J. Number. Methods Eng.* **1991**, *32*, 1115–1135. [CrossRef]
30. Kączkowski, Z. *Plates—Static Calculations*; Arkady: Warsaw, Poland, 1980. (In Polish)
31. Gilewski, W. *On the Criteria for Evaluation of Finite Elements—From Timoshenko Beam to Hencky-Bolle Plate*; Warsaw University of Technology Publishing House: Warsaw, Poland, 2005. (In Polish)
32. Gilewski, W. *Physical Shape Functions in the Finite Element Method*; Studies in Civil Engineering 81, Polish Academy of Sciences, Committee of Civil Engineering: Warsaw, Poland, 2013. (In Polish)
33. Hughes, T.J.R.; Cottrell, J.A.; Bazilievs, Y. Isogeometric analysis: CAD, finite elements, NURBS, exact geometry and mesh refinement. *Compos. Methods Appl. Mech. Eng.* **2005**, *194*, 4135–4195. [CrossRef]
34. *Abaqus version 6.13. Simula, ABAQUS User Subroutine Reference Guide*; Dassault Systemes Simula Corp.: Providence, RI, USA, 2013.
35. Boulbes, R.J. *Troubleshooting Finite-Element Modeling with Abaqus*; With Applications in Structural Engineering Analysis; Springer Nature Switzerland AG: Basel, Switzerland, 2020.
36. Maplesoft – Mathematics-Based Software and Services for Education, Engineering and Research. Available online: https://www.maplesoft.com/support/help (accessed on 1 February 2020).
37. Daouadji, T.H.; Tounsi, A.; Adda Beida, E.A. Analytical solution for bensing analysis of functionally graded plates. *Sci. Iran. Trans. B Mech. Eng.* **2013**, *20*, 516–523.

materials

MDPI

Article

FE Analyses of Hyperelastic Solids under Large Bending: The Role of the Searle Parameter and Eulerian Slenderness

Federico Oyedeji Falope [1,2], Luca Lanzoni [1,2] and Angelo Marcello Tarantino [1,2,*]

1 Department of Engineering Enzo Ferrari, DIEF, University of Modena and Reggio Emilia, via P. Vivarelli 10, 41125 Modena, Italy; federicooyedeji.falope@unimore.it (F.O.F.); luca.lanzoni@unimore.it (L.L.)
2 Centro di Ricerca Interdipartimentale Costruzioni e del Territorio, CRICT, via P. Vivarelli 10, 41125 Modena, Italy
* Correspondence: angelomarcello.tarantino@unimore.it; Tel.: +39-059-205-6117

Received: 18 February 2020; Accepted: 24 March 2020; Published: 1 April 2020

Abstract: A theoretical model concerning the finite bending of a prismatic hyperelastic solid has been recently proposed. Such a model provides the 3D kinematics and the stress field, taking into account the anticlastic effects arising in the transverse cross sections also. That model has been used later to extend the Elastica in the framework of finite elasticity. In the present work, Finite Element (FE) analyses of some basic structural systems subjected to finite bending have been carried out and the results have been compared with those provided by the theoretical model performed previously. In the theoretical formulation, the governing equation is the nonlinear local relationship between the bending moment and the curvature of the longitudinal axis of the bent beam. Such a relation has been provided in dimensionless form as a function of the Mooney–Rivlin constitutive constants and two kinematic dimensionless parameters termed Eulerian slenderness and compactness index of the cross section. Such parameters take relevance as they are involved in the well-known Searle parameter for bent solids. Two significant study cases have been investigated in detail. The results point out that the theoretical model leads to reliable results provided that the Eulerian slenderness and the compactness index of the cross sections do not exceed fixed threshold values.

Keywords: Finite elements; finite bending; 3D elasticity; Eulerian slenderness; compactness index; Searle parameter; Elastica

1. Introduction

The nonlinear bending theory of elastic bodies has attracted a lot of interest because of its relevance in many physical and engineering applications. As an example, in the past decades, the diffusion of robotic technologies has demanded the precise knowledge of the mechanical response of technological components subjected to large bending.

The soft robots [1], i.e., robots based on extremely compliant components, are used to produce pneumatic robots [2], to simulate artificial systems [3], animals [4], human hands [5] and other gripper devices [6]. For these kinds of high-tech applications, the mechanical role of the external load is played by the light [7], humidity [8] or electricity [9] to drive motion.

In the aforementioned and many other applications, the theory of elastic bending allows predicting strains and stresses in the deformed solid. For situations in which the displacement and strains are small, the classical linearized elasticity theory provides reliable results. In such a framework, a certain number of closed-form solutions are available for beams [10], shells and plates [11,12] under different loading and boundary conditions. A few closed-form solutions can be found also for the 3D elastic bodies, with special reference to symmetric layouts [13,14].

However, linear elasticity cannot be used to properly assess the mechanical response of bodies that exhibit large displacements and/or strains, like, for example, tentacle action or hand closure in soft robots. For these and other contexts the finite elasticity is much more appropriate than the linearized theory.

From a mathematical standpoint, a large number of challenging insights are involved in the finite theory. Nevertheless, also in the framework of nonlinear elasticity, a certain number of analytical solutions in the context of homogeneous deformations are available for some basic layouts, like prismatic bodies under axial dead loads [15–18] or shear loads [19].

Concerning finite bending, various studies can be found in Literature [20–22]. Such studies are typically approached through the semi-inverse method: Some geometrical assumptions about the kinematics allows obtaining the displacement field, which is definitely assessed by solving a boundary value problem provided by the equilibrium conditions. However, all the aforementioned works have been carried out by assuming plane strain or plane stress conditions, thus reducing the problem of the finite bending of solids to a 2D problem.

Recently, the Rivlin formulation [20] has been extended to a 3D framework by taking into account also the anticlastic deformation arising in the transverse cross sections of the solid subjected to uniform bending [23]. Later, the analysis has been extended to beams subjected to variable bending moment [24,25]. In that study, a nonlinear relation between the bending moment and the curvature of the longitudinal axis has been found for a compressible Mooney–Rivlin material. As a matter of fact, this relation represents a generalization of the well known Elastica [26] in the context of finite elasticity.

In the present work, based on the theoretical formulation reported in [23,24], some study cases are analytically investigated. In particular, the applications consist of a clamped beam subjected to a couple or a shear force acting at its free end. The nonlinear relation between the bending moment and the curvature of the longitudinal axis is written in a dimensionless form, thus highlighting the relevance of the Eulerian slenderness together with a second dimensionless parameter, which stands for the compactness of the transverse cross sections. Such parameters are coupled into the well known Searle parameter [27,28]. The theoretical results in terms of displacements of the longitudinal axis, stretch and stress distributions within the cross sections are compared with the results provided by Finite Element (FE) simulations. The comparison allows assessing the reliability of the theoretical model to predict accurately the mechanical response of beams under large bending based on fixed threshold values of the governing parameters.

The paper is organized as follows: In Section 2 a brief remark about the theoretical model [23] is provided, with particular emphasis on the basic assumptions concerning the kinematics. The governing equation in terms of moment-curvature relationship is then provided in dimensionless form based on the definition of the Eulerian slenderness and the compactness index of the cross sections. The main results obtained by the theoretical model and the FE simulations are compared in Section 3 through the investigation of two study cases. In that Section, the role played by the Eulerian slenderness and the compactness of the cross sections is discussed in detail and relevant threshold values of the governing parameters are found. Finally, conclusions are drawn in Section 4.

2. The Theoretical Model for the Finite Bending of Solids

2.1. Remarks on the Theoretical Model

In the present Section the basic assumptions of the theoretical model are briefly recalled [23]. Let us consider a prismatic body of length L, width B and height H, placed in a Cartesian reference system $\{O, X, Y, Z\}$. The solid is uniformly bent around the X-axis by imposing a prescribed rotation angle $2\alpha_0 = L/R_0$ at the end cross sections, as sketched in Figure 1.

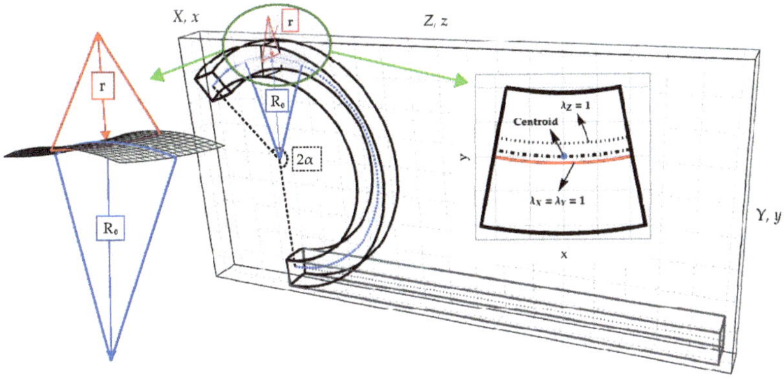

Figure 1. Prismatic hyperelastic solid under bending: Kinematic parameters, anticlastic surface and deformed configuration of the cross section.

Hereinafter R_0 denotes the radius of curvature of the centroidal line ($X = 0, Y = 0$) in the ZY-plane, which is constant along the longitudinal Z-axis, see Figure 1. The kinematics is such to distinguish the centroidal line from the longitudinal neutral fibre ($X = 0, Y = QM : \lambda_Z = 1$) and from the transverse neutral plane ($Y = QN : \lambda_X = \lambda_Y = 1$, being λ_J the stretch along the J-axis).

The kinematics is based on the following three basic hypotheses [23]:

1. the longitudinal fibres, parallel to Z-axis, after bending are deformed into arcs of circumferences (blue curves in Figure 1). As reported above, the longitudinal radius of curvature of the deformed centroidal fibre is denoted as R_0, whereas the longitudinal fibre with unitary transverse stretches ($\lambda_X = \lambda_Y = 1$) is characterized by the longitudinal radius of curvature R;
2. during bending, transverse cross sections belonging to XY planes preserve their planarity and exhibit the same deformation;
3. solid transverse fibres, parallel to X-axis, after bending are deformed into arcs of circumferences (red curves in Figure 1). The transversal fibre with unitary transverse stretches ($\lambda_X = \lambda_Y = 1$), is characterized by the anticlastic (or transverse) radius if curvature r.

Therefore, conversely to existing models on the 2D finite bending [20–22], in the theoretical formulation proposed in [23] a complete 3D description of the kinematics is provided. Due to bending, both the longitudinal and anticlastic curvatures of each fibre of the solid are considered. As a result, all components of the strain and stress fields are provided.

The second hypothesis is widespread in the framework of the linearized beam theory and it is known as Euler–Bernoulli assumption. As shown in [29,30], such an assumption is still valid at large displacements and strains. It is remarked that the proposed formulation neglects the deformation induced by shearing or axial loads.

From the third hypothesis about the anticlastic radius of curvature, it follows the invariance of the transverse stretch λ_X and λ_Y with respect to (w.r.t.) the X-variable. However, as shown in the following, such an assumption depends on the governing parameters.

Let us define $\beta_C = B/H$ as the *compactness index* of the cross sections. As already pointed out by Lamb [28] and experimentally by Searle [27], the cross sections of an elastic bent solid exhibit a variable curvature along the solid width. Such a variation increases in the neighborhood of the corners of the cross section.

The equilibrium condition, as shown in [23], is exactly fulfilled for the centroidal fiber in terms of local equilibrium equations, whilst it is satisfied in average at the boundaries. Therefore, moving away from the centroidal fiber, the equilibrium fulfillment loses its accuracy, but with negligible errors in the case of slender solids with compact cross sections [24].

The concept of slenderness can be referred to the Lagrangian or Eulerian configurations. In the classical beam theory [10], the Lagrangian slenderness is defined as the ratio $\beta_{LS} = L/\max(B, H)$ [10]. Generally, a beam can be considered slender if $\beta_{LS} \geq 10$ regardless the magnitude of strain or displacement fields involved in the deformation process (a definition of *Eulerian slenderness* β_{ES} will be provided in Section 2.2 based on the nonlinear moment-curvature relationship, showing that the bending problem is governed by two parameters: β_{ES} and β_C).

In the following we consider a hyperelastic prismatic solid characterized by a stored energy density function ω_{MR}, here assumed according to that of a compressible Mooney–Rivlin (MR) material [31]. The stored energy density function of an isotropic material is expressed as a function of the invariants I_1, I_2, I_3 of the right Cauchy–Green strain tensor $\mathbf{C} = \mathbf{F}^T\,\mathbf{F}$ as follows (for any details in terms of symbols and notation the Reader is referred to [23]):

$$\omega_{MR}(I_1, I_2, I_3) = a(I_1 - 3) + b(I_2 - 3) + c(I_3 - 1) - (a + 2b + c)\ln(I_3), \tag{1}$$

with

$$I_1 = \lambda_X^2 + \lambda_Y^2 + \lambda_Z^2, \quad I_2 = \lambda_X^2\lambda_Y^2 + \lambda_X^2\lambda_Z^2 + \lambda_Y^2\lambda_Z^2, \quad I_3 = \lambda_X^2\lambda_Y^2\lambda_Z^2,$$

where a, b and c denote the constitutive MR parameters.

Kinematics, equilibrium conditions and constitutive law allows assessing the displacement field $\mathbf{s}(X,Y,Z) = u\,\mathbf{i} + v\,\mathbf{j} + w\,\mathbf{k}$ as [23]

$$\mathbf{s} = \begin{bmatrix} -X + r\,e^{-\frac{X+QN}{r}} \sin\frac{X}{r} \\ -Y - R - QN + \left[R + r\left(1 - e^{-\frac{Y+QN}{r}}\cos\frac{X}{r}\right)\right]\cos\frac{Z}{R_0} \\ -Z + \left[R + r\left(1 - e^{-\frac{Y+QN}{r}}\cos\frac{X}{r}\right)\right]\sin\frac{Z}{R_0} \end{bmatrix}, \tag{2}$$

which is completely known once the following nonlinear system

$$\begin{cases} r\left[R_0^2(a+2b) - aR^2 + c\left(R^2 + R_0^2\right)\right] - 2R\left[R_0^2(a+3b) + c\left(R^2 + R_0^2\right)\right] = 0 \\ R_0 - R = r\left(1 - \cos\dfrac{B}{2r}\right) \\ QN = r\log\left(\cosh\dfrac{H}{2r}\right) \end{cases} . \tag{3}$$

It is remarked that Equation $(3)_1$ follows by imposing the equilibrium condition, $\mathrm{Div}\,\mathbf{T}_\mathrm{R} = \mathbf{0}$, along the Y and Z directions. Equation $(3)_2$ is obtained by imposing that the lateral surface of the bent solid is unloaded, i.e., the Piola-Kirchhoff stress vector, $\mathbf{t}_\mathrm{R} = \mathbf{T}_\mathrm{R}\mathbf{n}$, must be $\mathbf{0}$ on the contour of each cross section ($\mathbf{n}$ is the outward unit normal). Finally Equation $(3)_3$ comes from a simplifying assumption about stretch λ_Y along Y direction (for details, see [23]). Therefore system (3) is solved in the unknown kinematic parameters R, r and QN, and then the principal stretches can be evaluated as the roots of the diagonal components of the right Cauchy-Green strain tensor [23]

$$\lambda_X = \lambda_Y = e^{-\frac{QN+Y}{r}}, \quad \lambda_Z = \frac{r(1 - e^{-\frac{QN+Y}{r}}\cos\frac{X}{r}) + R}{R_0}. \tag{4}$$

Once the displacement field (2) is known, both the (first) Piola–Kirchhoff $\mathbf{T}_\mathrm{R}$ and Cauchy $\mathbf{T}$ stress tensors can be readily obtained as (for details about the expression of the Piola–Kirchhoff stress tensor, see Equations (30) and (56) of [23]. Analogous expressions for the Piola–Kirchhoff stress tensor are given also in [21,31])

$$\mathbf{T}_\mathrm{R} = \mathbf{R}\,\mathbf{S}, \quad \mathbf{T} = (\mathbf{F}^*)^{-1}\,\mathbf{T}_\mathrm{R}, \tag{5}$$

where symbol $^{(*)}$ stands for the cofactor and tensors $\mathbf{R}^T = \mathbf{R}^{-1}$ and $\mathbf{S}$ turn out to be

$$[\mathbf{R}] = \begin{bmatrix} \cos\frac{X}{r} & -\sin\frac{X}{r} & 0 \\ \cos\frac{Z}{R_0}\sin\frac{X}{r} & \cos\frac{X}{r}\cos\frac{Z}{R_0} & -\sin\frac{Z}{R_0} \\ \sin\frac{X}{r}\sin\frac{Z}{R_0} & \cos\frac{X}{r}\sin\frac{Z}{R_0} & \cos\frac{Z}{R_0} \end{bmatrix}, \qquad [\mathbf{S}] = \begin{bmatrix} S & 0 & 0 \\ 0 & S & 0 \\ 0 & 0 & S_Z \end{bmatrix},$$

with

$$S = 2e^{-\frac{3(QN+Y)}{r}}\left[-(a+2b+c)e^{\frac{4(QN+Y)}{r}} + ae^{\frac{2(QN+Y)}{r}} + b\right] + \frac{2e^{-\frac{5(QN+Y)}{r}}\left[be^{\frac{2(QN+Y)}{r}} + c\right]\left[(r+R)e^{\frac{QN+Y}{r}} - r\cos\frac{X}{r}\right]^2}{R_0^2},$$

$$S_Z = \frac{2e^{-\frac{5(QN+Y)}{r}}\left[(r+R)e^{\frac{QN+Y}{r}} - r\cos\frac{X}{r}\right]}{R_0}\left\{ae^{\frac{4(QN+Y)}{r}} + 2be^{\frac{2(QN+Y)}{r}} + c - \frac{R_0^2(a+2b+c)e^{\frac{6(QN+Y)}{r}}}{\left[(r+R)e^{\frac{QN+Y}{r}} - r\cos\frac{X}{r}\right]^2}\right\}.$$

Relations (1-5) will be used in Section 3 to compare the FE results with the theoretical predictions in terms of deformed configurations and stretch and stress distributions within the cross section.

2.2. Generalization to Variable Bending Moment

The theoretical model [23] was extended to the cases of nonuniform bending of slender beams in [24] (in [24], the deformed configuration of the centroidal fibre is described by the curvilinear abscissa s as sketched in Figure 2. Therefore, both the bending moment and the radii of curvature R and r turn out to be functions of the deformed beam axis, i.e., $m_x = m_x(s)$, $R_0 = R_0(s)$ and $r = r(s)$. Likewise [24], here R_0 is referred to the longitudinal neutral fibre corresponding to $Y = 0$, for which $\lambda_Z = 1$. In other words, the simplifications $R_0 = R$ and $QN = 0$ assumed in [24] is adopted here. Therefore, in the present work, the longitudinal axis of the beam corresponds to the fibre at $X = 0, Y = 0, Z = Z$). In particular, the nonlinear relation between the internal bending moment (moment due to the Eulerian stress T_3 over the transverse cross section [24]) and the longitudinal radius of curvature R_0 is given as

$$m_x(s) = \frac{E_{MR} I_X}{R_0(s)} + \frac{10}{3}\left[\frac{6B^2}{R_0(s)H}\right]^3 \sum_{i=1}^{2} g_i + O(R_0^{-5}), \tag{6}$$

where s is the curvilinear coordinate along the deformed configuration of the longitudinal axis, $I_X = BH^3/12$ is the second moment of inertia of the (undeformed) cross section and

$$g_i = \frac{H^{2(i+2)}}{B^{2i+1}} E_{MR,i},$$

are parameters in which $E_{MR,i}$ are defined as follows:

$$E_{MR,1} = 5\nu_{MR}\left\{a(6\nu_{MR}+3) + b[4\nu_{MR}(5-2\nu_{MR})+6] + c[2\nu_{MR}(7-6\nu_{MR})+3]\right\},$$
$$E_{MR,2} = 3\left\{4\nu_{MR}^3(8b+23c) - \nu_{MR}^2(7a+62b+79c) - 9\nu_{MR}(a+2b+c) - 6(a+2b+c)\right\}.$$

It is worth nothing that the transition from finite to linearized elastic theory reported in [23] has led to some relations between the MR constitutive parameters and the usual elastic moduli, i.e., the Young modulus E_{MR} and Poisson ratio ν_{MR}, according to

$$E_{MR} = \frac{4(a+b)(a+4b+3c)}{a+3b+2c}, \qquad \nu_{MR} = \frac{b+c}{a+3b+2c}. \tag{7}$$

In order to investigate the reliability of the theoretical model varying the geometry of the beam, let us rewrite Equation (6) in the following dimensionless form:

$$
\begin{aligned}
\overline{m}_x(s) &= \beta_{ES}^{-1}\Bigg\{1-\frac{\beta_{ES}^{-2}}{4E_{\rm MR}}\Bigg\{\beta_C^2\Bigg\{\frac{\nu_{\rm MR}}{3}a(6\nu_{\rm MR}+3)+b[4\nu_{\rm MR}(5-2\nu_{\rm MR})+6]+c[2\nu_{\rm MR}(7-6\nu_{\rm MR})+3]\Bigg\}+\\
&\quad \frac{4\nu_{\rm MR}^3(8b+23c)-\nu_{\rm MR}^2(7a+62b+79c)-9\nu_{\rm MR}(a+2b+c)-6(a+2b+c)}{5}\Bigg\}\Bigg\}+O(\beta_{ES}^{-5}),\\
&= \beta_{ES}^{-1}\left[1+\sum_{n=1}^{N}\beta_{ES}^{-2n}(\rho_{n0}+\rho_n\beta_C^2)\right]+O(\beta_{ES}^{-(2N+3)}),
\end{aligned} \tag{8}
$$

in which $\overline{m}_x(s) = 12m_x(s)/E_{\rm MR}BH^2$ denotes the dimensionless bending moment, whilst $\beta_{ES} = R_0/H$ is the Eulerian slenderness which, together with the compactness index $\beta_C = B/H$, governs the problem. Parameters ρ_{n0} and ρ_n involves only the constitutive parameters. The form of such parameters for $N = 2$ turns out to be

$$
\begin{aligned}
\rho_{10} &= \frac{4(8b+23c)\nu_{\rm MR}^3-(7a+62b+79c)\nu_{\rm MR}^2-9(a+2b+c)\nu_{\rm MR}-6(a+2b+c)}{5},\\
\rho_1 &= \frac{\nu_{\rm MR}\{(6\nu_{\rm MR}+3)a+[4\nu_{\rm MR}(5-2\nu_{\rm MR})+6]b+[2\nu_{\rm MR}(7-6\nu_{\rm MR})+3]c\}}{3},\\
\rho_{20} &= \frac{1}{840}\Big[2(166b+1441c)\nu_{\rm MR}^5-(31a+1382b+4891c)\nu_{\rm MR}^4-135(a+2b+c)\nu_{\rm MR}^3-\\
&\quad 390(a+2b+c)\nu_{\rm MR}^2-450(a+2b+c)\nu_{\rm MR}-180(a+2b+c)\Big],\\
\rho_2 &= \frac{\nu_{\rm MR}}{240}\Big[-8(10b+39c)\nu_{\rm MR}^4+2(15a+190b+343c)\nu_{\rm MR}^3+\\
&\quad 75(a+2b+c)\nu_{\rm MR}^2+120(a+2b+c)\nu_{\rm MR}+60(a+2b+c)\Big].
\end{aligned}
$$

Relation (8) represents the series expansion of Equation (3.1) of [32] w.r.t. the Eulerian slenderness, thus highlighting the relevance of such a parameter. The Equation (8) truncated at the leading order term, namely β_{ES}^{-1}, can be interpreted as the Elastica extended to the context of finite elasticity.

It is worth noticing that, in the Lamb theory for shells [28], that Author observed that *"for sufficiently small curvatures, i.e., so long as R is large compared to B^2/H the shell profile is close to a paraboloid"*. He found that this situation is encountered for low values of the dimensionless parameter $\sqrt[4]{3/4(1-\nu_{MR})}B/\sqrt{R_0H}$, being ν_{MR} the Poisson ratio. As this parameter increases, nonuniform curvature occurs in the cross sections, especially in the neighboring of the corners. Later, based on the Lamb work, Searle [27] experimentally observed that for a high value of the so-called Searle parameter β (it is $\beta^2 = B^2/R_0H$), the variation of the deformation field in the transverse cross sections significantly increases. In particular, for high values of β the profile of the cross sections of a bent plate is characterized by an almost flat region in the inner part of the cross section and low values of the anticlastic radius close to the corners. This effect gives rise to the "curl effect". The condition $\beta = 20$ was experimentally inferred by Searle in [27] as a threshold value to distinguish among bodies with compact cross sections (beams) and plates.

Based on the above positions, a straightforward relation between the Searle parameter and the Eulerian slenderness can be established by means of the compactness index β_C according to $\beta^2 = \beta_C^2/\beta_{ES}$. As shown in the following, the reliability of the theoretical model to properly describe the mechanical behavior of beams under large bending depends on both the parameters β_{ES} and β_C.

2.3. The Numerical Procedure

The implementation of the theoretical procedure consists of two main steps: The determination of the deformed axis of the beam by means of Equation (8), which is solved iteratively, and then the assessment of the deformed configuration of the beam axis. This allows the complete description of the 3D kinematics of the bent beam, accounting for the deformation of the transverse cross sections.

It is remarked that, according to [24], in the present work, R_0 is the radius of curvature of the longitudinal neutral fibre $X = 0, Y = 0$ (therefore $R = R_0$ and $QN = 0$).

The theoretical model is here implemented for two study cases: A cantilever beam subjected to constant (first study case) and variable (second study case) bending moment. Let us denote with n the number of discretizations of the beam axis, and with $s_i^{(j)}$ the ith node at the iteration jth. Each element (the elements are here assumed equispaced for simplicity, i.e., $\Delta s = L/n = \|s_{i+1} - s_i\|$) preserves its original length during deformation, as sketched in Figure 2.

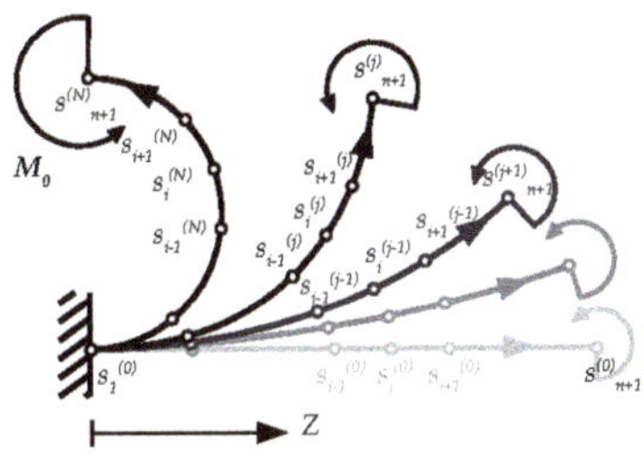

Figure 2. Discretization of the beam axis: Location of nodes at different iterations.

The iteration number corresponding to the achievement of the convergence criterion is denoted by N.

Therefore, starting form a prescribed guess solution (tipically, the solution provided by the linearized theory), the radius of curvature R_0 of the longitudinal axis of the beam is known at a given node $s_i^{(j)}$ through relation (6). Once the radius of curvature is known, the curvature $\chi_0(s) = 1/R_0(s)$, provides the rotation $\theta(s)$ at each node as

$$\theta(s_i^{(j)}) = \theta(s_1^{(j)}) + \frac{\Delta s}{2} \sum_{k=2}^{i} \left[\chi(s_k^{(j)}) + \chi(s_{k-1}^{(j)}) \right], \tag{9}$$

and then the displacement field follows according to:

$$v(s_i^{(j)}) = v(s_1^{(j)}) - \frac{\Delta s}{2} \sum_{k=2}^{i} \left[\sin \theta_{k-1}^{(j)} + \sin \theta_k^{(j)} \right], \tag{10}$$

$$w(s_i^{(j)}) = w(s_1^{(j)}) - \frac{\Delta s}{2} \sum_{k=2}^{i} \left[2 - \cos \theta_{k-1}^{(j)} - \cos \theta_k^{(j)} \right]. \tag{11}$$

Summing up, the present approach is based on the 1D solution of the governing Equation (8), followed by the determination of the displacement field of the transverse cross sections. In detail, the numerical procedure used in Section 3 consists of the following steps:

- starting from a trial solution corresponding to the linearized theory, the bending moment $m_x(s_i^{(1)})$ is obtained at the first iteration;
- for each node, Equation (6) is solved in the unknown radius of curvature $R_0(s_i^{(1)})$;
- then, the rotation and displacements fields are assessed from (9)–(11) for each node, providing the Eulerian coordinates of the beam axis, i.e., $[z_i^{(2)}, y_i^{(2)}] = [Z_i + w(s_i^{(1)}), Y_i + v(s_i^{(1)})]$;
- two convergence criteria have been adopted: The relative error between two subsequent iterations in terms of displacement norm and the potential energy;
- in case of not convergent results, the deformed configuration (not convergent) is used as guest solution for assessing a further bending moment distribution, $m_x(s_i^{(2)})$. Therefore, the iterative procedure restarts from the second issue since both the convergence criteria are reached.

Given the elastic potential energy Π at iteration j-th defined as

$$\Pi^{(j)}(\mathbf{s}^{(j)}, \boldsymbol{\theta}^{(j)}) = \frac{\Delta s}{2} \sum_{i=1}^{n+1} \chi(s_i^{(j)})\, m(s_i^{(j)}) - \mu \begin{cases} M_0\, \theta(s_{n+1}^{(j)}), & \text{study case 1} \\ F_0\, w(s_{n+1}^{(j)}), & \text{study case 2,} \end{cases}$$

the relative errors between subsequent iterations in terms of energy and the displacement norms are defined as

$$\epsilon_{\Pi^{(j)}} = \frac{\Pi^{(j)} - \Pi^{(j-1)}}{\Pi^{(j)}}, \qquad \epsilon_{\mathbf{s}^{(j)}} = \frac{\|\mathbf{s}^{(j)}\| - \|\mathbf{s}^{(j-1)}\|}{\|\mathbf{s}^{(j)}\|},$$

the solution at the iteration Nth is convergent if the following criterion holds true:

$$\max\left\{\epsilon_{\Pi^{(N)}}, \epsilon_{\|\mathbf{s}^{(N)}\|}\right\} < 10^{-8}. \tag{12}$$

3. Study Cases

Two study cases are considered here: A clamped beam subjected to a bending couple (case A_1 sketched in Figure 3a) and a clamped beam subjected to a dead shear force (case A_2 shown in Figure 3b) acting at the free end.

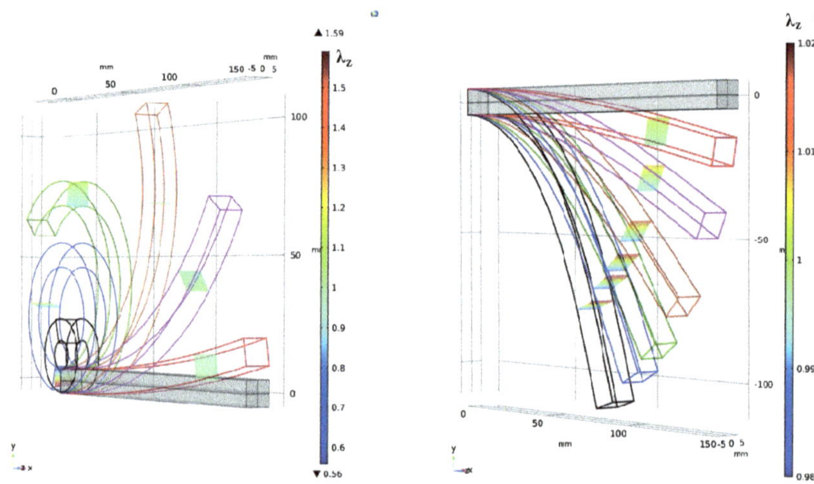

(**a**) A_1. (**b**) A_2.

Figure 3. Sketch of the study cases: Reference and deformed configurations provided by the FE solution.

For each application, three subcases related to three different sizes of the cross section have been considered according to Table 1. Note that both the thickness and length have been fixed, varying the width of the cross sections only.

Table 1. Dimensionless geometric parameters related the investigated subcases: $H = 10$ [mm], $\beta_C = B/H$ and $\beta_{LS} = L/\max(B,H)$.

Subcases	(a)	(b)	(c)
β_C	1	2	5
β_{LS}	15	7.5	3

In the following, notation $A_2(c)$ stands for study case A_2 with the cross section geometries corresponding to the subcase (c) and so on.

Subcase (a) represents a slender beam with square compact cross section, which properly fulfills the theoretical hypotheses. Subcase (b) represents an intermediate situation of a beam weakly slender, $\beta_{LS} < 10$, with a flat cross section ($\beta_C = 2$). To emphasize the effect induced by loss of slenderness, the subcase (c) resembling a plate ($\beta_{LS} = 3$ and $\beta_C = 5$) is considered also.

The following MR constitutive parameters have been adopted: $a = 1$, $b = 0.05$ and $c = 2.256$ [MPa] according to [32].

The external loads, M_0 and F_0 for cases A_1 and A_2 respectively, incremented by the load multiplier μ, are such to induce for $\mu = 1$ a deflection at the free end of $L/100$ according to the linearized theory.

For both the study cases, the theoretical model has been implemented by subdividing the longitudinal axis into 100 equispaced elements. The FE simulations involve high computational effort owing to the nonlinear nature of the problem. However, the FE solution provides higher accuracy in terms of equilibrium fulfillment w.r.t. the theoretical formulation, for which the equilibrium has been imposed in the weak form (it is remarked that in the theoretical formulation the equilibrium is satisfied exactly for the sole centroidal fiber).

Conversely to the theoretical model, the FE model accounts for the effect of both the axial or shear load on the deformation field. In other words, the theoretical formulation takes into account only the bending effects induced by the axial or shear forces. In the FE models, the external loads have been simulated indirectly by applying prescribed displacements in order to optimize the achievement of convergence. Therefore, prescribed rotation (case A_1) or transverse displacement (case A_2) have been imparted at the final cross section and to its centroid, respectively. As a consequence, a comparison between the theoretical predictions and FE solutions will be performed in terms of stress resultants over the cross sections. The principal stretch and stress distributions within the beam cross section will be discussed in the following sections.

3.1. Cantilever Beam Subjected to a Couple at Its Free End (Case A_1)

3.1.1. Theoretical Solution

A cantilever beam subjected to an external couple $M_0 = \mu E_{\mathrm{MR}} I_X / 50L$ acting at the free end, as sketched in Figure 4, is investigated in the present section. The values assumed by the load multiplier μ have been set according to Figure 4a. Due to the loading condition, the beam is subjected to a constant bending moment M_0 along its longitudinal axis.

The bending moment distribution w.r.t. the beam axis is obtained as $m_x(s_i) = M_0$, then the radius of curvature and rotation of the longitudinal axis are achieved by using relations (6)–(9). Finally, the deformed configuration is determined by using Equations (10) and (11) (see Figure 4b).

In order to investigate the effects of bending inside the cross section, stretches and stresses have been computed using Equations (2)–(5).

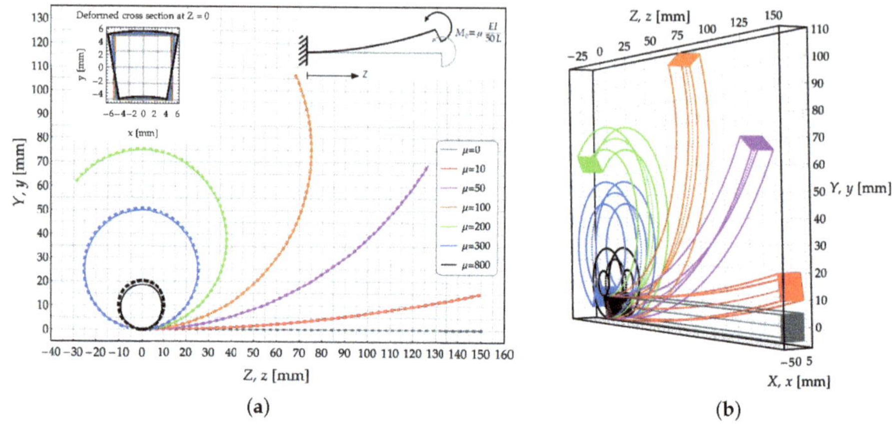

Figure 4. Study case A_1: Cantilever beam subjected to an external couple acting at its free end. (**a**) Deformed configuration of the longitudinal axis: Theoretical model (continuous lines) and FE model (dashed lines); (**b**) 3D skecth of the deformed configurations obtained by the theoretical solution.

3.1.2. The FE Model Solution

The FE simulations have been carried out by using COMSOL Multiphysics® v 5.5 software. The non-linear structural mechanics modules of the FE code allows simulating hyperelastic materials. The parametric approach of the FE code makes it possible to define the elastic strain energy function also for compressible materials. Therefore, the MR store energy function (1) has been directly implemented as a function of the principal invariants of the Cauchy-Green strain tensors.

Three FE models (see Figure 3) in agreement with the geometries listed in Table 1, have been modelled by using 4-nodes tetrahedron elements. In this way, each subcase is characterized by 131,282, 211,251 and 308,667 finite brick elements. The compressible MR law (1) is adopted in the FE simulation to compute the (second) Piola Kirchhoff stress tensor $\mathbf{T}_{\mathrm{R}}^{2nd} = \partial\omega_{MR}/\partial\mathbf{E}$ with $\mathbf{E} = (\mathbf{C} - \mathbf{I})/2$, being $\mathbf{C} = \mathbf{F}^T\mathbf{F}$ the right Cauchy-Green strain tensor and $\mathbf{I}$ the identity tensor.

Various tests about the optimal FE mesh have been required in order to obtain displacements at the centroid of the end cross section less than 1% w.r.t. more refined (but more time consuming) meshes.

The clamped end has been reproduced by restraining the out of plane displacement component for each node, namely $w(X, Y, 0) = 0$. In addition, the centroidal node at the clamped end has been fully restrained, i.e., $\mathbf{s}(0, 0, 0) = \mathbf{0}$. At the free end, the external couple has been reproduced by imposing a prescribed rotation along the X axis. Then, the problem has been handled by increasing step-by-step the prescribed rotation and evaluating a posteriori the corresponding bending moment resultant over the cross section.

3.1.3. Results and Comparison

The deformed configurations of the longitudinal axis are plotted in Figure 4a for different values of the of load multiplier μ. The growth of load multiplier μ is such to induce the beam rolling up on itself, with the simultaneous increase of the beam curvatures.

A first comparison between theoretical predictions and FE results is performed in terms of the bending moment at $Z = L$. In the FE simulations, the bending moment has been assessed by integrating the elementary moments due to the Cauchy stress T_3 w.r.t. the x axis over the deformed cross section.

As reported in Table 2, the relative error of the FE results w.r.t. the theoretical predictions, denoted as $\epsilon_{r,M_0(L)}$, assumes lower values in correspondence of moderate loads. Indeed, it changes in sign as the load multiplier increases starting from $\mu = 10$.

Table 2. Study case A_1: Relative error on the bending moment $\epsilon_{r,m_x(L)}$, Searle parameter $\beta^2 = B^2/R_0H = \beta_C^2/\beta_{ES}$ and weight of the fifth-order truncated term in Equation (8).

Load	Subcases								
Mult.	(a)			(b)			(c)		
μ	$\epsilon_{r,m_x(L)}$	β^2	$\rho_{20}+\rho_2\beta_C^2$	$\epsilon_{r,m_x(L)}$	β^2	$\rho_{20}+\rho_2\beta_C^2$	$\epsilon_{r,m_x(L)}$	β^2	$\rho_{20}+\rho_2\beta_C^2$
10	-3.1×10^{-3}	0.01	1.6×10^{-9}	-5.7×10^{-3}	0.05	-1.2×10^{-4}	2.8×10^{-1}	0.33	2.1×10^{1}
50	-2.4×10^{-3}	0.07	9.8×10^{-7}	-5.5×10^{-3}	0.27	$-8.\times10^{-2}$	2.7×10^{-1}	1.68	5.4×10^{2}
100	$2.\times10^{-4}$	0.13	1.5×10^{-5}	-3.9×10^{-3}	0.53	-1.7	2.6×10^{-1}	3.49	2.3×10^{3}
200	1.3×10^{-2}	0.26	2.3×10^{-4}	5.6×10^{-4}	1.07	5.3×10^{1}	2.3×10^{-1}	19.32	7.1×10^{4}
300	2.5×10^{-2}	0.38	$1.\times10^{-3}$	6.5×10^{-3}	1.62	7.4×10^{1}	2.3×10^{-1}	20.35	7.9×10^{4}
800	1.6×10^{-1}	0.87	2.3×10^{-2}	5.8×10^{-2}	4.71	3.1×10^{2}	3.6×10^{-1}	24.17	1.1×10^{5}

Both the Eulerian slenderness and the compactness of the cross section, represented here through the parameters β_{ES} and β_C, affect the reliability of the theoretical model. To highlight this aspect, the values assumed by the Searle parameter β^2 have been listed in Table 2 for each subcase and load increment. In addition, in the same table the values of the fifth-order term in Equation (8), namely $\rho_{20}+\rho_2\beta_C^2$, have been reported also. Since the calculations have been carried out by truncating the moment-curvature relationship (8) to the third-order term, the evaluation of term $\rho_{20}+\rho_2\beta_C^2$ allows assessing the approximation of the performed analysis.

The Searle parameters, here written as the product of the Eulerian slenderness (which varies with the load multiplier) and the compactness of the cross sections (which is independent of the load multiplier), increases as the load multiplier increases due to the decrease of β_{ES}.

Note that the growth of the Eulerian slenderness decreases the relevance of the high order terms in relation (6). In particular, terms $\rho_{20}+\rho_2\beta_C^2$ listed in Table 2 put in light the relevance of the higher-order terms w.r.t. the first one in the moment-curvature expression. The comparison between theoretical predictions and FE results allows assessing the reliability of the theoretical formulation varying the parameters β_{ES} and β_C.

Results provided by the theoretical model and the FE simulations, in terms of principal stretch distributions inside the cross section at $Z = 0$ for $\mu = 100$, are shown in Figure 5. The contour plots represent the theoretical results whereas the FE solutions are represented with solid isolines with boxed values. The same contour range has been used for representing the principal stretches.

Moving from compact to flat cross sections, i.e., as β_C increases (from left to right in Figure 5), the gap between the FE and theoretical results increases. The FE results indicate that as β_C increases the principal stretches variability w.r.t. the X axis (namely, along the cross section width) takes relevance. On the other hand, the comparison reported in Figure 5 highlights the reliability of the theoretical model to predict the kinematics of the bent beam since, in that case, the Searle parameter assumes low values, being $\beta^2 < 0.5$.

For small values of the Searle parameter, the theoretical and FE solutions in terms of transverse stretch λ_X are almost indistinguishable. In general, the theoretical solution is more reliable in the region close to the core of the cross section, where the theoretical formulation exactly fulfills the equilibrium condition, see Figures 5a,b,d,e,g,h. The gap between the theoretical and FE results slightly increases for the transverse principal stretch λ_Y w.r.t. stretch λ_X.

Conversely, rough theoretical predictions in terms of stretch occur for subcase $A_1(c)$ because of the relevant value assumed by the Searle parameter ($\beta^2 = 3.49$). In order to highlight the effect induced by loss of Eulerian slenderness on the reliability of the theoretical model, Figure 6 shows the increase of the gap between the principal stretches λ_X provided by the theoretical formulation and the FE simulations for subcase $A_1(b)$.

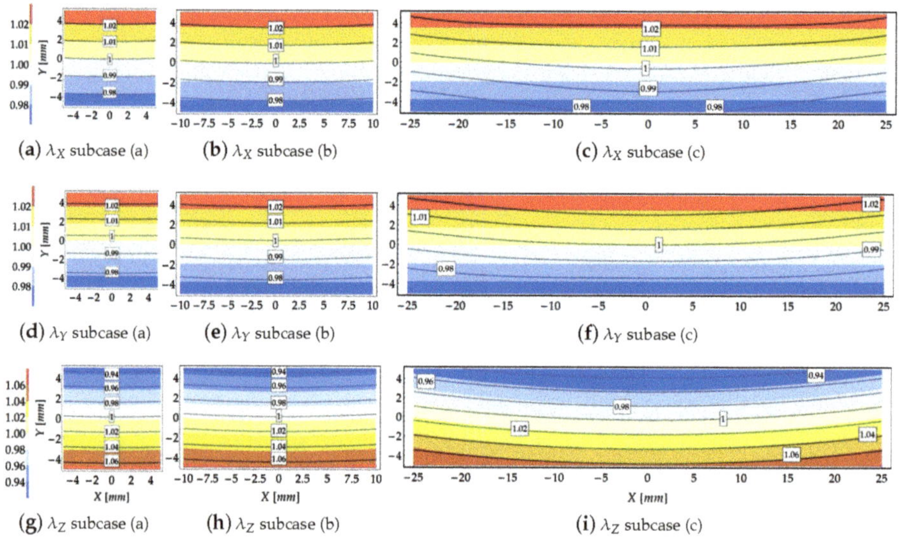

(a) λ_X subcase (a) (b) λ_X subcase (b) (c) λ_X subcase (c)

(d) λ_Y subcase (a) (e) λ_Y subcase (b) (f) λ_Y subase (c)

(g) λ_Z subcase (a) (h) λ_Z subcase (b) (i) λ_Z subcase (c)

Figure 5. Study case A_1 at load increment 3 ($\mu = 100$): Comparison in terms of stretches provided by the theoretical model (contour plot) and the FE solution (solid isolines with boxed value).

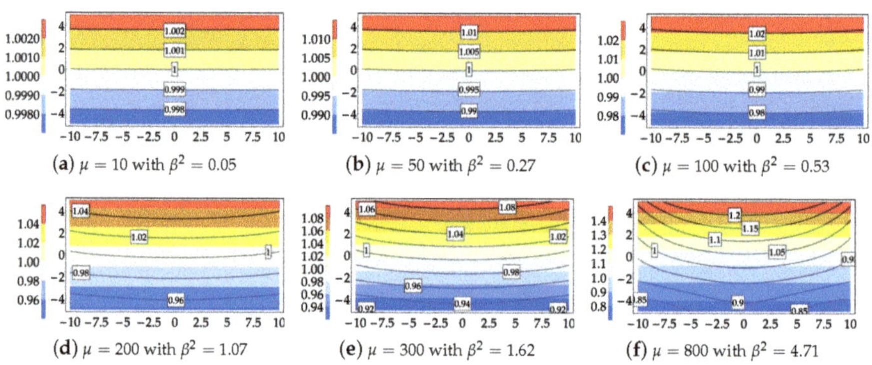

(a) $\mu = 10$ with $\beta^2 = 0.05$ (b) $\mu = 50$ with $\beta^2 = 0.27$ (c) $\mu = 100$ with $\beta^2 = 0.53$

(d) $\mu = 200$ with $\beta^2 = 1.07$ (e) $\mu = 300$ with $\beta^2 = 1.62$ (f) $\mu = 800$ with $\beta^2 = 4.71$

Figure 6. Study case $A_1(b)$: Transversal principal stretch λ_X. Theoretical model (contour plot) and FE simulation (solid isolines with boxed value) for each value of the load multiplier and related values of the Searle parameter.

Figure 6 clearly shows that around the third and fourth load increment a significant gap between the analytical and numerical solutions is encountered.

Another source of mismatch between the theoretical predictions and the FE solutions lies in the approximated nature of the moment-curvature relation (6). It must be remarked that the higher-order terms in the governing equation assume values comparable or greater than the leading order term. Such aspect will be investigated indetail in a forthcoming work.

The distribution of the principal stretch λ_Z within the cross section is displayed in Figure 5g–i. As expected, the isolines are almost equispaced in agreement with the Euler–Bernoulli beam theory, as confirmed also by the FE results. However, as the slenderness decreases (from left to right in Figures 5), the gap between analytical and numerical isolines increases. Note also that the longitudinal neutral fiber for which $\lambda_Z = 1$ is not horizontal, but it resembles an arc of circumference.

The distribution of the Eulerian stresses inside the cross sections is shown in Figure 7 for $\mu = 100$. Differently from Figure 5, for sake of graphical representation each contour plot of Figure 7 is provided with a proper contour legend owing to the wide range assumed by stresses.

As the compactness index β_C increases (from case $A_1(a)$ to case $A_1(c)$), the FE model highlights a sensible change in the distribution of the internal stresses due to "plate effects", neglected in the theoretical formulation. The FE results indicate that, for $\beta_C < 0.5$, the transverse principal stresses T_1 and T_2, assume values two or three orders lower than that of the principal stress T_3, Figure 7a,b,d,e. Note also that, for case $A_1(c)$, the transverse stress T_1 is, in average, higher than that for cases $A_1(a)$ and $A_1(b)$. However, despite the discordance between theoretical and numerical predictions, it is shown that the theoretical model reproduces a main stress T_3 of the same order of those provided by FE simulations.

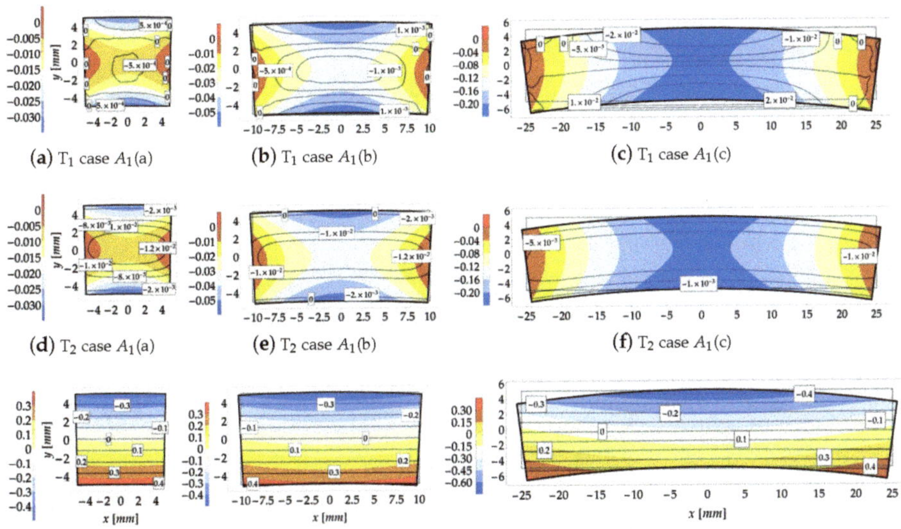

(**a**) T_1 case A_1(a) (**b**) T_1 case A_1(b) (**c**) T_1 case A_1(c)

(**d**) T_2 case A_1(a) (**e**) T_2 case A_1(b) (**f**) T_2 case A_1(c)

(**g**) T_3 case A_1(a) (**h**) T_3 case A_1(b) (**i**) T_3 case A_1(c)

Figure 7. Study case $A_1(b)$ at load increment 3 ($\mu = 100$): Stresses comparison [MPa] between the theoretical model (contour plot) and FE model (isolines with boxed value).

3.2. *Cantilever Subjected to a Shear Force Acting at Its Free end (Case A_2)*

3.2.1. Theoretical Solution

The study case A_2, related to a cantilever subjected to a transverse shear force acting at its free end is investigated in the present Section. A dead shear load $F_0 = 3EI\mu/100L^2$ is applied and incremented according to the load multipliers reported in Figure 8a. In that figure the theoretical model (solid lines) and FE solutions (dashed lines) are compared in terms of deformed configurations (see Figure 8b).

Conversely to case A_1, here the bending moment varies along the longitudinal axis of the beam and it depends on the deformed configuration. Furthermore, in the deformed configuration the shearing dead load produces also axial stresses, see Figure 8a. It is remarked that the theoretical model neglects the effects induced by shear and axial loads on the strain field.

As reported above, the theoretical model is implemented starting with the guess solution corresponding to the solution provided by the linearized elasticity. Once the deformed configuration is known, the bending moment follows from the equilibrium. Then, the obtained bending moment at each node is plugged into relation (6), whose solution provides the longitudinal radius of curvature and, in turn, the rotation and displacements fields according to Equations (9)–(11). At this point, the algorithm restarts until the obtained solution converges according to the criterion given in Equation (12).

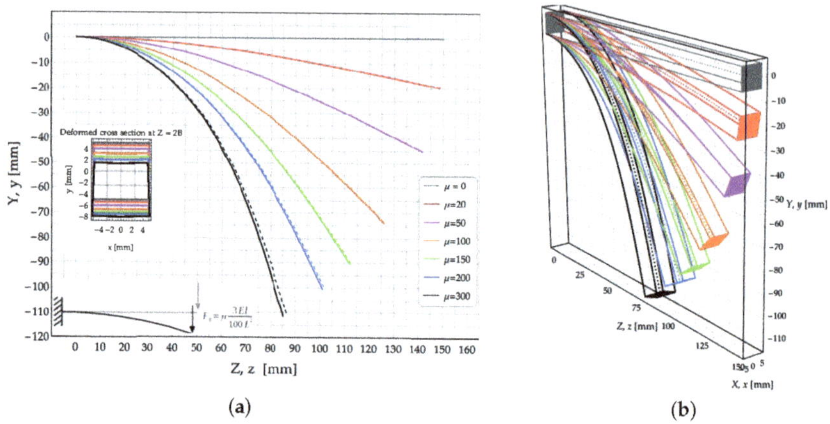

Figure 8. Study case A_2: Cantilever beam subjected to a transverse dead load acting at its free end. (**a**) Deformed configuration of the longitudinal axis: Theoretical model (continuous lines) and FE model (dashed lines); (**b**) 3D skecth of the deformed configurations obtained by the theoretical solution.

As expected, the convergence rate decreases as the load multiplier increases, as confirmed by the iteration number required to reach the convergence, namely $N = 13$, 20, 40, 55, 84 and 205, for each load increment, respectively, and it is not significantly affected by the beam geometry.

3.2.2. The FE Solution

The deformed configurations provided by FE simulations are shown in Figure 3b. The geometries listed in Table 1 have been simulated by using 4-nodes tetrahedron elements. Such subcases are characterized by 130569, 212161 and 284231 number of FEs.

Conversely to the study case A_1, the cross section at $Z = 0$ has been fully restrained (namely $\mathbf{s}(X, Y, 0) = \mathbf{0}$) to mitigate the noise induced by a single concentrated reactive force at the clamped end. Moreover, the comparison between the theoretical and FE predictions has been performed at $Z = 2B$ (the value of $2B$ has been taken here as an extinction length equals to $2\max\{B, H\}$.).

Like the previous study case, the external load is simulated by applying a displacement. In particular, the prescribed displacement values $v(0, 0, L)$ corresponding to the load increments considered in the theoretical model are listed in Table 3.

Table 3. Prescribed displacements $-v(0, 0, L)$ adopted in the FE simulations.

Load Multiplier μ		20	50	100	150	200	300
	(a)	19.643	45.230	73.921	90.333	100.303	111.447
Subcases	**(b)**	19.646	45.263	74.039	90.521	100.536	111.737
	(c)	19.666	45.494	74.917	91.970	102.423	115.291

As already observed, the FE models account for the deformations induced by both the axial and shear loads also. In order to quantify such contributes, the following two dimensionless measures of deformation can be introduced

$$\Delta L = \frac{1}{L}\int_0^L \lambda_Z(0, 0, Z)\, dZ - 1, \tag{13}$$

$$\Delta\gamma = \frac{2}{BHL}\int_{\mathcal{B}} \sqrt{|C_{YZ}(X, Y, Z)|}\, dV. \tag{14}$$

Such quantities, representative of the deformation induced by the axial and shear stresses, have been been reported in Figure 9 varying the load multiplier.

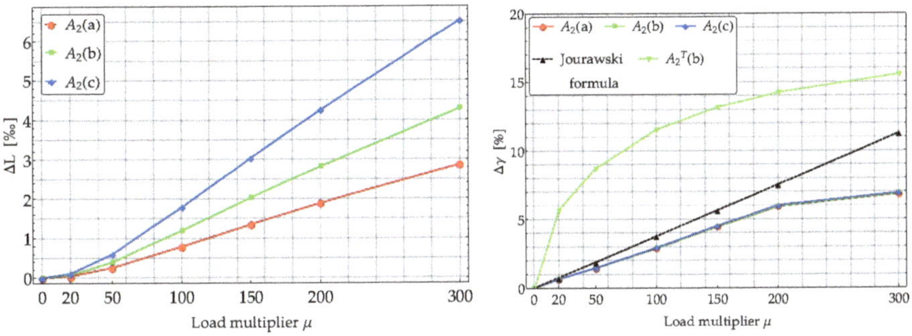

(**a**) Length variation. (**b**) Shear deformation.

Figure 9. Effects of axial and shear deformations about study case A_2.

The axial elongation ΔL (13) highlights the length variation associated whit the axial component of the external force. Such a variation turns out to be very small, also for high values of the load multiplier, as shown in Figure 9a. This contribution exhibits a different nonlinear growth with the load multiplier varying the subcases. It follows that the geometry of the beams significantly influences the value of ΔL for $\beta_C > 1$.

The (global) measure of the shear deformation (14) can be interpreted, according to the linearized elasticity, as average engineering shear strain in the YZ-planes and it quantifies the sliding occurring in the planes of the cross sections. The amount of shear $\Delta\gamma$ contributes negligibly to the rotation θ of the cross section. Based on Figure 9b, it could seem that $\Delta\gamma$ be much more relevant than ΔL. Nonetheless, the maximum value of $\Delta\gamma$ turns out to be only 7% of the final angle of rotation of the cross section due to bending, which is $\theta(L) = -73°$.

In Figure 9b $A_2^T(b)$ denotes a further subcase for which it has been assumed $H = 20$ mm and $B = 10$ mm. The average contribution of the shear strain seems to be almost independent of the geometry of the beam cross section, as predicted by linearized elasticity. Indeed, according to the well known Jourawski formula [33] in the framework of linearized elasticity, one has

$$\tau_{YZ}(Y, L) = \frac{F_0 S_X(Y)}{I_X B}, \tag{15}$$

in which τ_{YZ} is the shear stress, F_0 the external shear force acting at a given beam cross section and $S_X(Y)$ is the first order moment w.r.t. the X axis of the part of the cross section overlaying the fibre at distance Y form the X axis. Therefore, by assuming $C_{YZ} \cong \gamma_{YZ}$ it follows that $C_{YZ} \cong \tau_{YZ}/G$, with $G = E/2(1+\nu)$. Considering the values of the shear load F_0 reported in Figure 8a and applying the Jourawski formula one finds

$$\gamma_{YZ} \quad = \quad \mu \frac{3(\nu+1)}{400} \left(\frac{H}{L}\right)^2 = k\mu. \tag{16}$$

Therefore, Equation (16) does not depend explicitly by the width B of the cross section. Conversely, γ_{YZ} significantly depends on the heigth H of the cross section, as confirmed by curve of Figure 9b related to case $A_2^T(b)$.

3.2.3. Results and Comparison

For each load increment of case A_2, the deformed configurations of the longitudinal axis of the bent beam are reported in Figure 8a. In that figure, solid lines represent the analytical predictions whilst dashed lines reproduce the FE results. A sketch of the 3D configurations is provided in Figure 8b.

The relative errors of the FE results w.r.t. the theoretical predictions in terms of resultant of the shear stresses are listed in Table 4.

Table 4. Study case A_2: Relative error on the shear resultant $\epsilon_{r,F_0(L)}$, Searle parameter $\beta^2 = B^2/R_0H = \beta_C^2/\beta_{ES}$ and weight of the fifth-order truncated term $\rho_{20} + \rho_2\beta_C^2$ in Equation (8).

Load	Subcases								
Mult.	(a)			(b)			(c)		
μ	$\epsilon_{r,F_0}(L)$	β^2	$\rho_{20} + \rho_2\beta_C^2$	$\epsilon_{r,F_0}(L)$	β^2	$\rho_{20} + \rho_2\beta_C^2$	$\epsilon_{r,F_0}(L)$	β^2	$\rho_{20} + \rho_2\beta_C^2$
20	$-4. \times 10^{-3}$	0.02	1.4×10^{-8}	-1.3×10^{-2}	0.08	-5.2×10^{-4}	$-5. \times 10^{-2}$	0.22	9.4
50	-7.7×10^{-3}	0.05	4.3×10^{-7}	-3.2×10^{-2}	0.18	-1.6×10^{-2}	-1.4×10^{-1}	0.51	4.9×10^{1}
100	-1.6×10^{-2}	0.09	$4. \times 10^{-6}$	-7.8×10^{-2}	0.31	-1.5×10^{-1}	-3.9×10^{-1}	0.81	1.3×10^{2}
150	-2.4×10^{-2}	0.12	1.2×10^{-5}	-1.3×10^{-1}	0.39	-4.1×10^{-1}	-7.2×10^{-1}	0.96	1.8×10^{2}
200	-2.9×10^{-2}	0.14	2.2×10^{-5}	-1.7×10^{-1}	0.45	-7.5×10^{-1}	-1.1	1.01	$2. \times 10^{2}$
300	-3.1×10^{-2}	0.18	4.7×10^{-5}	-2.5×10^{-1}	0.52	-1.5	-2.2	1.09	2.3×10^{2}

As for the previous case, the relative errors concerning the reactive force provided by the FE solution and the theoretical model keeping fixed the displacement $v(0, 0, L)$ start from negative values that increase as the load multiplier increase. This means that the theoretical model provides stiffer response w.r.t. the numerical ones. Indeed the FE solutions accounts for the shear and axial compliance also.

The discussion provided in Section 3.1.3 about the reliability of the theoretical model based on the Searle parameter holds also for the case of variable bending moment, as the study case at hand. Indeed, the comparison provided in Figure 10 in terms of stretches displays good agreement between theoretical and numerical predictions for low values of the Searle parameter, see Table 4. In detail, for $\beta^2 \leq 0.5$ the analytical formulation provides extremely accurate results.

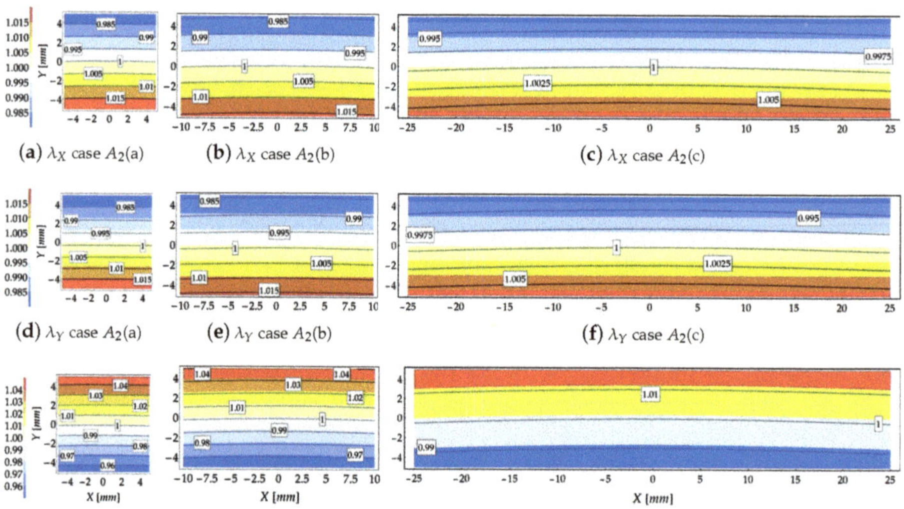

(a) λ_X case A_2(a) **(b)** λ_X case A_2(b) **(c)** λ_X case A_2(c)

(d) λ_Y case A_2(a) **(e)** λ_Y case A_2(b) **(f)** λ_Y case A_2(c)

(g) λ_Z case A_2(a) **(h)** λ_Z case A_2(b) **(i)** λ_Z case A_2(c)

Figure 10. Study case A_2 at increment 3 ($\mu = 100$): Comparison in terms of stretches provided by the theoretical model (contour plot) and the FE solution (solid isolines with boxed value).

Note that for the study case A_2, the values of the Eulerian slenderness are lower than those related to case A_1, as listed in Table 2. It means that, despite of the large values of the displacements, the parameter $\beta_{ES} = H/R_0$, is smaller. Conversely to case A_1, in case A_2 the bending moment at the clamped side increases slowly as the load multiplier increases owing to the reduction of the external load arm.

For subcases $A_2(a)$ and $A_2(b)$, for which $\beta^2 < 0.5$ (see Table 4), the theoretical and numerical results in terms of stretch distributions within the cross section agree well, as shown in Figure 10. For these cases, as shown in Figures 10a,b,d,e,h,g the theoretical model is able to grasp both the stretches magnitude and their distribution within the cross section. In particular, for case $A_2(a)$ the stretches predicted by the theoretical formulation and those provided by the FE code are almost coincident.

Moving from the subcase (a) to subcase (c) the analytical model reduces its accuracy, leading to an underestimation of the stretches as compared with those furnished by the FE simulations. However, since $\beta^2 < 0.5$ (cases $A_2(a)$ and $A_2(b)$ of Figure 10), the theoretical model preserves its reliability to predict the stretches. For the subcase $A_2(c)$, for which $\beta^2 = 0.81$, the theoretical formulation loses its reliability, as confirmed by Figures 10c,f,i.

As the compactness index β_C increases (from left to right in Figure 11), the FE results show that the transverse principal stress T_2 becomes comparable w.r.t. principal stress T_3.

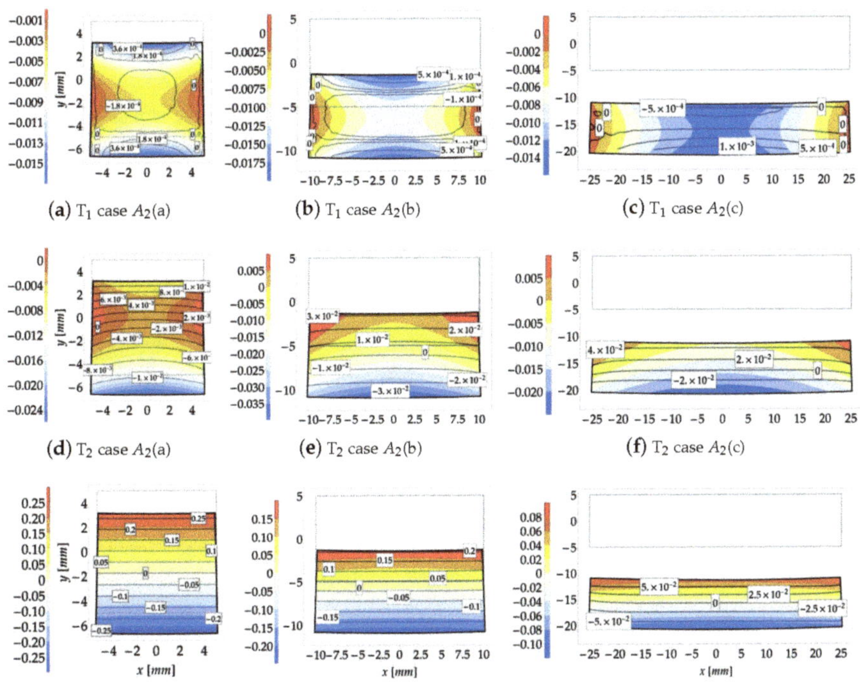

(**a**) T_1 case A_2(a) (**b**) T_1 case A_2(b) (**c**) T_1 case A_2(c)

(**d**) T_2 case A_2(a) (**e**) T_2 case A_2(b) (**f**) T_2 case A_2(c)

(**g**) T_3 case A_2(a) (**h**) T_3 case A_2(b) (**i**) T_3 case A_2(c)

Figure 11. Study case A_2 at increment 3 ($\mu = 100$): Stresses comparison [MPa] between the theoretical model (contour plot) and FE model (isolines with boxed value).

As already observed, conversely to the FE solution, the reliability of the theoretical formulation sensibly depends on the geometry of the cross sections. Indeed, for subcases in which $\beta^2 < 0.5$ (see Figure 11c,f), the analytical formulation loses its accuracy to predict stretches and stresses. Moreover, the FE solution indicates that stress T_1 assumes negligible values w.r.t. the other principal stresses.

In particular, the principal stress T_1 is about one or two orders lower than T_2, and three orders lower than T_3 when $\beta_C < 0.5$.

As expected, the principal stress T_3 is accurately predicted by the theoretical model when $\beta^2 < 0.5$, as displayed in Figures 11f,h,i.

4. Conclusions

The finite bending of a homogeneous hyperelastic beam made of a compressible Mooney–Rivlin material has been investigated here. Two study cases have been analyzed in detail: A cantilever subjected to a prescribed couple and a cantilever subjected to an external shear load acting at its free end. For each study case, the main analytical results provided by the theoretical model [23] have been compared with those obtained by FE simulations.

The governing Equation (6), has been rearranged here in dimensionless form, allow identifying two dimensionless governing parameters here termed Eulerian slenderness $\beta_{ES} = R_0/H$ and an index about the compactness of the cross section, i.e., $\beta_C = B/H$. Such parameters are involved in the Searle parameter β [27,28] according to the expression $\beta^2 = \beta_C^2/\beta_{ES}$.

The comparison between the theoretical and numerical predictions in terms of the deformed configurations of the bent body together with stretch and stress distributions within the cross section, has been provided and discussed. Such comparison allows assessing the accuracy of the theoretical model to predict the mechanical response of beams under large bending. It is shown that the analytical model provides accurate results provided that the following conditions about the governing parameters hold true: $\beta_C < 0.5$ and $\beta^2 < 0.5$ (namely, $\beta_{ES} < 0.125$).

Once the aforementioned threshold values are overcome, the deformed configuration of the cross sections differs from an arc of circumference as hypothesized in the theoretical formulation. This fact is due to relevant transverse deformations according to the theory of elastic plates.

The study cases here considered allows confirming the negligible effects of the shear and axial loads compared with the bending effects. Furthermore, the present study allows validating the simplifying hypotheses involved in the theoretical formulation [23] also for beams subjected to variable bending moment.

The obtained moment-curvature relationship extends the classical Elastica to the framework of finite elasticity. In a forthcoming work, the authors will compare the results provided by Equation (8) with those furnished by the Elastica with the aid of experimental tests on rubber-like materials.

Author Contributions: F.O.F.: Conceptualization, Methodology, Writing. L.L.: Data curation, investigation. A.M.T.: Supervision, Reviewing and Editing. All authors have read and agreed to the published version of the manuscript.

Funding: This research received no external funding.

Acknowledgments: Financial support from the Italian Ministry of Education, University and Research (MIUR) in the framework of the Project PRIN2015 (code 2015JW9NJT) *"Advanced mechanical modeling of new materials and structures for the solution of 2020 Horizon challenges"* is gratefully acknowledged.

Conflicts of Interest: The authors declare no conflicts of interest.

Abbreviations

The following abbreviations are used within the manuscript:

w.r.t.	with respect to
MR	Mooney–Rivlin

References

1. Schmitt, F.; Piccin, O.; Barbé, L.; Bayle, B. Soft robots manufacturing: A review. *Front. Robot. AI* **2018**, *5*, 84. [CrossRef]

2. Martinez, R.V.; Branch, J.L.; Fish, C.R.; Jin, L.; Shepherd, R.F.; Nunes, R.M.; Suo, Z.; Whitesides, G.M. Robotic tentacles with three-dimensional mobility based on flexible elastomers. *Adv. Mater.* **2013**, *25*, 205–212. [CrossRef]
3. Mac Murray, B.C.; An, X.; Robinson, S.S.; van Meerbeek, I.M.; O'Brien, K.W.; Zhao, H.; Shepherd, R.F. Poroelastic foams for simple fabrication of complex soft robots. *Adv. Mater.* **2015**, *27*, 6334–6340. [CrossRef]
4. Wen, L.; Wang, T.; Wu, G.; Liang, J. Quantitative thrust efficiency of a self-propulsive robotic fish: Experimental method and hydrodynamic investigation. *IEEE/ASME Trans. Mechatron.* **2012**, *18*, 1027–1038. [CrossRef]
5. Polygerinos, P.; Correll, N.; Morin, S.A.; Mosadegh, B.; Onal, C.D.; Petersen, K.; Cianchetti, M.; Tolley, M.T.; Shepherd, R.F. Soft robotics: Review of fluid-driven intrinsically soft devices; manufacturing, sensing, control, and applications in human-robot interaction. *Adv. Eng. Mater.* **2017**, *19*, 1700016. [CrossRef]
6. Ilievski, F.; Mazzeo, A.D.; Shepherd, R.F.; Chen, X.; Whitesides, G.M. Soft robotics for chemists. *Angew. Chem. Int. Ed.* **2011**, *50*, 1890–1895. [CrossRef] [PubMed]
7. Wang, H.; Ma, J.; Ren, Z.; Gong, Z.; Hao, Y.; Wang, T.; Wen, L. Fiber-reinforced soft robotic anthropomorphic finger. In Proceedings of the 2016 International Conference on Robotics and Automation Engineering (ICRAE), Jeju-Do, Korea, 24–29 August 2016; IEEE: Piscataway, NJ, USA, 2016, pp. 1–5.
8. de Haan, L.T.; Verjans, J.M.; Broer, D.J.; Bastiaansen, C.W.; Schenning, A.P. Humidity-responsive liquid crystalline polymer actuators with an asymmetry in the molecular trigger that bend, fold, and curl. *J. Am. Chem. Soc.* **2014**, *136*, 10585–10588. [CrossRef] [PubMed]
9. Zhao, J.; Abbas, A. A low-cost soft coiled sensor for soft robots. In Proceedings of the ASME 2016 Dynamic Systems and Control Conference, Minneapolis, MN, USA, 12–14 October 2016.
10. Timoshenko, S.; Timoshenko, S.; Goodier, J. *Theory of Elasticity*; McGraw-Hill Book Company Inc.: New York, NY, USA,1951.
11. Timoshenko, S.P.; Woinowsky-Krieger, S. *Theory of Plates and Shells*; McGraw-Hill: New York, NY, USA, 1959.
12. Lekhnitskii, S.G. *Anisotropic Plates*; Technical Report; Foreign Technology Div Wright-Patterson Afb Oh: Dayton, OH, USA, 1968.
13. Lekhnitskii, S.; Fern, P.; Brandstatter, J.J.; Dill, E. Theory of elasticity of an anisotropic elastic body. *Phys. Today* **1964**, *17*, 84. [CrossRef]
14. Barber, J.R. *Elasticity*; Springer: Dordrecht, The Netherlands, 1992.
15. Tarantino, A.M. Homogeneous equilibrium configurations of a hyperelastic compressible cube under equitriaxial dead-load tractions. *J. Elast.* **2008**, *92*, 227. [CrossRef]
16. Lanzoni, L.; Tarantino, A.M. Damaged hyperelastic membranes. *Int. J. Non-Linear Mech.* **2014**, *60*, 9–22. [CrossRef]
17. Lanzoni, L.; Tarantino, A.M. Equilibrium configurations and stability of a damaged body under uniaxial tractions. *Zeitschrift für angewandte Mathematik und Physik* **2015**, *66*, 171–190. [CrossRef]
18. Lanzoni, L.; Tarantino, A.M. A simple nonlinear model to simulate the localized necking and neck propagation. *Int. J. Non-Linear Mech.* **2016**, *84*, 94–104. [CrossRef]
19. Rashid, B.; Destrade, M.; Gilchrist, M.D. Mechanical characterization of brain tissue in simple shear at dynamic strain rates. *J. Mech. Behav. Biomed. Mater.* **2013**, *28*, 71–85. [CrossRef] [PubMed]
20. Rivlin, R. Large elastic deformations of isotropic materials. V. The problem of flexure. *Proc. R. Soc. Lond. Ser. A. Math. Phys. Sci.* **1949**, *195*, 463–473.
21. Ogden, R.W. *Non-Linear Elastic Deformations*; Courier Corporation: Chelmsford, MA, USA, 1997.
22. Bruhns, O.T.; Xiao, H.; Meyers, A. Finite bending of a rectangular block of an elastic Hencky material. *J. Elast. Phys. Sci. Solids* **2002**, *66*, 237.
23. Lanzoni, L.; Tarantino, A.M. Finite anticlastic bending of hyperelastic solids and beams. *J. Elast.* **2018**, *131*, 137–170. [CrossRef]
24. Lanzoni, L.; Tarantino, A.M. The bending of beams in finite elasticity. *J. Elast.* **2020**, *139*, 91–121. doi:10.1007/s10659-019-09746-8. [CrossRef]
25. Lanzoni, L.; Tarantino, A. Mechanics of high-flexible beams under live loads. *J. Elast.* **2020**, 1–26. IN PRESS doi:10.1007/s10659-019-09759-3. [CrossRef]
26. Bigoni, D. *Extremely Deformable Structures*; Springer: Basel, Switzerland, 2015; Volume 562.
27. Searle, G.F.C. *Experimental Elasticity: A Manual for the Laboratory*; Cambridge University Press: Cambridge, UK, 1908.

28. Lamb, H. On the flexure of a flat elastic spring. *Philos. Mag. J. Theor. Exp. Appl. Phys.* **1891**, *31*, 182–188.
29. Falope, F.; Lanzoni, L.; Tarantino, A. Bending device and anticlastic surface measurement of solids under large deformations and displacements. *Mech. Res. Commun.* **2019**, *97*, 52–56. [CrossRef]
30. Falope, F.; Lanzoni, L.; Tarantino, A.M. The bending of fully nonlinear beams. Theoretical, numerical and experimental analyses. *Int. J. Eng. Sci.* **2019**, *145*, 103167. [CrossRef]
31. Ciarlet, P.G.; Geymonat, G. Sur les lois de comportement en élasticité non linéaire compressible. *CR Acad. Sci. Paris Sér. II* **1982**, *295*, 423–426.
32. Tarantino, A.M.; Lanzoni, L.; Falope, F.O. *The Bending Theory of Fully Nonlinear Beams*; Springer: Basel, Switzerland, 2019.
33. Jourawski, D. Sur le Résistance d'un Corps Prismatique et d'une Pièce Composée en Bois ou on Tôle de Fer à une Force Perpendiculaire à leur Longeur. *Annales des Ponts et Chaussées* **1856**, *12*, 328–351.

materials

MDPI

Article

Homogenization and Equivalent Beam Model for Fiber-Reinforced Tubular Profiles

Daniel Gnoli [1], Sajjad Babamohammadi [2] and Nicholas Fantuzzi [1,*]

1 DICAM Department, University of Bologna, 40136 Bologna, Italy; daniel.gnoli@studio.unibo.it
2 Gruppo COSMI, 48122 Ravenna, Italy; sajjad.babamohammadi@gruppocosmi.com
* Correspondence: nicholas.fantuzzi@unibo.it; Tel.: +39-0512093494

Received: 14 April 2020; Accepted: 28 April 2020; Published: 30 April 2020

Abstract: The current work presents a study on hollow cylinder composite beams, since hollow cylinder cross-sections are one of the principal geometry in many engineering fields. In particular, the present study considers the use of these profiles for scaffold design in offshore engineering. Composite beams cannot be treated as isotropic ones due to couplings mainly present among traction, torsion, bending and shear coefficients. This research aims to present a simple approach to study composite beams as they behave like isotropic ones by removing most complexities related to composite material design (e.g., avoid the use of 2D and 3D finite element modeling). The work aims to obtain the stiffness matrix of the equivalent beam through an analytical approach which is valid for most of the laminated composite configurations present in engineering applications. The 3D Euler–Bernoulli beam theory is considered for obtaining the correspondent isotropic elastic coefficients. The outcomes show that negligible errors occur for some equivalent composite configurations by allowing designers to continue using commercial finite element codes that implement the classical isotropic beam model.

Keywords: pultruded beams; effective stiffness matrix; FRP; hollow circular beams; finite element method

1. Introduction

Hollow cylinders are one of the most common elements in offshore engineering, e.g., risers, pipes and generally used for scaffolding systems. Among vast materials to produce these kinds of elements, all of them have one mutual defect which is an important factor in an offshore environment; that is corrosion. The most common material used in offshore structures is steel, which although it is stiff and light (compared to reinforced concrete) but it can be easily corroded, which makes maintenance activities very expensive. On the other hand, plastics, a non-environmentally friendly material, can be in this context environmental-friendly. This is due to the fact that fewer materials are involved due to less frequent maintenance, thus, less resources. Steel structural elements should be changed frequently in an offshore environment to maintain offshore structural integrity. Steel is used also in onshore civil applications when being lightweight is an important prerequisite at the design phase. One of the plastics' problem is the low stiffness and strength compared to steel. To overcome this problem fibers can be added in order to make the so-called Fiber Reinforced Plastics (FRPs). FRPs can be produced by different methods which leads to different stiffness and strength. However, composite materials do not behave like well-known isotropic ones. Therefore over the years, several methodologies have been utilized to analyze composite materials and structures simply and accurately.

Regarding hollow circular cross-sections, the stress distribution of anisotropic configurations has been led by Lekhnitskii [1] which was a starting point for further investigations by other scientists and engineers. Based on this, Jolicoer and Cardou [2] studied hollow cylinders under bending, tensile and torsions. The hollow cylinders behavior under hygrothermal and mechanical loads was investigated by Kollar et al. [3,4]. They followed displacement-based approach in their works, as well as Bhaskar and Varadan [5], Xia et al. [6,7], Calhoglu et al. [8] and Bakian et al. [9]. Verijenko et al. [10] studied the behavior of laminated cylinder under internal and external pressures. Roque and Ferreira [11] studied plates and shell deformation by means of Reddy's theory. Salahifar and Mohareb [12] studied shell cylinders under harmonic forces. Laminated tubes were also considered by Tarn et al. [13] to present a state space approach to torsion, extension, bending, shearing and pressure. Bai et al. [14] investigated the buckling behavior of thermoplastic pipes under combined bending and tension loads. Derisi et al. [15] and Shadmehri et al. [16] studied the composite tubes under bending. Effect of pressure, shear and torsion on anisotropic materials was analyzed by Ting [17,18]. Dynamic axial compression on thin wall circular tubes was investigated by Uchikawa et al. [19]. Non-classical effect on hollow cylinders was studied by Silvestre [20] which led us to present a formula on generalized beam theory. Kardomateas [21] investigated orthotropic shells under internal and external pressure. Wang et al. [22] presented a study on compressive behaviors of FRP tubular beams. In addition, other works [23–27] have been presented on the investigation of different laminated composites subjected to several loadings. Different investigations have been done on multi-layered hollow cylinders [28,29]. Khalili et al. [30] analyzed the dynamic behavior of multilayered composite plates. Experimental data and laboratory tests were provided by Ascione et al. [31], Boscato and Russo [32], Philippidis and Vassilopoulos [33] and Quadrino et al. [34] to analyze the behavior of composite laminates. Ascione et al. [31] presented a bearing design formula, Boscato and Russo [32] presented dynamic parameters, Philippidis and Vassilopoulos [33] presented the effect of off-axis loading o fatigue and static behavior, whereas Quadrino et al. [34] investigated the local behavior of commercial I-shaped tubes. Furthermore, Madenci et al. [35] used both experimental and theoretical data to investigate the effect of flexure and Xin et al. [36] in-plane compression and shear on GFRPs. Experimental studies also have been done by Mayookh et al. [37] on the flexural creep of unidirectional bars and by Zhang et al. [38] on correlation between mechanical properties of FRPs. A homogeneous approach was considered by Sun and Li [39] to investigate the load-deformation of thick laminate and Sun et al. [40] to study the stress analysis of hollow cylinders. Other current applications of FRP tubular profiles have been presented in for retaining walls [41], column jacketing [42] and reinforcement of concrete slabs [43].

Homogenization is a method to investigate the macroscopic behavior of a material by considering a replacement of that material to an equivalent homogeneous one [44–46]. Thick hollow beams were investigated by Kim and White [47]. In addition, Yildiz and Sarikanat [48] used finite element analysis to investigate the material properties of multilayered hollow composites; whereas, Ferreira [49] used global meshless approximation to analyze them. Yazdani Sarvestani et al. [50] investigated the thick orthotropic cantilever under transverse loading and the effect of shear on stress distributions of thick composites [51]. Theory on governing equations of anisotropic thin-walled members was presented by Berdichevski et al. [52]. Kollar and Pluzsik [53] studied thin-walled composite beams and presented a theory to calculate the stiffness matrix of closed sections. Jung and Lee [54] investigated thin-walled I-beam composites. Lateral buckling of thin-walled composites was investigated by Lee [55]. Dynamic response of FRPs have been investigated by Boscato [56]. Corotational method was used by de Miranda et al. [57] and Ruggerini et al. [58] to investigate post-buckling and non-linear GBT buckling analysis, respectively.

Lately, Babamohammadi et al. [59] presented research based on a homogenization method of composite beams which lead to a simple design procedure of frame structures made of FRP beams. The present research develops further and expands such previous work by including an in-depth analysis of the mechanical behavior of hollow composite beams with several standards and not-standard lamination schemes as well as critical discussion and remarks on possible applications of the present methodology for practical engineering purposes.

2. Motivation

The mechanical behavior of isotropic beams is well-known in engineering practices and it is used for designing any kind of frame structure. However, if the beam is made up of fiber-reinforced composites its behavior cannot be attributed to one of the isotropic materials.

Stiffness matrices of hollow cylindrical composite structures can be determined by employing a finite element program. In this work, ABAQUS software [60] has been utilized for hollow profiles of 4 mm thickness which is the standard dimension of scaffolding beam components. Several stacking sequences are analyzed with different arrangements of the reinforcing fibers to investigate different coupling effects given by the composite configuration in comparison with the classical isotropic model.

The stiffness matrix of a two-node Euler–Bernoulli isotropic beam with compact cross-section, in the 3D space, has six degrees of freedom per node as 12×12 matrix $[K] = k_{i,j}$ for $i, j = 1, 2, \ldots, 12$. However, it is not necessary to define 144 parameters. If the cross-section is compact and with a double-symmetry (as in the present case) it is sufficient to compute six indipendent parameters. Below definitions of the stiffness matrix components are given.

$$[K] = \begin{bmatrix} k_{1,1} & 0 & 0 & 0 & k_{1,5} & 0 & k_{1,7} & 0 & 0 & 0 & k_{1,11} & 0 \\ & k_{2,2} & 0 & k_{2,4} & 0 & 0 & 0 & k_{2,8} & 0 & k_{2,10} & 0 & 0 \\ & & k_{3,3} & 0 & 0 & 0 & 0 & 0 & k_{3,9} & 0 & 0 & 0 \\ & & & k_{4,4} & 0 & 0 & 0 & k_{4,8} & 0 & k_{4,10} & 0 & 0 \\ & & & & k_{5,5} & 0 & k_{5,7} & 0 & 0 & 0 & k_{5,11} & 0 \\ & & & & & k_{6,6} & 0 & 0 & 0 & 0 & 0 & k_{6,12} \\ & & & & & & k_{7,7} & 0 & 0 & 0 & k_{7,11} & 0 \\ & & & & & & & k_{8,8} & 0 & k_{8,10} & 0 & 0 \\ & & & & & & & & k_{9,9} & 0 & 0 & 0 \\ & & & & & & & & & k_{10,10} & 0 & 0 \\ & & & & & & & & & & k_{11,11} & 0 \\ \text{sym} & & & & & & & & & & & k_{12,12} \end{bmatrix} \quad (1)$$

that can be written in compact matrix form as

$$[K] = \begin{bmatrix} [k_1] & [k_2] & -[k_1] & [k_2] \\ & [k_3] & [k_2] & [k_4] \\ & & [k_1] & -[k_2] \\ \text{sym} & & & [k_3] \end{bmatrix} \quad (2)$$

where

$$\begin{aligned} [k_1] &= \begin{bmatrix} \frac{12EI}{L^3} & 0 & 0 \\ 0 & \frac{12EI}{L^3} & 0 \\ 0 & 0 & \frac{EA}{L} \end{bmatrix}, \quad [k_2] = \begin{bmatrix} 0 & \frac{6EI}{L^2} & 0 \\ -\frac{6EI}{L^2} & 0 & 0 \\ 0 & 0 & 0 \end{bmatrix} \\ [k_3] &= \begin{bmatrix} \frac{4EI}{L} & 0 & 0 \\ 0 & \frac{4EI}{L} & 0 \\ 0 & 0 & \frac{GJ}{L} \end{bmatrix}, \quad [k_4] = \begin{bmatrix} \frac{2EI}{L} & 0 & 0 \\ 0 & \frac{2EI}{L} & 0 \\ 0 & 0 & -\frac{GJ}{L} \end{bmatrix}. \end{aligned} \quad (3)$$

E, G are classical elastic properties and A, I, L are cross-section area, the moment of inertia and length of the beam. Definitions for the sitffness components $k_{i,j}$ for $i, j = 1, 2, \ldots, 12$ in Equation (1) are straightforwardly given by comparison with definitions in Equation (3). It is clear that for isotropic beams only six parameters are sufficient to define the whole stiffness matrix of the beam.

As far as structures made of composite materials are concerned, there are still no relations available that define the stiffness matrix within a specific theoretical framework. However, it can be demonstrated that orthotropic beams (with fibers parallel to the beam's principal axis) have the same nonzero elements in the stiffness matrix as the equivalent isotropic one. Thus, if homogenized mechanical properties can be carried out for composites hollow beams the approach for isotropic beams can be transferred to composite configurations. In this regard, a study was carried out by Reddy [61] wherein the bending, vibration and buckling plate problems of laminated composite plates were compared by increasing the number of plies in cross- and angle-ply configurations. Reddy showed that by increasing number of laminae in cross-ply orientation, the mechanical behavior of the plate tends to be equivalent to the same plate in orthotropic configuration (e.g., single-ply with orientation 0). When angle-ply plates are studied, the mechanical characteristic is asymptotic to behavior between (45–45) and (45–45) s and does not change when the number of layers is large.

The latter has been generalized to hollow circular beams in order to see if such asymptotic behavior could be used to simply model these beams as equivalent isotropic or orthotropic ones.

3. Theoretical Background

Taking the studies of Reddy [61] on laminated composite plates as a reference, in the following, sections cross- and angle-ply laminated hollow circular beams are investigated. The geometric and mechanical properties considered are listed in Table 1. Note that geometric properties selected are the ones typical of beams used in scaffolding systems. Optimization of cross-section properties is not analyzed because out of the scope of the present work. Mechanical properties taken here are typical engineering constants of Carbon FRP (CFRP). This study is not limited to this geometry and/or mechanical properties but aims to present a generalized framework that is able to work in every context.

Table 1. Geometric and mechanical properties.

Geometric		Mechanical	
Average radius	27 mm	E_1	145,849.69 MPa
Length	1000 mm	E_2	11,030 MPa
Laminate thickness	4 mm (Fixed)	ν_{12}	0.28
Layer thickness	Variable	G_{12}	6209.89 MPa
Area	678.58 mm^2	G_{13}	6209.89 MPa
Polar inertia	497,400 mm^4	G_{23}	3860.5 MPa
Inertia	248,774.1516 mm^4		

The stiffness matrix of the composite beam has been computed using ABAQUS as described in [59] using a 3D shell model and laminated shell elements of first-order (Mindlin theory). The cylindrical geometry is restrained at the two ends through rigid links which simulate the two end nodes of the equivalent beam. Stiffness matrix components (following the structure given by Equation (2)) are determined by setting alternatively one unitary displacement and by retrieving the corresponding boundary forces. For each unitary displacement, a stiffness matrix row is carried out. Since the equivalent beam has two nodes with 6 degrees of freedom each, the stiffness matrix result to be 12×12. In addition, equivalent elastic properties via an analytical cross-section homogenization by Sun et al. [40] is also considered. The present methodology is based on the ensemble of the latter. Interested readers are asked to refer to the cited works for further details on such procedures that are not reported below for the sake of conciseness.

3.1. *Cross-Ply Laminates*

A typical circular hollow profile with cross-ply configuration is depicted in Figure 1. The stiffness matrix of composite beams with a different number of plies is presented below with geometric and mechanical properties listed in Table 1. Configurations are alternatively symmetric and antisymmetric (with the only exception of the first two which are (0) and (90)). All the nonzero stiffness matrix components are compared with the same given by an orthotropic configuration (0). Figure 2a–e display such comparison and a summary in terms of relative error is depicted in Figure 2f where the error tends to be −40% by increasing the number of plies.

Clearly, the global trend of all stiffness constants deviates from the initial orthotropic behavior and becomes stable when 3 or 4 layers are considered. Thus, it can be established that the study of Reddy [61] for cross-ply plates does not apply to hollow circular cylinders but it is closer to the behavior of the angle-ply plates. In detail, it can be noted that $k_{1,1}$, $k_{3,3}$, $k_{1,5}$ and $k_{4,4}$, decrease until reaching an almost constant value. Specifically, symmetrical configurations reach this value from above, whereas the others from the bottom. The variation in torsional stiffness $k_{6,6}$ is negligible since it does not deviate from the initial configuration, thus torsional stiffness does not improve by introducing more cross-ply layers. In conclusion, it is possible to confirm that by increasing the number of layers in a cross-ply lamination scheme, keeping a constant thickness, about 40% loss of stiffness is generated, while the torsional stiffness remains almost constant.

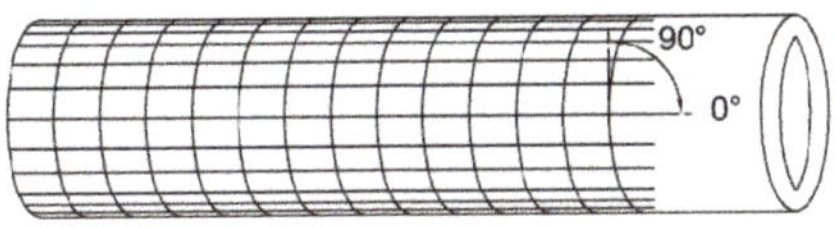

Figure 1. Cross-ply laminates.

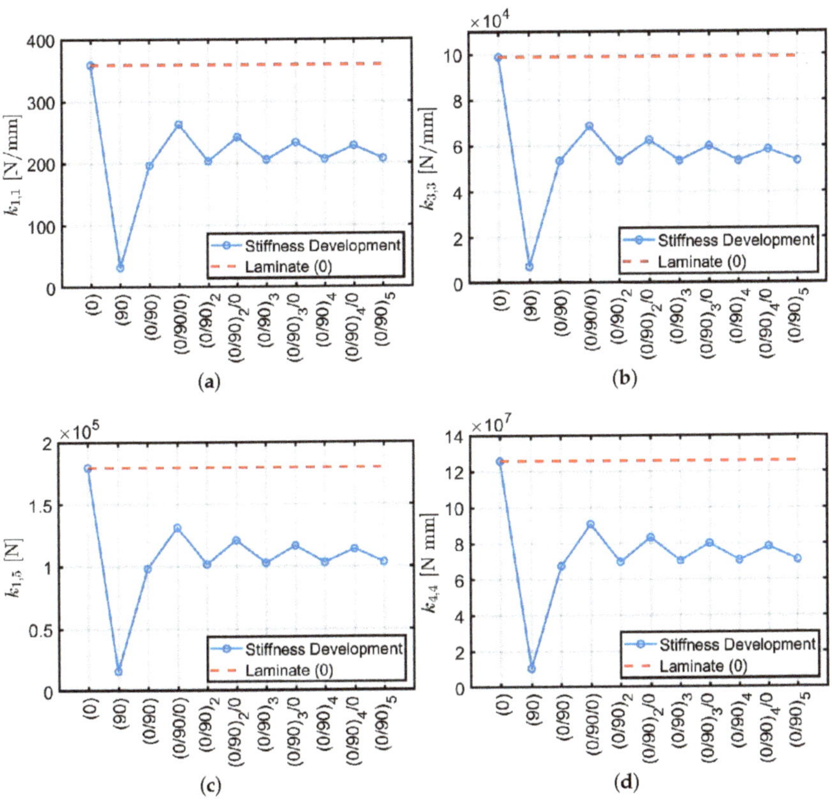

Figure 2. *Cont.*

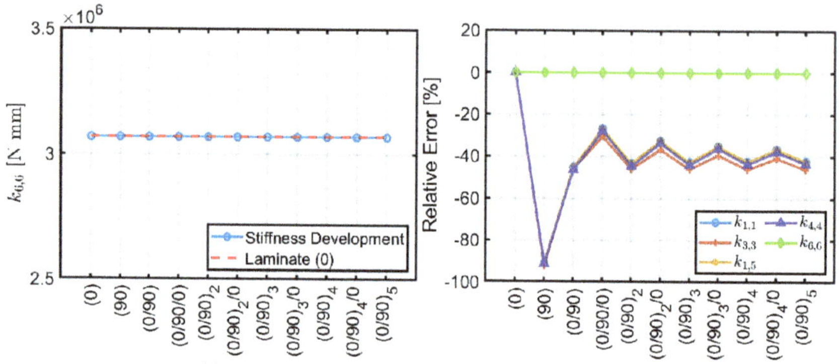

(e) (f)

Figure 2. Stiffness trends with respect to the number of plies (**a**–**e**) and relative error with respect to an orthotropic configuration (**f**).

The validity of the finite element model is tested by comparing it with the results of the analytical approach by Sun et al. [40] which provides homogenized modulus of elasticity E_z and tangential modulus $G_{\theta z}$ for a composite cross-section in planar strain. For instance, by purging the stiffnesses $k_{3,3}$ (axial stiffness) and $k_{6,6}$ (torsional stiffness) computed with ABAQUS by L/A and L/J an equivalent elastic and shear moduli can be retrieved. It is recalled that $k_{3,3} = EA/L$ and $k_{6,6} = GJ/L$ for isotropic Euler–Bernoulli beam with compact cross-section as shown in Equation (3). The values obtained by both methods are shown in Figure 3 where they almost coincide. This demonstrates that ABAQUS and Sun et al. [40] method give the same results for the axial and torsional stiffnesses $k_{3,3}$ and $k_{6,6}$. Even though the two models have a different mathematical background they provide the same result because axial and torsional stiffnesses do not depend on beam slenderness as shear and bending behaviors do. ABAQUS model is introduced because Sun et al. [40] method does not carry out shear stiffness and it is not able to predict accurately bending behavior of composite beams [59]. Therefore one of the aims of the present work is to reply to this problem.

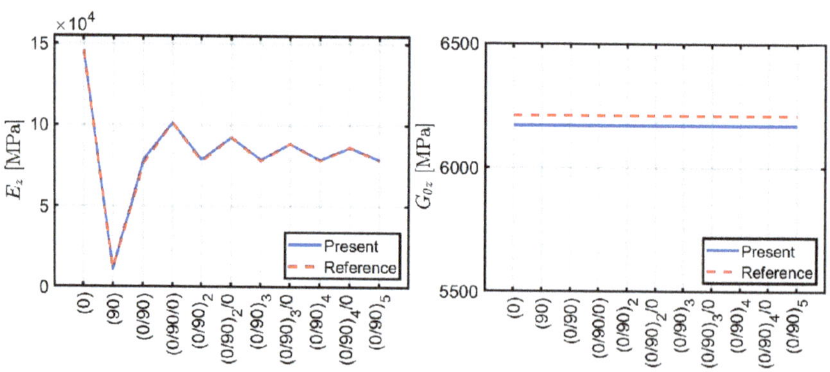

(a) (b)

Figure 3. Equivalent moduli using ABAQUS and Sun et al. [40] approach: (**a**) normal E_z; (**b**) shear $G_{\theta z}$.

For the sake of conciseness, variation with respect to different mechanical or geometric properties is not reported. However, it has been observed by the authors that by varying the mechanical properties or geometric ratios (e.g., E_1, A and L) the results presented below are not affected (more details can be found in [62]). Thus, the data given represent a general FRP hollow beam and this study can be easily adapted to different materials or geometry.

3.2. Angle-Ply Laminates

Angle-ply lamination ($\pm45/\ldots$) is used for structures that have to resist mainly against torsion or internal pressure. A typical configuration is depicted in Figure 4. The same geometric properties of the previous case are considered in the following and the number of plies is increased starting from the single-ply (45). Trends of the stiffness values for each configuration are assessed. Considering a constant thickness of 4 mm, Figure 5 shows the plots of the stiffness matrix coefficients and the relative error with respect to the scheme of initial lamination (45).

By examining the graphs obtained (Figure 5) it is possible to remark that compared to the cross-ply configuration, the stiffnesses $k_{1,1}$, $k_{1,5}$ and $k_{4,4}$, tend to increase to a constant value regardless of the lamination scheme, whether symmetrical or not. The axial stiffness $k_{3,3}$ approximately keeps a constant value. For the torsional stiffness $k_{6,6}$, a substantial increase is noted. Since the angle of the reinforcing fibers differs from the longitudinal (0) and transverse (90) directions, coupling stiffnesses appear due to shear-bending ($k_{1,4}$) and axial-torsional ($k_{3,6}$). This agrees with Reddy's studies on angle-ply reinforced plates: the structural strength has an asymptotic behavior by increasing the number of plies in the stacking sequence. In conclusion, with reference to Figure 5h it is possible to observe a variation of about 40% of the stiffness values $k_{1,1}$, $k_{1,5}$ and $k_{4,4}$ analogous to the previous cross-ply case, whereas $k_{3,3}$ has approximately a constant trend, practically is characterized by zero relative error. It is remarked that coupling stiffnesses $k_{1,4}$ and $k_{3,6}$ tend to zero by increasing the number of plies (100% of relative error with respect to (45) configuration). Thus, for these stacking sequences, the composite tends to have an orthotropic behavior with very small couplings.

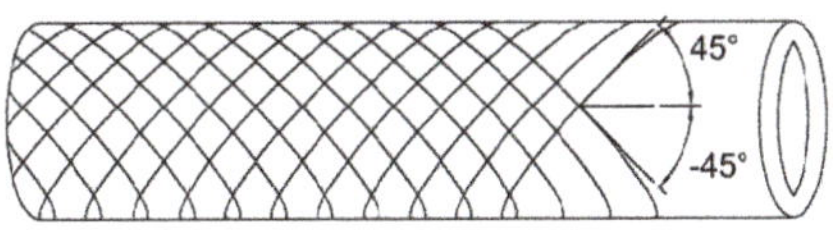

Figure 4. Angle-ply laminates.

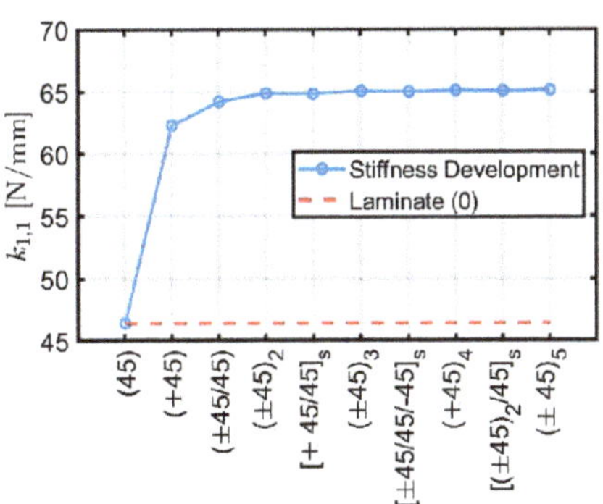

(a)

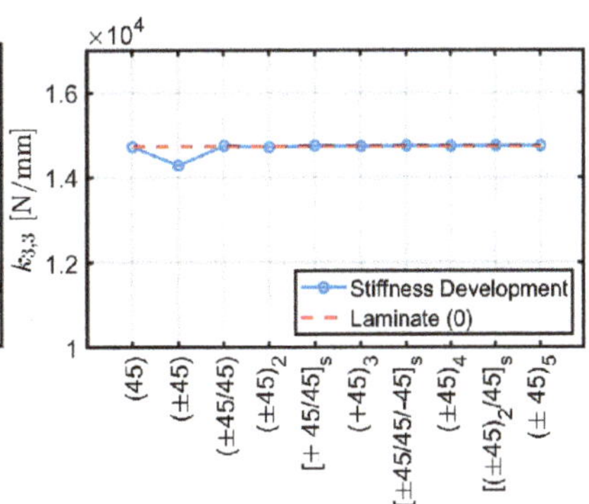

(b)

Figure 5. *Cont.*

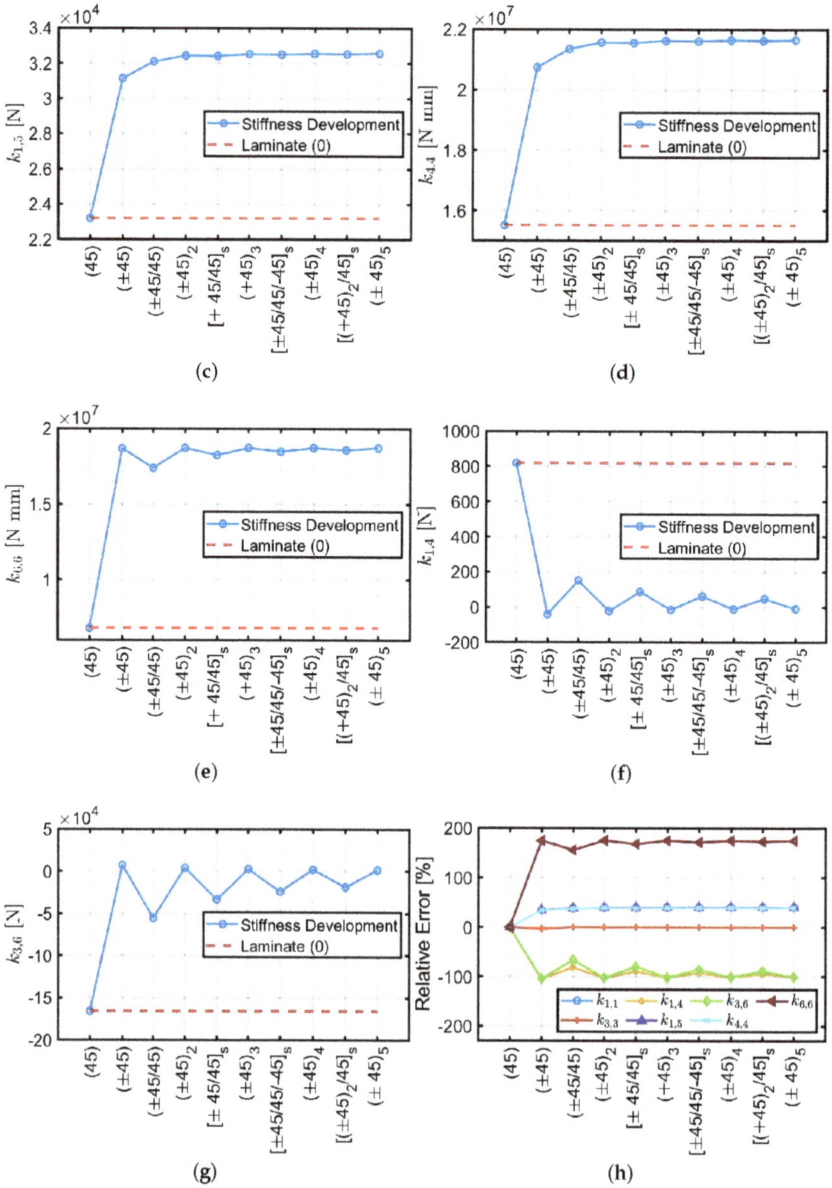

Figure 5. Stiffness trends with respect to number of plies (**a**–**g**) and relative error with respect to orthotropic configuration (**h**).

Analogously to the previous section a comparison between ABAQUS and an analytical formulation [40] is performed for angle-ply configurations. Stiffness terms $k_{3,3}$ and $k_{6,6}$ are carried out via ABAQUS and purged by multiplication of L/A and L/J, respectively in order to derive E_z and $G_{\theta z}$. Such comparison is represented in Figure 6. It can be observed that both methods give the same results in terms of $k_{3,3}$ and $k_{6,6}$ as in the cross-ply case presented in the previous section.

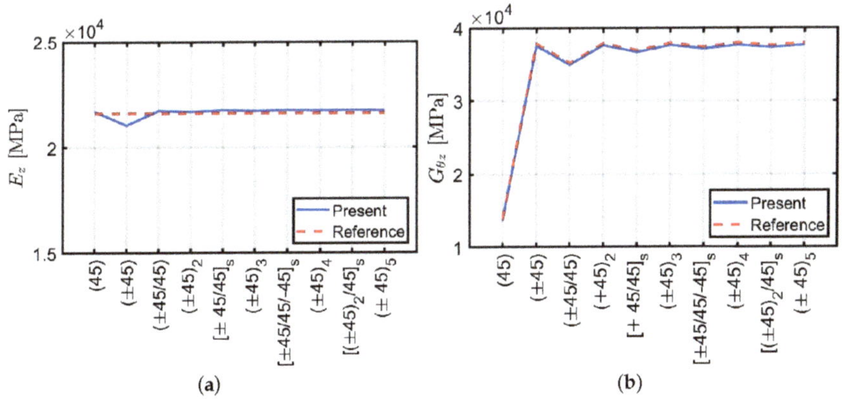

Figure 6. Equivalent moduli using ABAQUS and Sun et al. [40] approach: (**a**) normal E_z; (**b**) shear $G_{\theta z}$.

In conclusion, the present ABAQUS model is able to approximate accurately the homogenized normal E_z and shear $G_{\theta z}$ moduli by considering axial $k_{3,3}$ and torsional $k_{6,6}$ stiffness components, respectively.

4. Beam Slenderness Effect

As aforementioned, shear and bending stiffness depend on beam slenderness and they are not predicted accurately by Sun et al. [40] method. Therefore, the aim of the present section is to show such behaviors for several laminates. The geometry of the cross-section is kept constant and beam length is increased from 1 to 5 m. For cross-$(0/90/\ldots)$ and angle-ply $(\pm 45/\ldots)$ laminations the number of plies is increased as in the previous section. Moreover, quasi-isotropic configuration $(0/\pm 45/90)_s$ is taken into consideration as well as other particular stacking sequences taken from the literature. Main stiffness components are represented in a dimensionless form with respect to the correspondent stiffness component given by the Euler–Bernoulli theory. For instance shear stiffness $k_{1,1}$ is multiplied by $L^3/(12E_zI)$, bending stiffness $k_{4,4}$ by $L/(4E_zI)$, shear-bending stiffness $k_{1,5}$ by $L^2/(6E_zI)$, axial stiffness $k_{3,3}$ by L/E_zA and torsional stiffness $k_{6,6}$ by $L/(G_{\theta z}J)$. In addition, the extra shear-bending term $k_{4,11}$ which is not classical is compared to its main stiffness contribution on the main diagonal $k_{4,4}$. The homogenized moduli E_z and $G_{\theta z}$ are computed by inverting axial and torsional stiffnesses, respectively, as shown in the previous section.

4.1. Cross-Ply Laminates

Results obtained are depicted in Figure 7 where, as expected, axial $k_{3,3}$ and torsional $k_{6,6}$ rigidity do not depend on beam slenderness. In addition, as already expected no bending coupling is observed, thus $k_{4,11} = 0$. Moreover, the E_z and $G_{\theta z}$ values coincide with the ones given by Sun et al. [40] formulation. On the contrary shear $k_{1,1}$, bending $k_{4,4}$ and shear-bending $k_{1,5}$ moduli tend to follow classical formulas (e.g., $k_{1,1} = 12E_zI/L^3$, $k_{4,4} = 4E_zI/L$) only for large slenderness. In particular, there is a more accelerated trend for symmetrical configurations.

It should be remarked that the calculation of the stiffness matrix by using classical formulas varies with beam slenderness. This error leads to lower stiffness and consequently, it should be considered in the design phase by the introduction of a correction factor. For symmetric lamination schemes and with slenderness values $L/h > 30$–35 the error is less than 5%.

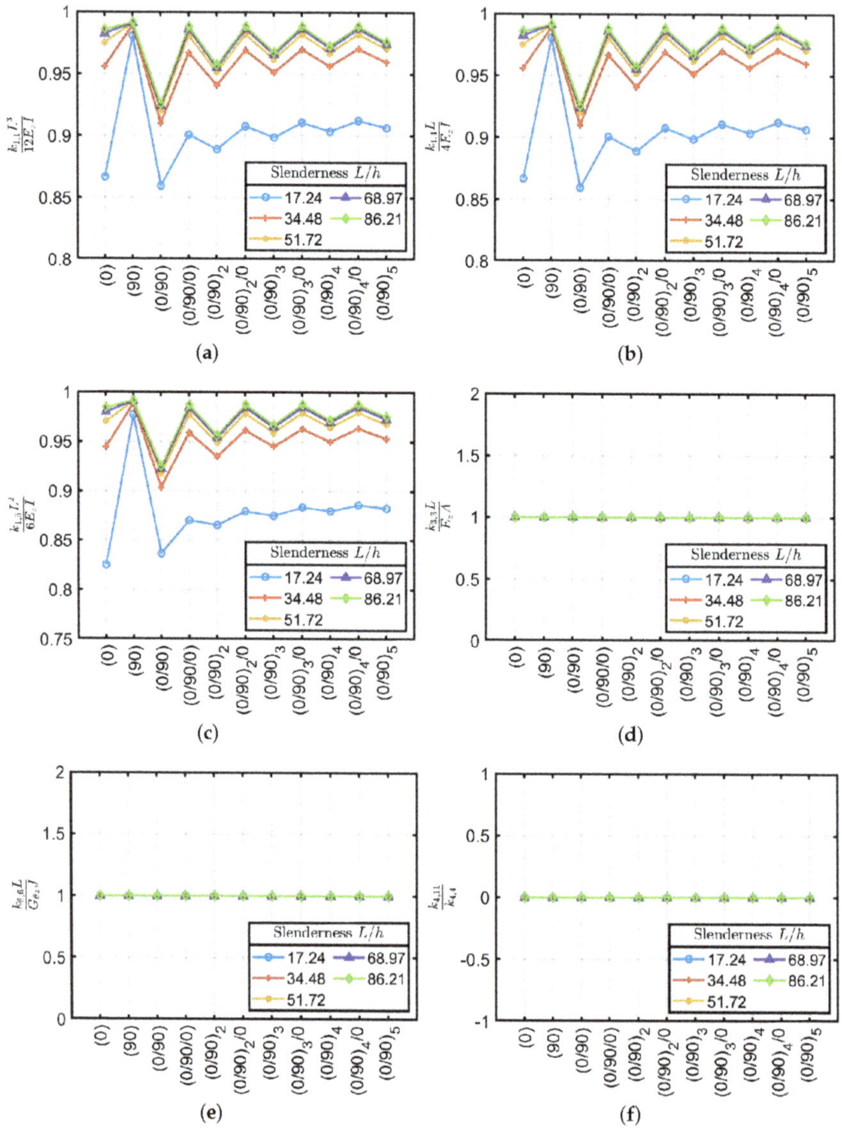

Figure 7. Stiffness coefficients **(a)** $\frac{k_{1,1}L^3}{12E_zI}$, **(b)** $\frac{k_{4,4}L}{4E_zI}$, **(c)** $\frac{k_{1,5}L^2}{6E_zI}$, **(d)** $\frac{k_{3,3}L}{E_zA}$, **(e)** $\frac{k_{6,6}L}{G_{\theta z}J}$ and **(f)** $k_{4,11}/k_{4,4}$ as a function of slenderness L/h for cross-ply laminates.

4.2. Angle-Ply Laminates

For the angle-ply configurations ($\pm45/\ldots$) the values compared between the stiffnesses generated by ABAQUS and the Euler–Bernoulli theory for isotropic materials were depicted in Figure 8. For the present configuration, as in the previous cross-ply one, the axial $k_{3,3}$ and torsional $k_{6,6}$ stiffness does not depend on the slenderness and the equivalent stiffness carried out are close to the ones obtained via Sun et al. [40] approach.

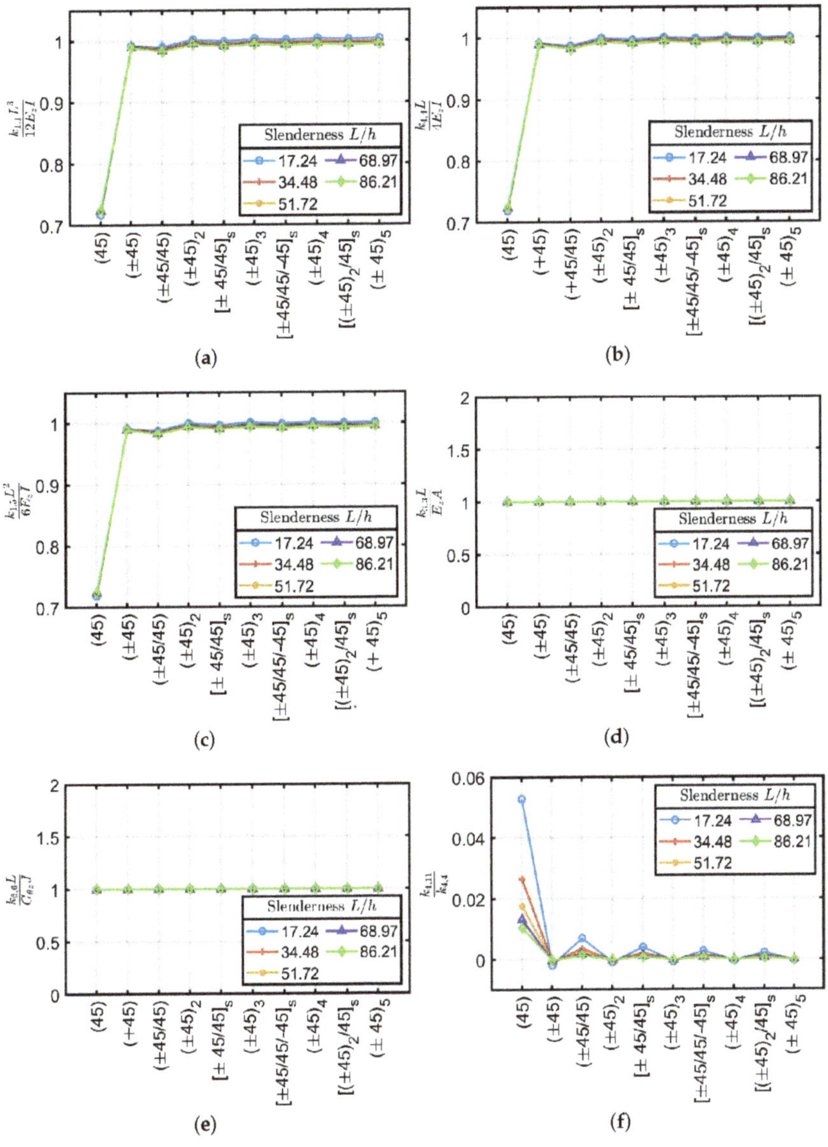

Figure 8. Stiffness coefficients **(a)** $\frac{k_{1,1}L^3}{12E_zI}$, **(b)** $\frac{k_{4,4}L}{4E_zI}$, **(c)** $\frac{k_{1,5}L^2}{6E_zI}$, **(d)** $\frac{k_{3,3}L}{E_zA}$, **(e)** $\frac{k_{6,6}L}{G_{\theta z}J}$ and **(f)** $k_{4,11}/k_{4,4}$ as a function of slenderness L/h for angle-ply laminates.

By considering shear and bending stiffnesses $k_{1,1}$, $k_{4,4}$ and $k_{1,5}$ the difference between the (45) laminate and the others can be easily noted. The single-ply configuration has approximately 30% of error with respect to classical formulas. By increasing the number of plies, on the contrary, the computed values are in agreement with the classical values if they were made of an equivalent isotropic material where the average error is below 2%. Antisymmetric lamination schemes have better behavior by excluding the scheme (±45) which shows a slightly higher stiffness (about 1%) than the reference ones. It is remarked that the angle-ply ($\pm45/\ldots$) configuration has a very small variation with respect to beam slenderness. On the contrary, a large variation was observed for the cross-ply cases.

Finally, the angle-ply configuration ($\pm45/\ldots$) has an extra coupling between bending moments ($k_{4,11}$) that is not present in the cross-ply configurations. This coupling tends to decrease by increasing the beam slenderness. Compared to the main bending stiffness $k_{4,4}$, this is minimal and can be considered negligible, except for the first lamination scheme (45) which has a difference of about 5%, consequently, they can be neglected for design purposes. The scheme (45) is feasible but not used in applications as it generates delamination and durability problems of the composite. Usually, the laminations are composed of several crossed layers, it is therefore legitimate, given the proposed results, to neglect the component $k_{4,11}$ in practical applications.

4.3. Quasi-Isotropic Configuration

Quasi-isotropic laminates are made of three or more orthotropic plies of identical thickness and material. The stack considers for each ply $+\theta$ a corresponding one with $-\theta$ over the middle surface arranged with the same sequence order as depicted in Figure 9. In order to obtain an element with good shear, bending, traction and torsion properties a quasi-isotropic configuration with a combination of cross-($0/90/\ldots$) and angle-ply ($\pm45/\ldots$) has been considered as $(0/\pm45/90)_s$. The main feature of this typology is that the coupling stiffnesses $k_{1,4}$ and $k_{3,6}$ are zero. These laminates are widely used in practical applications, because they offer homogeneous stiffness and strength in all directions, and also have a good performance against crack propagation.

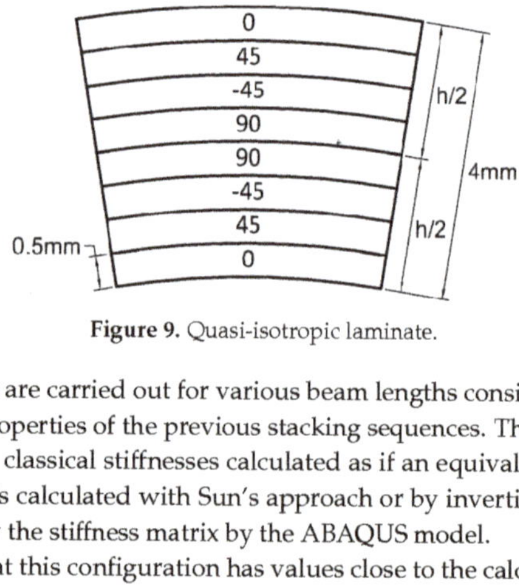

Figure 9. Quasi-isotropic laminate.

Stiffness components are carried out for various beam lengths considering the same geometric section and mechanical properties of the previous stacking sequences. The variations of the stiffness values with respect to the classical stiffnesses calculated as if an equivalent isotropic material with $E = E_z$ is considered. E_z is calculated with Sun's approach or by inverting the main component of axial stiffness $k_{3,3}$ given by the stiffness matrix by the ABAQUS model.

It is also observed that this configuration has values close to the calculated stiffness as if it was a homogenized isotropic material. As a matter of fact, stiffnesses $k_{1,1}$, $k_{4,4}$ and $k_{1,5}$ are all below 3% compared to the reference values. Furthermore, as the slenderness increases, this error decreases until it reaches a constant value of less than 1%.

The homogenized elastic E_z and shear $G_{\theta z}$ moduli are obtained from axial $k_{3,3}$ and torsional $k_{6,6}$ stiffnesses by multiplying them by A/L and J/L, respectively. Thus, the ratios between $k_{3,3}$ and $k_{6,6}$ and the same calculated via Sun et al. [40] procedure are the same.

It has been shown also for the quasi-isotropic configuration $(0/\pm45/90)_s$, that stiffness ratios are almost constant with respect to the beam slenderness. In the current configuration, the stiffness due to the coupling between the moments ($k_{4,11}$) tends to increase until it reaches a constant value as the element's slenderness increases; the values generated with respect to the bending stiffness ($k_{4,4}$) are, however, minimal and negligible (see Figure 10).

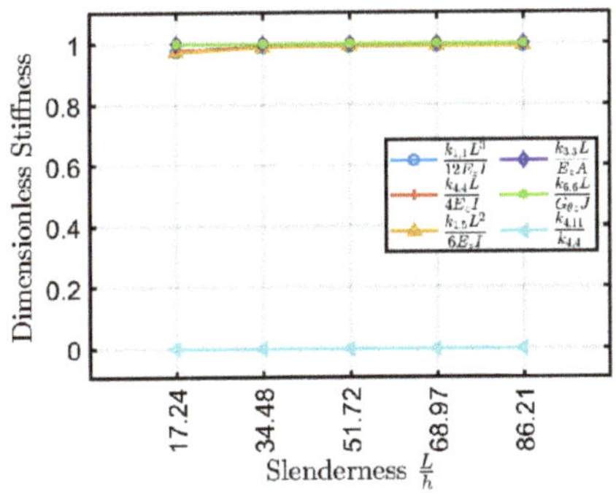

Figure 10. Stiffness coefficients $\frac{k_{1,1}L^3}{12E_zI}$, $\frac{k_{4,4}L}{4E_zI}$, $\frac{k_{1,5}L^2}{6E_zI}$, $\frac{k_{3,3}L}{E_zA}$, $\frac{k_{6,6}L}{G_{\theta z}J}$ and $k_{4,11}/k_{4,4}$ as a function of slenderness L/h for quasi-isotropic laminate.

4.4. Bouligand Laminates

In this section, analyses are carried out by using bio-inspired Bouligand lay-ups of 5° at each subsequent layer as presented by Mencattelli and Pinho [63] on laminated plates with ultra-thin thickness. It has been shown that by reducing the variation of the angles between the layers (e.g., the pitch $\Delta\theta$) to the values that imitate the natural microstructure, it is possible to obtain at the same time a strong increase in the energy dissipated and in the maximum load-bearing capacity.

Bouligand unit considers rotation of 180° inside each unit, however, 1/4 of a Bouligand unit has been considered here as the main unit. Thus, the following configurations and nomenclature will be used.

- $L.1 \equiv (0/5/\ldots/40/45)_s$
- $L.2 \equiv (0/5/\ldots/85/90)_s$
- $L.3 \equiv (0/5/\ldots/85/90/-85/\ldots/-5/0)_4$
- $L.4 \equiv (0/5/\ldots/85/90/-85/\ldots/-5/0)_{2s}$

A graphical representation of the four sequences considered are given in Figure 11. $L.1$ and $L.2$ are 1/4 and 1/2 of a Bouligand unit, respectively. $L.3$ and $L.4$ are an antisymmetric and a symmetry scheme inspired by Mencattelli and Pinho [63], respectively.

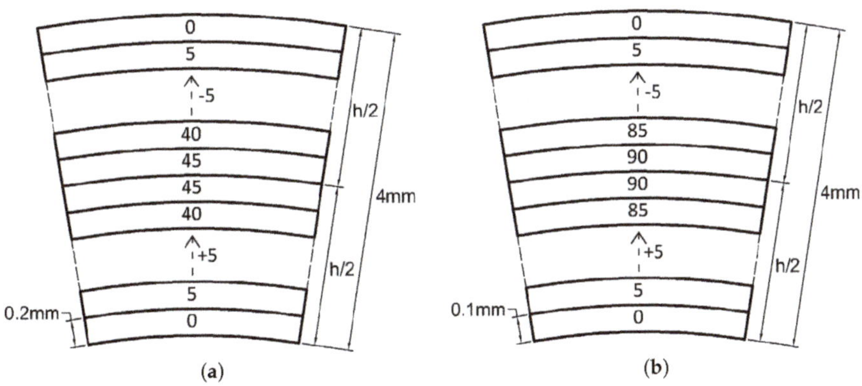

Figure 11. *Cont.*

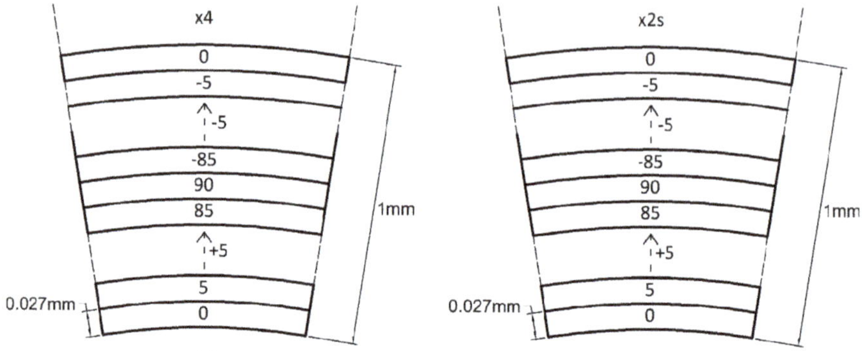

Figure 11. (**a**) *L*.1 laminate, (**b**) *L*.2 laminate, (**c**) *L*.3 laminate and (**d**) *L*.4 laminate.

By keeping the same geometric section and mechanical properties of the previous cases, the following analyses are carried out on element lengths from 1 to 5 m. In order to correctly isolate the effect of the pitch angle of the reinforcement fibers $\Delta\theta$, Mencattelli and Pinho [63] have decided to eliminate the couplings in the stiffness matrix by creating symmetric laminates. Therefore, *L*.4 adopts a symmetrical lamination scheme consisting of 2 symmetric modules (each module is defined by a complete set of the fiber angles, or 360°) with 146 layers in total:

$$\begin{aligned}(0/5/\ldots/85/90/-85/\ldots/-5/0/5/\ldots/85/90/-85/\ldots/-5/0)_s = \\ = (0/5/\ldots/85/90/-85/\ldots/-5/0)_{2s}\end{aligned} \tag{4}$$

Stiffness variation as a function of the slenderness with respect to the classical stiffness relations are depicted in Figure 12. By using a symmetrical lamination scheme, in addition to the cancellation of the coupling stiffnesses, the variation between the stiffnesses $k_{1,1}$, $k_{4,4}$, $k_{1,5}$, $k_{3,3}$ and $k_{6,6}$ is further reduced with respect to the classical definitions. Likewise in this case, the relationship between the coupling stiffness $k_{4,11}$ and the bending stiffness $k_{4,4}$ is negligible.

By comparing the obtained results, it is shown that the substantial differences between the selected lamination schemes are caused by the coupling stiffnesses $k_{1,4}$, $k_{3,6}$ and $k_{4,11}$. These coupling terms have higher values for laminates with few layers, even if symmetrical, and will tend to fade out by increasing the number of layers, thus tend to a balanced configuration. Although the first two configurations (*L*.1, *L*.2) are symmetrical, fairly pronounced coupling stiffnesses are generated. The validation of this behavior has been done through the Sun [40] model according to which coupling values are generated which are also different from zero and high between tension and torsion. Regarding Figure 13, it is possible to note that the laminate *L*.4 studied by Mencattelli and Pinho [63] has average axial, shear and bending stiffness values, high torsional and zero coupling stiffness values with respect to the others. For this reason, *L*.4 has been selected for the following comparisons with respect to the others presented.

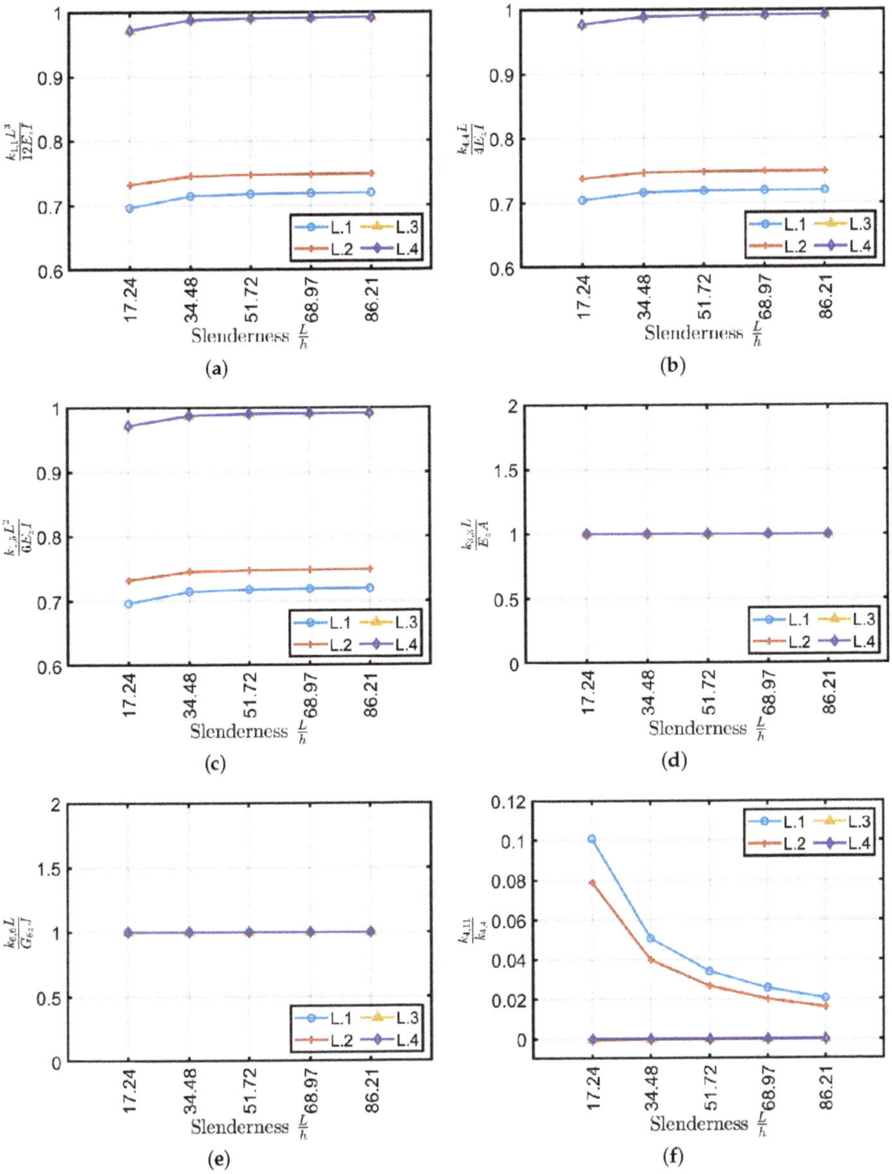

Figure 12. Stiffness coefficients as a function of slenderness for $L.1$, $L.2$, $L.3$ and $L.4$ **(a)** $\frac{k_{1,1}L^3}{12E_zI}$, **(b)** $\frac{k_{4,4}L}{4E_zI}$, **(c)** $\frac{k_{1,5}L^2}{6E_zI}$, **(d)** $\frac{k_{3,3}L}{E_zA}$, **(e)** $\frac{k_{6,6}L}{G_{\theta z}J}$ and stiffness ratio **(f)** $k_{4,11}/k_{4,4}$ for Bouligand laminates.

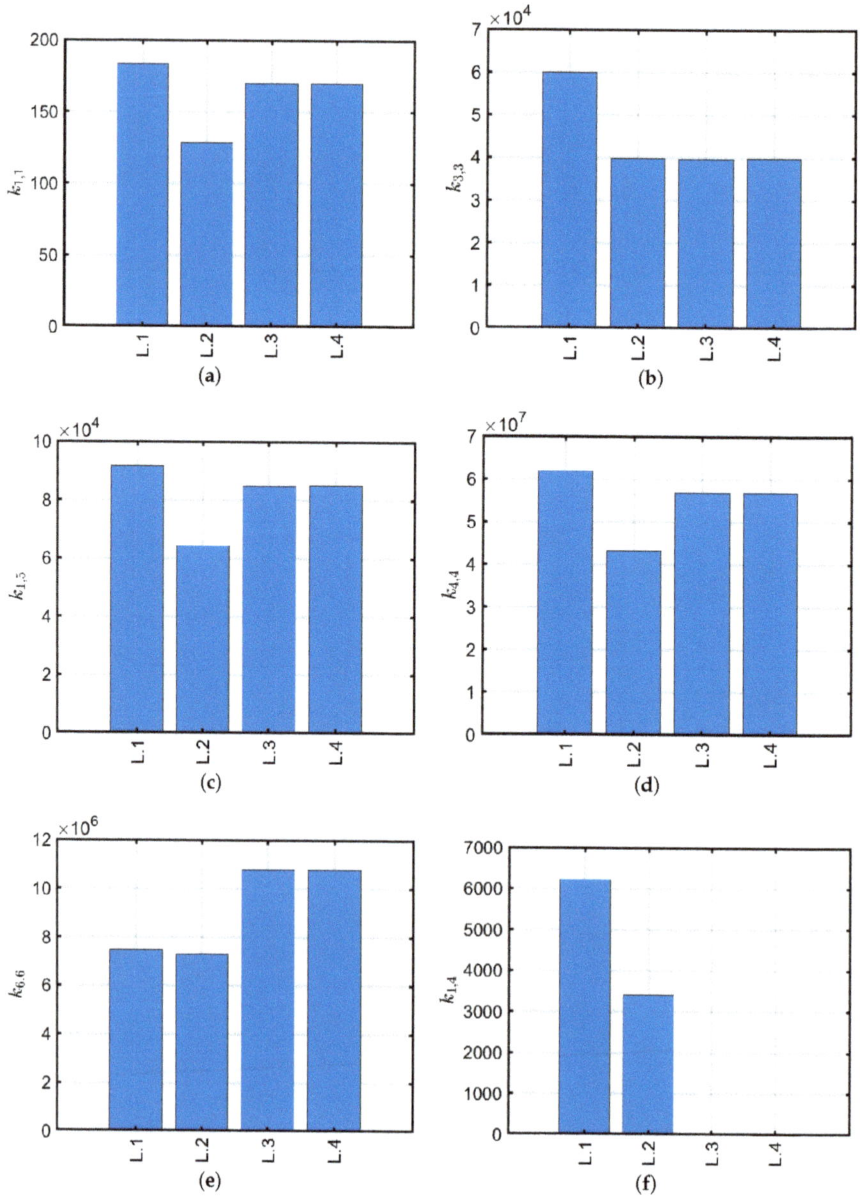

Figure 13. *Cont.*

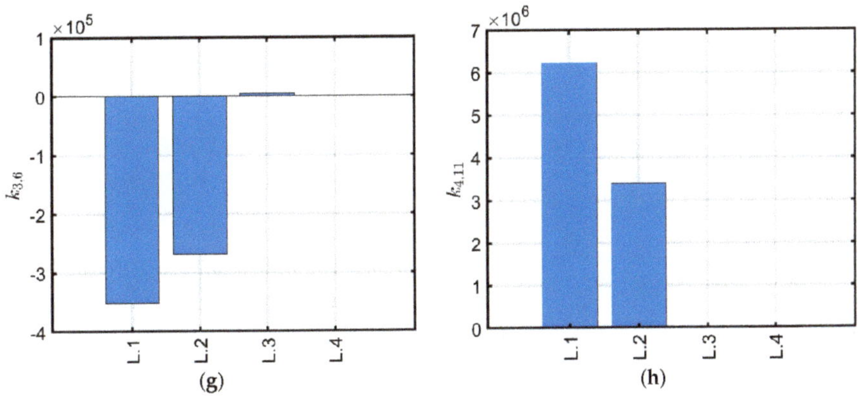

Figure 13. Stiffness components (**a**–**h**) of the considered Bouligand laminates for $L = 1$ m.

5. Comparison

Following the analyses carried out, an overall view is given on the behavior of the various lamination schemes, evaluating the variation of the stiffnesses of the laminates with respect to the orthotropic scheme (0) and the coupling stiffnesses based on their relevance with respect to the correspondent main stiffness on the main diagonal of the stiffness matrix.

Figure 14 shows the comparison for the main stiffnesses involved in the present study with respect to the reference (0) configuration. Note that angle-ply configurations have the highest stiffnesses compared to the others, but they also have shear-bending coupling $k_{1,4}, k_{4,11} \neq 0$ and traction-torsion coupling $k_{3,6} \neq 0$. Whereas, quasi-isotropic and *L*.4 configurations show negligible couplings and lower stiffness values analogous to angle-ply configurations. From the obtained results the following comments arise:

- All the stiffnesses tends to have a constant value by increasing beam slenderness.
- The orthotropic scheme (0) is very efficient if subjected to shear, axial and bending actions, but not for torsional ones. However, it has a significant variation while increasing slenderness. Furthermore, it suffers from delamination, because the fibers being all parallel, tend to form a preferential fracture plane.
- The angle-ply scheme $(\pm 45)_5$ works conversely with respect to the orthotropic configuration (0). It has high torsional stiffness, but low axial, shear and bending ones. For the present configuration, the shear-bending coupling is negligible. It is possible to use Sun et al. [40] formulae to homogenize equivalent properties because its properties do not depend on the beam slenderness.
- The angle-ply scheme $(\pm 30)_5$ represents an intermediate solution between the orthotropic (0) and the angle-ply $(\pm 45)_5$ configurations. It has a lower axial, shear and bending stiffness of 50–60% approximately and torsional stiffness of 80% with respect to (0). The present stack, as well as $(\pm 45)_5$, has very low shear-bending coupling stiffness values which can be neglected and invariability of stiffness constants with respect to slenderness.
- The quasi-isotropic laminate $(0/\pm 45/90)_s$ has stiffness values similar to the laminate $(\pm 30)_5$, except for the torsional stiffness which decreases by approximately 30%. Being a quasi-isotropic laminate, it has a negligible shear-bending coupling stiffness. In this case, the constants of the stiffness matrix vary slightly with increasing slenderness.
- The *L*.4 laminate, studied by Mencattelli and Pinho [63], with the only exception of the orthotropic configuration (0), presents slightly higher values for axial, shear and bending stiffnesses, but lower values for torsional stiffness with respect to the others. Coupling effects are minimal, and a

very small dependency on the beam slenderness for the main stiffness components $k_{1,1}$, $k_{4,4}$, $k_{1,5}$, $k_{3,3}$ and $k_{6,6}$ is observed.

In conclusion, the best axial, shear and bending properties are achieved by the configuration (0) however such selection has very small torsional stiffness and low durability and reliability. Therefore, quasi-isotropic and $L.4$ configurations are suggested for having good mechanical performances in almost all conditions. These configurations do not present coupling terms so they can be easily characterized.

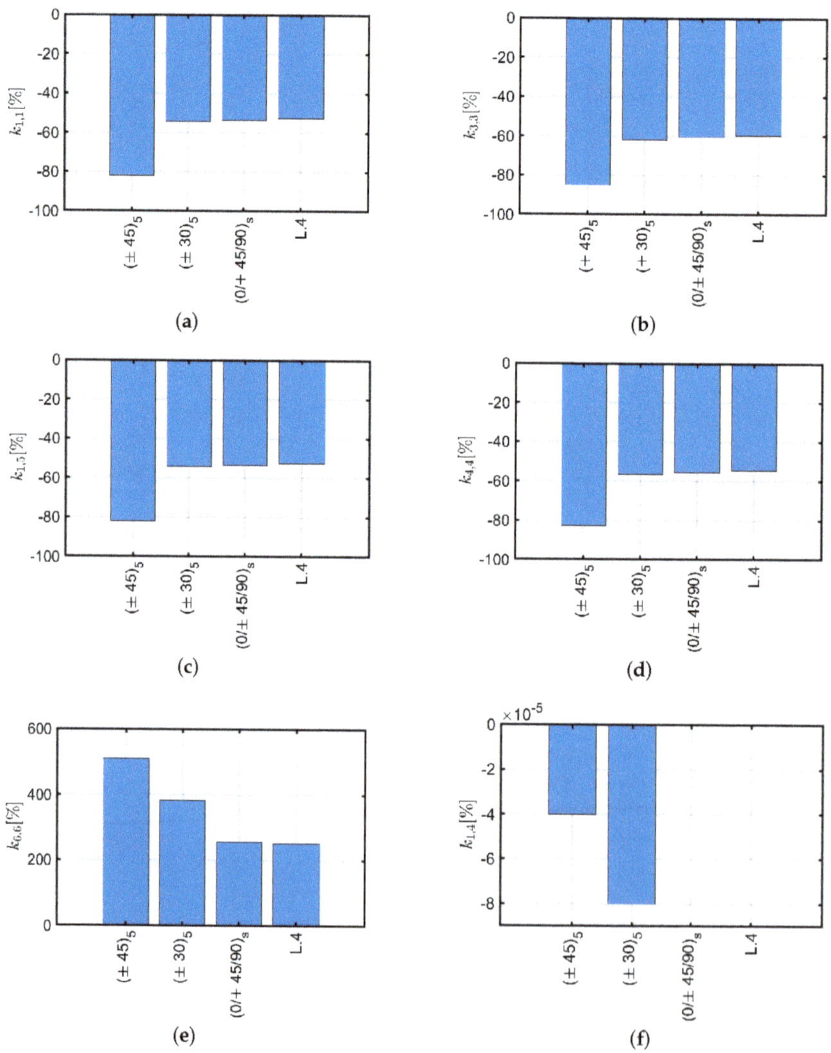

Figure 14. *Cont.*

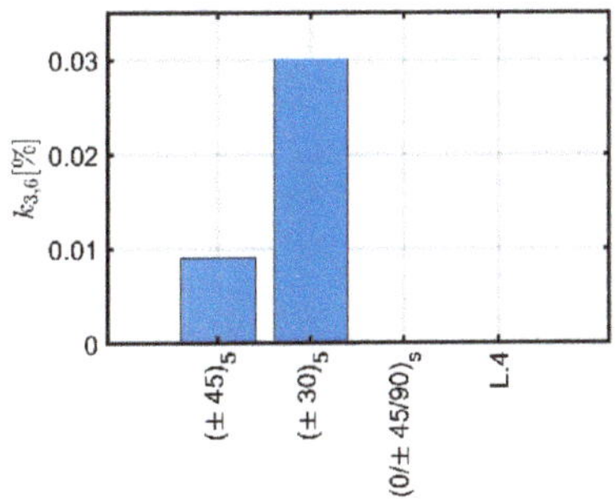

(g)

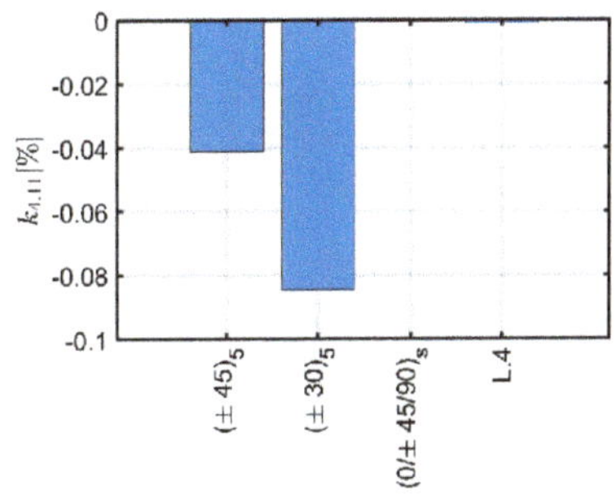

(h)

Figure 14. Comparison among stiffness components (**a**–**h**) for angle-ply, quasi-isotropic and *L*.4 laminates with $L = 1$ m respect to the orthotropic scheme (0).

6. Laminates with Highly Positive and Negative Poisson Values

In the previous section, the most common composite materials such as fiberglass/epoxy or graphite/epoxy have been considered. Such composites have a stiffness ratio $E_1/E_2 = 8$–16. In particular, the material considered in the previous analyses is characterized by $E_1/E_2 \approx 13$ (see Table 1). However, there is another class of highly orthotropic composite materials which are defined by $E_1/E_2 = 5000$–$50,000$ while using RP6410 or RP6442 as a polyurethane matrix, depending on the volume fraction of the fibers. Mechanical properties of these two constituents are listed in Table 2. The equivalent orthotropic coefficients for the present laminates are given in Table 3 which shows a stiffness ratio of approximately 42,800 and 10,530 respectively. Peculiarities and features of these laminates applied to flat plates have been described by Peel [64].

Table 2. Mechanical properties of laminates with high stiffness ratio [64].

Materials	E [MPa]	ϵ [%]
RP6410	1.65	3.30
RP6442	7.00	5.25
Graphite	276,000	-

Table 3. Mechanical properties of: Graphite/RP6410 and Graphite/RP6442 [64].

	E_1 [MPa]	E_2 [MPa]	$\nu_{12} = \nu_{13}$	ν_{23}	$G_{12} = G_{13} = G_{23}$ [MPa]
Graphite/RP6410	120,000	2.85	0.41	0.502	0.949
Graphite/RP6442	126,600	12.02	0.41	0.504	4

This high stiffness ratio allows having a laminate with a high positive or negative engineering Poisson ratio of ν_{xy}. A high Poisson is given by a balanced lamination and a negative Poisson is given by a symmetrical scheme. According to the studies by Peel [64], unbalanced lamination schemes $[\theta/\alpha]_s$ have very small (highly negative) Poisson ratios. In particular, the laminate $(15/5)_s$ (represented in Figure 15) has a Poisson ratio of about −34 for the Graphite/RP6410 material. Its engineering constants are listed in Table 4 alongside with those defined by the CFRP material.

Table 4. Engineering constants for $(15/5)_s$ using material from Table 2 and CFRP using material from Table 1.

Graphite/RP6410		CFRP	
E_x	4498.23 MPa	E_x	94,509.78 MPa
E_y	3.02 MPa	E_y	11,234.40 MPa
ν_{xy}	−34.04	ν_{xy}	0.342
ν_{yx}	−0.023	ν_{yx}	0.041
G_{xy}	23.33 MPa	G_{xy}	7395.17 MPa

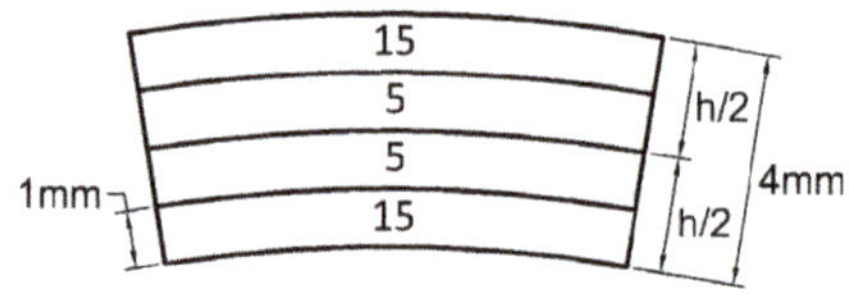

Figure 15. The $(15/5)_s$ laminate.

The stiffness coefficients are carried out below for the materials listed in Table 4. The comparison between the two types of material is addressed on the same tubular profile of the previous cases. The coefficients of the stiffness matrix are evaluated and their variations with respect to beam slenderness are show in Figure 16. Both configurations have a small variation with respect to beam slenderness. The main difference is observed in the values of the stiffness coefficients $k_{1,1}$, $k_{4,4}$ and $k_{1,5}$ that compared to the same stiffnesses calculated by Euler–Bernoulli theory using homogenization a difference of approximately 90% in the case of Graphite/RP6410 and approximately 30% in the case of CFRP occurs. Regarding the coupling coefficients, both materials have significant values due to the unbalanced lamination scheme. In particular, Graphite/RP6410 will have lower coupling values due to the poor resistance in the transverse direction of the material, but by observing Figure 16f, a higher effect than the $k_{4,11}$ given by CFRP.

In the following, four configurations investigated numerically and experimentally by Peel [64] are used for studying the stiffness of tubular profiles. Stiffness ratios and equivalent Poisson ratios are listed in Table 5. The sequences are $(15_2/30_2)_s$ with negative Poisson ratio (NP1 and NP2) and $(\pm 15/\pm 30)_s$ with a high Poisson ratio (HP1 and HP2). The enumeration 1 and 2 identifies the material used Graphite/RP6410 and Graphite/RP6442, respectively. The same nomenclature is used in Figure 17. The analysis was carried out by dividing the combinations according to the lamination schemes and comparing the two materials.

The laminate $(15_2/30_2)_s$ (NP1 and NP2) and $(\pm 15/\pm 30)_s$ (HP1 and HP2) are presented in Figure 18. The results for $(15_2/30_2)_s$ obtained were expected also considering the same given in Figure 16 where a smaller Poisson ratio was considered. The error given for the shear and bending behavior is approximately 98% on the contrary traction and torsion work well as in all the previous cases. Regarding the influence of the coupling coefficient $k_{4,11}$ with respect to the bending stiffness $k_{4,4}$, both materials behave in the same way, that is, tends to decrease by increasing beam slenderness. Such ratio goes from a value of about 40% ($L/h = 12.24$) to 10% ($L/h = 86.21$) so it cannot be neglected.

A further comparison can be made with the previous lamination scheme $(15/5)_s$ presented in the Figure 16, which, observing the Graphite/RP6410, is characterized by a minor error towards the calculation of the stiffness coefficients with the Euler–Bernoulli formulation, but by a much greater influence of the coupling stiffness $k_{4,11}$, by about double.

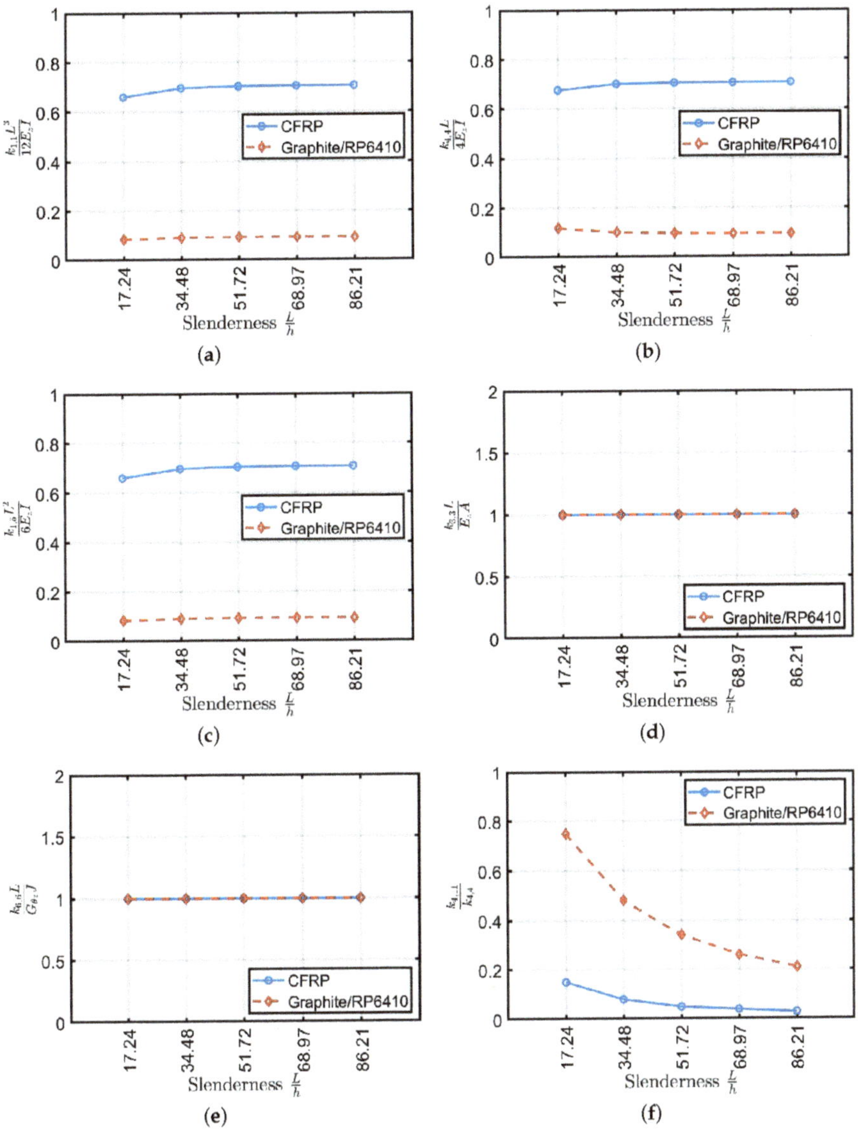

Figure 16. Stiffness coefficients as a function of slenderness for $(15/5)_s$ (**a**–**e**) and stiffness ratio $k_{4,11}/k_{4,4}$ (**f**).

Table 5. Stacking sequencies presented by Peel [64] stiffness and equivalent Poisson ratios.

Laminates	Material	E_1/E_2	ν_{xy}
NP1 $(15_2/30_2)_s$	Graphite/RP6410	42,800	−6.38
HP1 $(\pm15/\pm30)_s$	Graphite/RP6410	42,800	3.73
NP2 $(15_2/30_2)_s$	Graphite/RP6442	10,530	−6.15
HP2 $(\pm15/\pm30)_s$	Graphite/RP6442	10,530	3.72

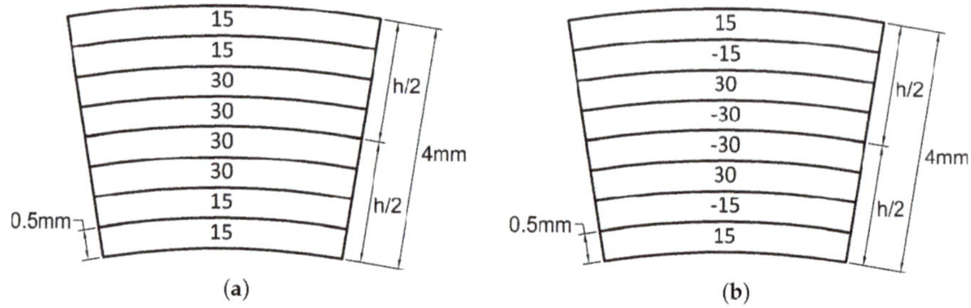

Figure 17. (**a**) The $(15_2/30_2)_s$ laminate, (**b**) $(\pm 15/\pm 30)_s$ laminate.

The laminate $(\pm 15/\pm 30)_s$ (HP1 and HP2) is made up of highly orthotropic composite materials and a balanced lamination scheme. Assessment is depicted in the Figure 18. Compared to the previous laminate with a negative Poisson ratio, the balanced laminates generate stiffness values that are close to those obtained with the Euler–Bernoulli formulation for an isotropic material. This is expected by any balanced laminate, even in the case of highly orthotropic composite materials. The values of the coupling stiffnesses are negligible as a matter of fact $k_{4,11}/k_{4,4}$ is almost zero.

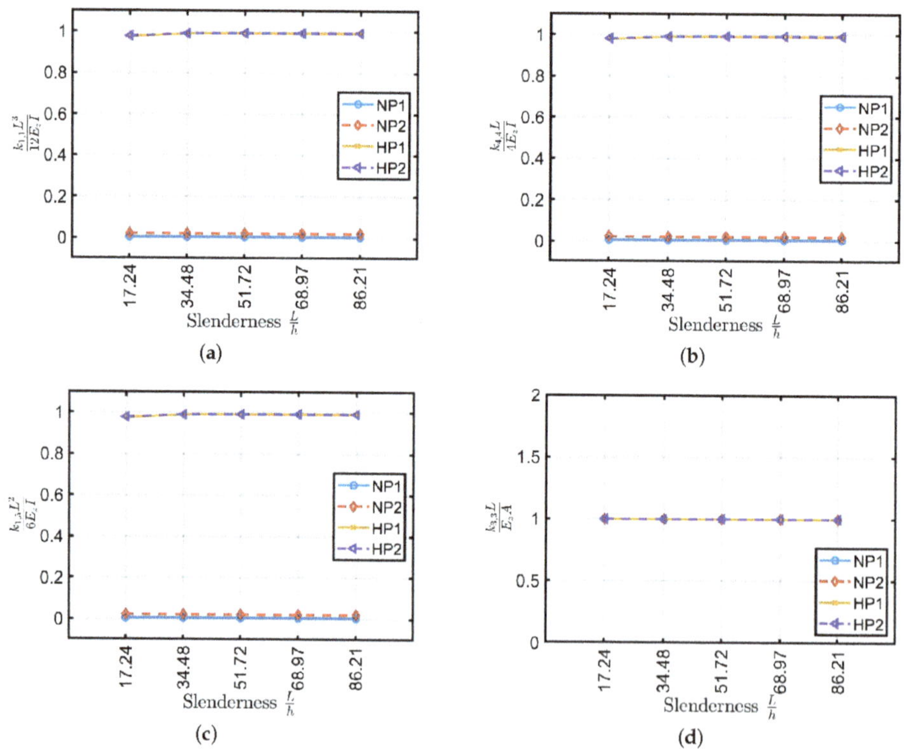

Figure 18. *Cont.*

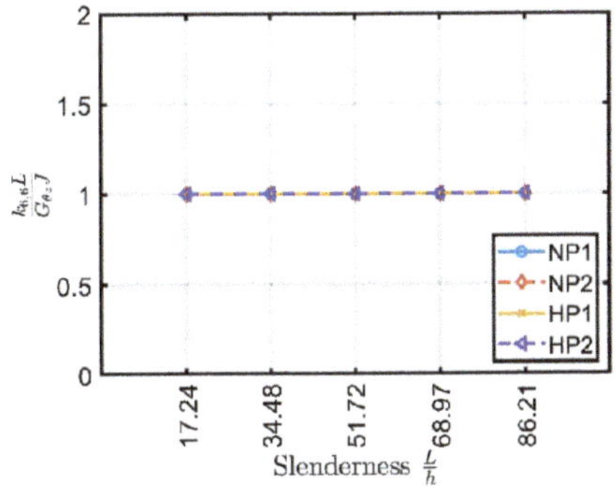

(e)

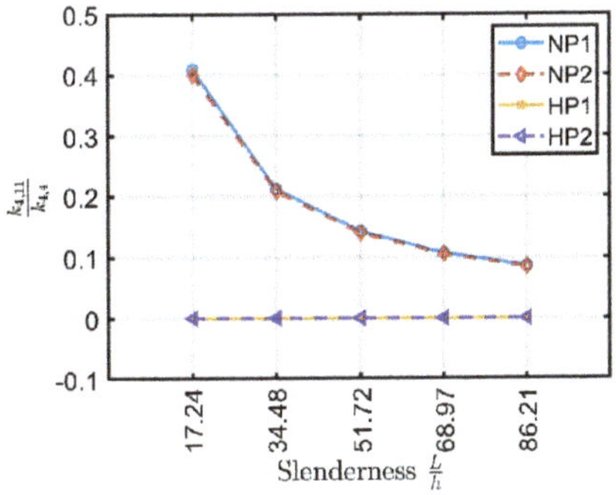

(f)

Figure 18. Stiffness coefficients as a function of slenderness for $(15_2/30_2)_s$ and $(\pm15/\pm30)_s$ (**a**–**e**) and stiffness ratio $k_{4,11}/k_{4,4}$ (**f**).

7. Validity and Limitations

After investigating several lamination schemes with standard and high stiffness ratios (E_1/E_2) according to beam slenderness and by comparing it to an equivalent homogenized isotropic Euler–Bernoulli beam model, the following remarks can be reported:

- cross-ply laminates, except laminate (90) have lower values, approximately 10% for $L = 1$ m and highly variable as a function of slenderness with respect to the equivalent orthotropic model;
- (45) scheme has lower values than the reference of about 30% and invariant with respect to beam slenderness;
- (30) and $(\pm30/30)$ scheme have lower values than the reference of about 55% and 5%, respectively, and they are invariant with respect to beam slenderness;
- unbalanced laminates with a fiber pitch angle of $\Delta\theta = 5°$, $L.1$ and $L.2$ have values lower than the reference of approximately 30% and 25%, respectively;
- $(15/5)_s$ scheme has lower values than the reference of approximately 30% for the CFRP configuration and 90% for the Graphite/RP6410;
- $(15_2/30_2)_s$ scheme has values very large errors of about 100% and it is invariant with respect to beam slenderness.

If cross-ply laminates are not considered, it is possible to observe that the calculation of the coefficients of the stiffness matrix is possible by using isotropic definitions of the Euler–Bernoulli theory (Equation (3) for balanced laminates and for angle-ply $(\pm\alpha/\ldots)$ ones. It is remarked that angle-ply must have more than one ply in the stack, because as they increase the number of plies they tend to behave approximately as balanced with negligible coupling stiffness values.

The problem arises when cross-ply are also considered, because for low slenderness values, although balanced, they have stiffness values that deviate of about 10% from those defined from Euler–Bernoulli's theory.

In order to analyze the couplings which might occur in the present laminates, the ABAQUS shell model is considered as a reference. Composite plate/shell Mindlin theory considers at the constitutive level the matrices $[A]$, $[B]$ and $[D]$ which represent the membrane, the bending-membrane coupling, and the bending stiffness matrices, respectively.

It is well-known that the membrane-bending stiffness matrix ($[B]$) is zero for symmetric laminates and it is not zero for antisymmetric cross-ply (in particular $B_{11} = -B_{22}$ applies) and angle-ply (in particular $B_{16} \neq B_{26} \neq 0$ applies) configurations. Since both symmetric and antisymmetric schemes present errors with respect to the reference configuration, it is clear that such behavior is not due to only the membrane-bending coupling matrix $[B]$. In other words, balanced laminates have $A_{1,6} = A_{2,6} = 0$ as well as cross-ply, however the latter presents larger errors in the definitions of $k_{1,1}$, $k_{4,4}$ and $k_{1,5}$ with respect to the same computed with balanced laminates.

A condition that allows using the analytical model of Sun et al. [40] and subsequently calculate the stiffness of the various beams via the Euler–Bernoulli theory for isotropic materials is sought. The authors investigated the following stiffness ratios $A_{1,6}/A_{6,6}$ $A_{2,6}/A_{6,6}$, $A_{1,6}/A_{1,1}$, $A_{1,6}/A_{2,2}$, $D_{1,6}/D_{6,6}$,$D_{2,6}/D_{6,6}$, $D_{1,6}/D_{1,1}$, $D_{1,6}/D_{2,2}$. Differences among laminates and their configurations are mainly given by the presence of normal-shear couplings A_{ij} and D_{ij} when $ij = 16$ or 26. In particular, it is relevant to analyze the ratio between the latter with the main stiffness terms when $ij = 11, 22$ or 66. For the sake of conciseness, not all various combinations are reported, because only a few results are important to the present problem. Among the aforementioned list of ratios the only ones that result to be relevant were $A_{1,6}/A_{6,6}$ and $A_{2,6}/A_{6,6}$ (depicted in Figure 19). Such ratios are considered when the trend was similar to the correspondent error in the stiffness matrix components. To this aim, the variation of the error for $k_{1,1}$ has been taken as a reference for a beam of $L = 1$ m and reported in Figure 20. Extensive details and plots regarding such ratios can be found in [62].

Among the ratios considered, leaving aside the cross-ply laminates, the one that has a trend similar to the error of $k_{1,1}$ is $A_{1,6}/A_{6,6}$. This is why a deeper investigation is provided to determine the validity of the stiffness calculation $k_{1,1}$ with the Euler–Bernoulli theory. Therefore, angle-ply laminates with lamination scheme of the type $(\alpha/-\alpha/\alpha)$ with the orientation of the laminae α defined with a difference of 2.5° between plies are considered. In this way, it would be possible to determine the minimum orientation angle of the laminates for which the calculation of the stiffness with the Euler–Bernoulli formulas for an isotropic material is correct. The analysis was carried out by analyzing the stiffness coefficients $\overline{Q}_{ij}$ which represent the variables of A_{ij} since the thickness of each ply is assumed constant.

$$\frac{A_{1,6}}{A_{6,6}} = \frac{\sum_{k=1}^{3}\int_{\zeta_k}^{\zeta_{k+1}}\overline{Q}_{1,6}^{(k)}d\zeta}{\sum_{k=1}^{3}\int_{\zeta_k}^{\zeta_{k+1}}\overline{Q}_{6,6}^{(k)}d\zeta} = \frac{2\overline{Q}_{1,6}^{(\alpha)}\frac{1}{3} - \overline{Q}_{1,6}^{(-\alpha)}\frac{1}{3}}{3\overline{Q}_{6,6}^{(\pm\alpha)}\frac{1}{3}} = \frac{\overline{Q}_{1,6}}{3\overline{Q}_{6,6}} \tag{5}$$

The lamination schemes analyzed and the values obtained are represented in Figure 21a. It is expected to find a trend for $k_{1,1}$ according to the value depicted in Figure 21a. In other words, the laminate $(\pm 5/5)$ should present an error in the calculation of the stiffness with the Euler–Bernoulli theory similar or slightly greater than the laminate $(\pm 30/30)$. On the contrary, Figure 21b shows that error on $k_{1,1}$ decreases by increasing α. It can be concluded that the error committed while stiffness is calculated using the Euler–Bernoulli theory is not a function of the ratio $\overline{Q}_{1,6}/(3\overline{Q}_{6,6})$. Moreover, such a problem might be due to not uniform torsional and cross-section distortion effects which cannot be evaluated with the present model and must be verified via 3D modeling or higher-order beam formulation. However, the present methodology can be accurately utilized for angle-ply configurations when $\alpha > 30°$.

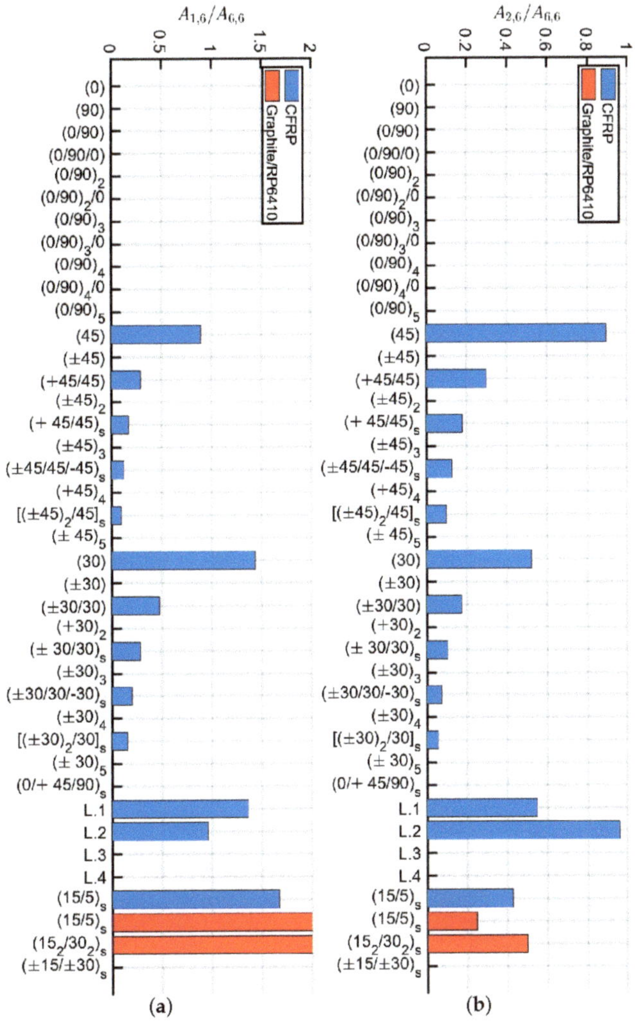

Figure 19. Ratios (**a**) $A_{1,6}/A_{6,6}$ and (**b**) $A_{2,6}/A_{6,6}$ for different laminates.

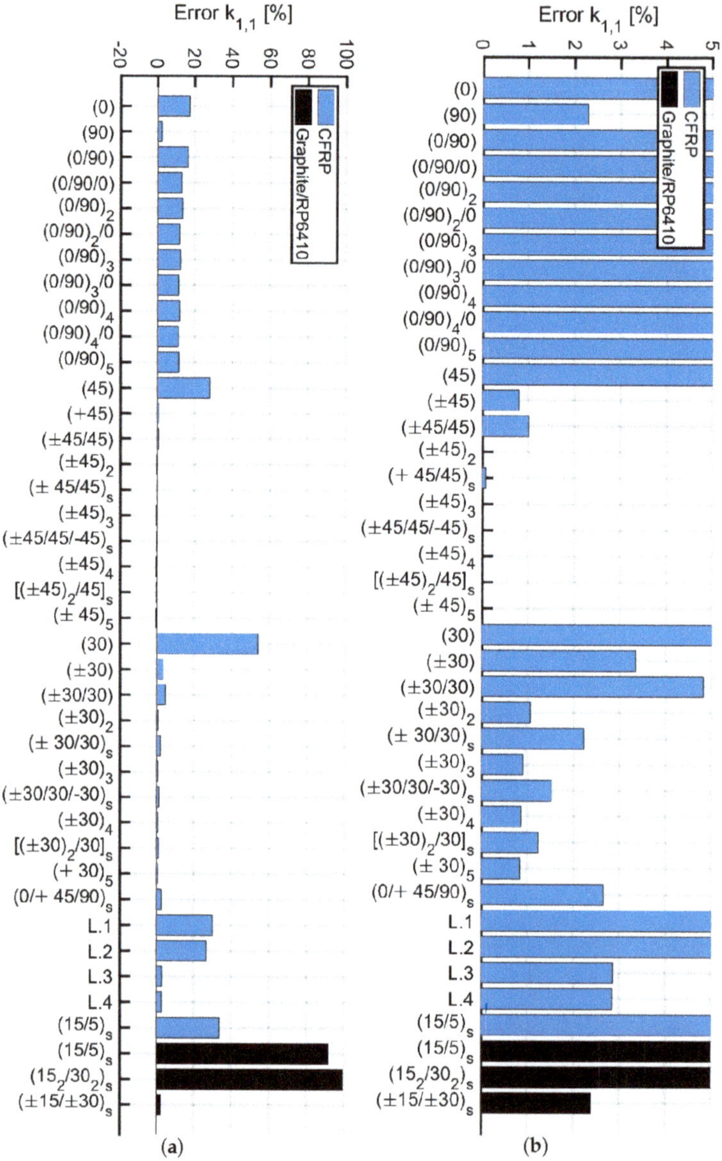

Figure 20. Percentage of error computing $k_{1,1}$ with the present methodology: (**a**) overall error; (**b**) the plot limited up to 5% of error.

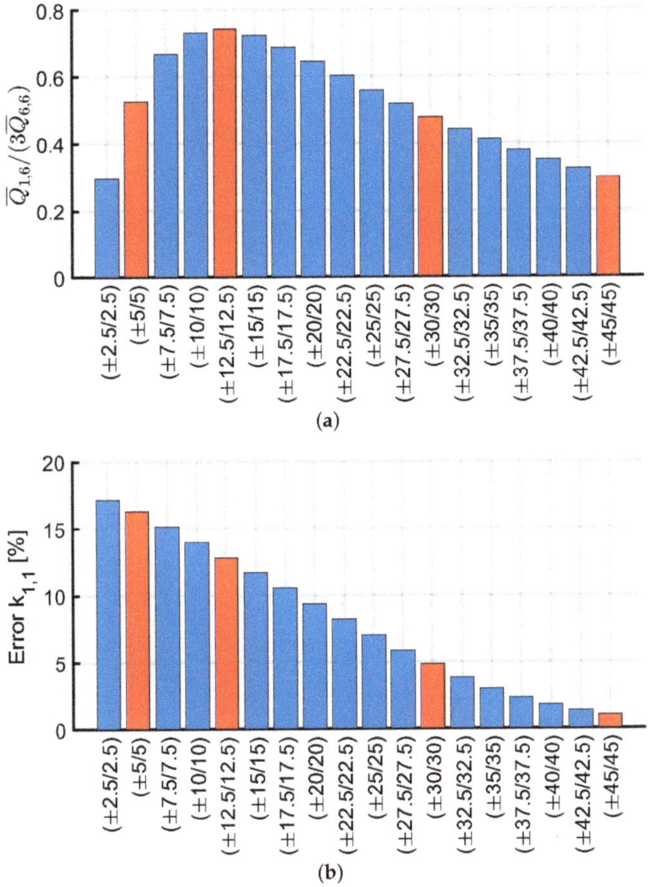

Figure 21. Trend by increasing α for laminates $(\alpha/-\alpha/\alpha)$ of: **(a)** $\overline{Q}_{1,6}/(3\overline{Q}_{6,6})$; **(b)** error on $k_{1,1}$.

8. Conclusions

In conclusion, it has been verified by increasing the number of plies in cross- and angle-ply configurations that hollow circular beam behavior does not tend to be like the one of an orthotropic model (0).

Subsequently, an analytical method is proposed for the calculation of the stiffness matrix components $k_{i,j}$ for $i,j = 1,2,\ldots,12$ of an equivalent 3D Euler–Bernoulli beam model. The present method resulted to be a valid alternative for investigating composite hollow beams in a simplified manner without the need for complex 3D shell modeling. The methodology demonstrated to work well for balanced laminates with $A_{1,6}/A_{6,6} < 50\%$ which corresponds to an error $< 5\%$ on the shear and bending stiffness components. By accurately analyzing angle-ply $(\pm\alpha/\alpha)$ configurations it has been demonstrated that a condition on the ply angle is $\alpha > 12.5^\circ$ in order to have acceptable results.

On the contrary, the condition $A_{1,6}/A_{6,6} < 50\%$ is not valid for cross-ply, that have $A_{1,6} = 0$ and they present an error around 15% for small slenderness. These errors might be due to some non-uniform cross-section torsional effects that cannot be predicted with the present modeling.

A balanced bio-inspired Bouligand-type laminate has been observed to be a good candidate for the present approach because other than being completely defined by the present analytical procedure such scheme demonstrated to have high ductility, dissipated energy, and high load-carrying capability.

The present method is valid also for investigating hollow beams with orthotropic materials with high stiffness ratio E_1/E_2 that have a high Poisson ratio, which corresponds to a balanced laminate. On the contrary, symmetric laminates with the same materials that have a negative Poisson ratio cannot be modeled accurately (because their lamination scheme is not balanced).

Author Contributions: Background investigation and analysis, S.B. and N.F.; Code Validations, S.B. and N.F. software, S.B. and N.F.; conceptualization, N.F.; methodology, N.F.; validation, D.G., S.B. and N.F.; formal analysis, D.G.; investigation, D.G.; resources, S.B. and N.F.; data curation, D.G. and N.F.; writing—original draft preparation, D.G. and S.B.; writing—review and editing, N.F.; visualization, N.F.; supervision, N.F.; project administration, N.F. All authors have read and agreed to the published version of the manuscript.

Funding: This research received no external funding.

Acknowledgments: The authors acknowledge "Fondazione Flaminia" (Ravenna, Italy) for supporting the present research.

Conflicts of Interest: The authors declare no conflict of interest.

References

1. Leknitskii, S.G. *Theory of Elasticity of an Anisotropic Body*; Mir Publishers: Moscow, Russia, 1981.
2. Jolicoeur, C.; Cardou, A. Analytical solution for bending of coaxial orthotropic cylinders. *J. Eng. Mech.* **1994**, *120*, 2556–2574. [CrossRef]
3. Kollár, L.; Patterson, J.M.; Springer, G.S. Composite cylinders subjected to hygrothermal and mechanical loads. *Int. J. Solids Struct.* **1992**, *29*, 1519–1534. [CrossRef]
4. Kollár, L.; Springer, G.S. Stress analysis of anisotropic laminated cylinders and cylindrical segments. *Int. J. Solids Struct.* **1992**, *29*, 1499–1517. [CrossRef]
5. Bhaskar, K.; Varadan, T. Exact elasticity solution for laminated anisotropic cylindrical shells. *J. Appl. Mech.* **1993**, *60*, 41–47. [CrossRef]
6. Xia, M.; Takayanagi, H.; Kemmochi, K. Analysis of multi-layered filament-wound composite pipes under internal pressure. *Compos. Struct.* **2001**, *53*, 483–491. [CrossRef]
7. Xia, M.; Kemmochi, K.; Takayanagi, H. Analysis of filament-wound fiber-reinforced sandwich pipe under combined internal pressure and thermomechanical loading. *Compos. Struct.* **2001**, *51*, 273–283. [CrossRef]
8. Çallıoğlu, H.; Ergun, E.; Demirdağ, O. Stress analysis of filament-Wound composite cylinders under combined internal pressure and thermal loading. *Adv. Compos. Lett.* **2008**, *17*, 096369350801700102. [CrossRef]
9. Bakaiyan, H.; Hosseini, H.; Ameri, E. Analysis of multi-layered filament-wound composite pipes under combined internal pressure and thermomechanical loading with thermal variations. *Compos. Struct.* **2009**, *88*, 532–541. [CrossRef]
10. Verijenko, V.E.; Adali, S.; Tabakov, P.Y. Stress distribution in continuously heterogeneous thick laminated pressure vessels. *Compos. Struct.* **2001**, *54*, 371–377. [CrossRef]
11. Roque, C.; Ferreira, A. New developments in the radial basis functions analysis of composite shells. *Compos. Struct.* **2009**, *87*, 141–150. [CrossRef]
12. Salahifar, R.; Mohareb, M. Finite element for cylindrical thin shells under harmonic forces. *Finite Elem. Anal. Des.* **2012**, *52*, 83–92. [CrossRef]
13. Tarn, J.Q.; Wang, Y.M. Laminated composite tubes under extension, torsion, bending, shearing and pressuring: A state space approach. *Int. J. Solids Struct.* **2001**, *38*, 9053–9075. [CrossRef]
14. Bai, Y.; Ruan, W.; Cheng, P.; Yu, B.; Xu, W. Buckling of reinforced thermoplastic pipe (RTP) under combined bending and tension. *Ships Offshore Struct.* **2014**, *9*, 525–539. [CrossRef]
15. Derisi, B.; Hoa, S.V.; Xu, D.; Hojjati, M.; Fews, R. Mechanical behavior of Carbon/PEKK thermoplastic composite tube under bending load. *J. Thermoplast. Compos. Mater.* **2011**, *24*, 29–49. [CrossRef]
16. Shadmehri, F.; Derisi, B.; Hoa, S. On bending stiffness of composite tubes. *Compos. Struct.* **2011**, *93*, 2173–2179. [CrossRef]
17. Tings, T. New solutions to pressuring, shearing, torsion and extension of a cylindrically anisotropic elastic circular tube or bar. *Proc. R. Soc. Lond. Ser. A Math. Phys. Eng. Sci.* **1999**, *455*, 3527–3542. [CrossRef]
18. Ting, T. Pressuring, shearing, torsion and extension of a circular tube or bar of cylindrically anisotropic material. *Proc. R. Soc. Lond. Ser. A Math. Phys. Eng. Sci.* **1996**, *452*, 2397–2421.

19. Uchikawa, Y.; Itabashi, M.; Kawata, K. On crashworthiness of FRP thin-walled circular tubes under dynamic axial compression. *Adv. Compos. Mater.* **1997**, *6*, 239–252. [CrossRef]
20. Silvestre, N. Non-classical effects in FRP composite tubes. *Compos. Part B Eng.* **2009**, *40*, 681–697. [CrossRef]
21. Kardomateas, G.A. Elasticity solutions for sandwich orthotropic cylindrical shells under external/internal pressure or axial force. *AIAA J.* **2001**, *39*, 713–719. [CrossRef]
22. Wang, Y.; Chen, G.; Wan, B.; Han, B. Compressive Behavior of Circular Sawdust-Reinforced Ice-Filled Flax FRP Tubular Short Columns. *Materials* **2020**, *13*, 957, doi:10.3390/ma13040957. [CrossRef] [PubMed]
23. Zhang, C.; Hoa, S.V.; Liu, P. A method to analyze the pure bending of tubes of cylindrically anisotropic layers with arbitrary angles including 0 or 90. *Compos. Struct.* **2014**, *109*, 57–67. [CrossRef]
24. Sarvestani, H.Y.; Sarvestani, M.Y. Free-edge stress analysis of general composite laminates under extension, torsion and bending. *Appl. Math. Model.* **2012**, *36*, 1570–1588. [CrossRef]
25. Chen, T.; Chung, C.T.; Lin, W.L. A revisit of a cylindrically anisotropic tube subjected to pressuring, shearing, torsion, extension and a uniform temperature change. *Int. J. Solids Struct.* **2000**, *37*, 5143–5159. [CrossRef]
26. Chouchaoui, C.; Ochoa, O. Similitude study for a laminated cylindrical tube under tensile, torsion, bending, internal and external pressure. Part I: governing equations. *Compos. Struct.* **1999**, *44*, 221–229. [CrossRef]
27. Zhang, C.; Hoa, S.V. A limit-based approach to the stress analysis of cylindrically orthotropic composite cylinders (0/90) subjected to pure bending. *Compos. Struct.* **2012**, *94*, 2610–2619. [CrossRef]
28. Guz, I.A.; Menshykova, M.; Paik, J.K. Thick-walled composite tubes for offshore applications: an example of stress and failure analysis for filament-wound multi-layered pipes. *Ships Offshore Struct.* **2017**, *12*, 304–322. [CrossRef]
29. Menshykova, M.; Guz, I. Stress analysis of layered thick-walled composite pipes subjected to bending loading. *Int. J. Mech. Sci.* **2014**, *88*, 289–299. [CrossRef]
30. Khalili, S.; Dehkordi, M.B.; Carrera, E. A nonlinear finite element model using a unified formulation for dynamic analysis of multilayer composite plate embedded with SMA wires. *Compos. Struct.* **2013**, *106*, 635–645. [CrossRef]
31. Ascione, F.; Feo, L.; Maceri, F. An experimental investigation on the bearing failure load of glass fibre/epoxy laminates. *Compos. Part B Eng.* **2009**, *40*, 197–205. [CrossRef]
32. Boscato, G.; Russo, S. Free vibrations of pultruded FRP elements: Mechanical characterization, analysis, and applications. *J. Compos. Constr.* **2009**, *13*, 565–574. [CrossRef]
33. Philippidis, T.P.; Vassilopoulos, A. Fatigue of composite laminates under off-axis loading. *Int. J. Fatigue* **1999**, *21*, 253–262. [CrossRef]
34. Quadrino, A.; Penna, R.; Feo, L.; Nisticò, N. Mechanical characterization of pultruded elements: Fiber orientation influence vs web-flange junction local problem. Experimental and numerical tests. *Compos. Part B Eng.* **2018**, *142*, 68–84. [CrossRef]
35. Madenci, E.; Özkılıç, Y.O.; Gemi, L. Experimental and Theoretical Investigation on Flexure Performance of Pultruded GFRP Composite Beams with Damage Analyses. *Compos. Struct.* **2020**, *242*, 112162. [CrossRef]
36. Xin, H.; Mosallam, A.; Liu, Y.; Xiao, Y.; He, J.; Wang, C.; Jiang, Z. Experimental and numerical investigation on in-plane compression and shear performance of a pultruded GFRP composite bridge deck. *Compos. Struct.* **2017**, *180*, 914–932. [CrossRef]
37. Mayookh Lal, H.; Xian, G.; Thomas, S.; Zhang, L.; Zhang, Z.; Wang, H. Experimental Study on the Flexural Creep Behaviors of Pultruded Unidirectional Carbon/Glass Fiber-Reinforced Hybrid Bars. *Materials* **2020**, *13*, 976. [CrossRef]
38. Zhang, S.; Xing, T.; Zhu, H.; Chen, X. Experimental Identification of Statistical Correlation between Mechanical Properties of FRP Composite. *Materials* **2020**, *13*, 674. [CrossRef]
39. Sun, C.T.; Li, S. Three-dimensional effective elastic constants for thick laminates. *J. Compos. Mater.* **1988**, *22*, 629–639. [CrossRef]
40. Sun, X.S.; Tan, V.B.C.; Chen, Y.; Tan, L.B.; Jaiman, R.K.; Tay, T.E. Stress analysis of multi-layered hollow anisotropic composite cylindrical structures using the homogenization method. *Acta Mech.* **2014**, *225*, 1649–1672. [CrossRef]
41. Ferdous, W.; Almutairi, A.D.; Huang, Y.; Bai, Y. Short-term flexural behaviour of concrete filled pultruded GFRP cellular and tubular sections with pin-eye connections for modular retaining wall construction. *Compos. Struct.* **2018**, *206*, 1–10. [CrossRef]

42. Mohammed, A.A.; Manalo, A.C.; Ferdous, W.; Zhuge, Y.; Vijay, P.; Pettigrew, J. Experimental and numerical evaluations on the behaviour of structures repaired using prefabricated FRP composites jacket. *Eng. Struct.* **2020**, *210*, 110358. [CrossRef]
43. Al-Rubaye, M.; Manalo, A.; Alajarmeh, O.; Ferdous, W.; Lokuge, W.; Benmokrane, B.; Edoo, A. Flexural behaviour of concrete slabs reinforced with GFRP bars and hollow composite reinforcing systems. *Compos. Struct.* **2020**, *236*, 111836. [CrossRef]
44. Babuška, I. Homogenization approach in engineering. In *Computing Methods in Applied Sciences and Engineering*; Springer: Berlin/Heidelberg, Germany, 1976; pp. 137–153.
45. Sanchez-Palencia, E. Homogenization method for the study of composite media. In *Asymptotic Analysis II*; Springer: Berlin/Heidelberg, Germany, 1983; pp. 192–214.
46. Sanchez-Palencia, E. Homogenization in mechanics. A survey of solved and open problems. *Rend. Sem. Mat. Univ. Politec. Torino* **1986**, *44*, 1–45.
47. Kim, C.; White, S.R. Analysis of thick hollow composite beams under general loadings. *Compos. Struct.* **1996**, *34*, 263–277. [CrossRef]
48. Yildiz, H.; Sarikanat, M. Finite-element analysis of thick composite beams and plates. *Compos. Sci. Technol.* **2001**, *61*, 1723–1727. [CrossRef]
49. Ferreira, A. Thick composite beam analysis using a global meshless approximation based on radial basis functions. *Mech. Adv. Mater. Struct.* **2003**, *10*, 271–284. [CrossRef]
50. Yazdani Sarvestani, H.; Hoa, S.V.; Hojjati, M. Stress analysis of thick orthotropic cantilever tubes under transverse loading. *Adv. Compos. Mater.* **2017**, *26*, 335–362. [CrossRef]
51. Yazdani Sarvestani, H.; Hojjati, M. Effects of lay-up sequence in thick composite tubes for helicopter landing gears. *Proc. Inst. Mech. Eng. Part G J. Aerosp. Eng.* **2017**, *231*, 2098–2110. [CrossRef]
52. Berdichevsky, V.; Armanios, E.; Badir, A. Theory of anisotropic thin-walled closed-cross-section beams. *Compos. Eng.* **1992**, *2*, 411–432. [CrossRef]
53. Kollar, L.P.; Pluzsik, A. Analysis of thin-walled composite beams with arbitrary layup. *J. Reinf. Plast. Compos.* **2002**, *21*, 1423–1465. [CrossRef]
54. Jung, S.N.; Lee, J.Y. Closed-form analysis of thin-walled composite I-beams considering non-classical effects. *Compos. Struct.* **2003**, *60*, 9–17. [CrossRef]
55. Lateral buckling analysis of thin-walled laminated composite beams with monosymmetric sections. *Eng. Struct.* **2006**, *28*, 1997–2009. [CrossRef]
56. Boscato, G. Comparative study on dynamic parameters and seismic demand of pultruded FRP members and structures. *Compos. Struct.* **2017**, *174*, 399–419. [CrossRef]
57. de Miranda, S.; Madeo, A.; Melchionda, D.; Patruno, L.; Ruggerini, A. A corotational based geometrically nonlinear Generalized Beam Theory: buckling FE analysis. *Int. J. Solids Struct.* **2017**, *121*, 212–227. [CrossRef]
58. Ruggerini, A.; Madeo, A.; Gonçalves, R.; Camotim, D.; Ubertini, F.; de Miranda, S. GBT post-buckling analysis based on the Implicit Corotational Method. *Int. J. Solids Struct.* **2019**, *163*, 40–60. [CrossRef]
59. Babamohammadi, S.; Fantuzzi, N.; Lonardi, G. Mechanical assessment of hollow-circular FRP beams. *Compos. Struct.* **2019**, *227*, 111313. [CrossRef]
60. *2011 ABAQUS Version 6.11: Analysis User's Manual*; Dassault Systemes Simulia Corp.: Providence, RI, USA, 2011.
61. Reddy, J.N. *Mechanics of Laminated Composite Plates and Shells: Theory and Analysis*; CRC Press: Boca Raton, FL, USA, 2004.
62. Gnoli, D. Studio di Profili Tubolari in FRP: Omogeneizzazione e Modello Trave Equivalente. Master's Thesis, University of Bologna, Bologna, Italy, 2020.
63. Mencattelli, L.; Pinho, S.T. Ultra-thin-ply CFRP Bouligand bio-inspired structures with enhanced load-bearing capacity, delayed catastrophic failure and high energy dissipation capability. *Compos. Part A Appl. Sci. Manuf.* **2019**, *129*, 105655. [CrossRef]
64. Peel, L.D. Exploration of high and negative Poisson's ratio elastomer-matrix laminates. *Phys. Status Solidi (b)* **2007**, *244*, 988–1003. [CrossRef]

materials

MDPI

Article

Rigid Finite Element Method in Modeling Composite Steel-Polymer Concrete Machine Tool Frames

Paweł Dunaj [1,*], Krzysztof Marchelek [1], Stefan Berczyński [1] and Berkay Mizrak [2]

1 Department of Mechanical Engineering and Mechatronics, West Pomeranian University of Technology, 71-310 Szczecin, Poland; krzysztof.marchelek@zut.edu.pl (K.M.); stefan.berczynski@zut.edu.pl (S.B.)

2 Mechanical Engineering Department, Yildiz Technical University, 34349 Beşiktaş/İstanbul, Turkey; info@berkaymizrak.com

* Correspondence: pawel.dunaj@zut.edu.pl

Received: 12 June 2020; Accepted: 13 July 2020; Published: 15 July 2020

Abstract: At the stage of designing a special machine tool, it is necessary to analyze many variants of structural solutions of frames and load-bearing systems and to choose the best solution in terms of dynamic properties, in particular considering its resistance to chatter. For this reason, it is preferred to adopt a low-dimensional calculation model, which allows the user to reduce the necessary calculation time while maintaining a high accuracy. The paper presents the methodology of modeling the natural frequencies, mode shapes, and receptance functions of machine tool steel welded frames filled with strongly heterogenous polymer concrete, using low-dimensional models developed by the rigid finite elements method (RigFEM). In the presented study, a RigFEM model of a simple steel beam filled with polymer concrete and a frame composed of such beams were built. Then, the dynamic properties obtained on the basis of the developed RigFEM models were compared with the experimental results and the 1D and 3D finite element models (FEM) in terms of accuracy and dimensionality. As a result of the experimental verification, the full structural compliance of the RigFEM models (for beam and frame) was obtained, which was manifested by the agreement of the mode shapes. Additionally, experimental verification showed a high accuracy of the RigFEM models, obtaining for the beam model a relative error for natural frequencies of less than 4% and on average 2.2%, and for the frame model at a level not exceeding 11% and on average 5.5%. Comparing the RigFEM and FEM models, it was found that the RigFEM models have a slightly worse accuracy, with a dimensionality significantly reduced by 95% for the beam and 99.8% for the frame.

Keywords: rigid finite element method; composite; steel-polymer concrete; machine tool; multibody system

1. Introduction

One of the alternative technologies for manufacturing frames forming machine tool support systems is the technology of welding steel bodies. The basic components of such bodies are hollow steel profiles made of sheet metal or solid rolled profiles. Those profiles can be filled with a specially selected composite material whose task it is to increase the dissipation of vibration. Then, such a body is a hybrid body and it can effectively compete with cast iron body technology.

The use of hybrid body technologies (i.e., those resulting from the combination of different materials in order to exploit their properties synergically) in machine tool load-bearing systems manufactured individually or in small batches can provide specific benefits in terms of a significant reduction in the design time, particularly in terms of the calculations needed to shape their structure and the desired static and dynamic properties [1]. It is particularly important to shape the desired resistance of the designed machine tool frame to chatter [2–7]. The problems related to the design and modeling of frames of technological machines include a significant number of issues, which translate into a rich scientific literature. Analyzing the review publications on the currently proposed solutions, a large variety in their design can be noticed [8–14]. The static and dynamic properties of composite machine tool load-bearing systems are most often assessed using the finite element model (FEM) [15–22].

In [23], the authors present a structural solution of a gantry milling machine tool bed composed of a welded steel frame filled with polymer concrete. This solution was aimed at improving the dynamic properties of the machine tool. In relation to the fact that the required stiffness of the body was ensured by the welded frame and its good damping properties were provided primarily by the filling material, six different structural variants were proposed. The subsequent variants differed in the wall thickness of the steel frame and, consequently, in the volume of the filled internal spaces of the elements forming the frame. To assess the impact of the changes in the wall thickness of the static stiffness, natural frequencies, and mode shapes, the finite element models of the proposed variants were analyzed using the Ansys 6.0 software. The results of the model calculations were compared with the experimental ones, both for the values of the estimated loss factor and the natural frequencies; the relative errors were obtained at the level of up to 20%.

In [24], the authors present a solution for the construction of the column body of the milling machine, consisting of a cast iron frame, concrete, and elastomeric material, aimed at improving the damping properties of the structure. The body design process was supported by the analysis of its FEM model. The experimental verification of the results of the FEM modeling showed discrepancies in the values of natural frequencies at a level not exceeding 2.3% of the relative error.

In [25], the authors discuss a solution consisting of replacing the steel body elements of a table milling machine with composite parts made of aluminum, steel, polymer concrete, and carbon fibre-based composites. The models were built using the ANSYS software. Cubic elements characterized by eight nodes (SOLID45) were used to discretize the aluminum, steel, and polymer concrete parts. A layered version of the mentioned finite element (SOLID46) was used for the discretization of the parts made of the carbon fibre-based composite. A modal analysis of the body constraint in the places of foundation of the actual structure was carried out for the constructed model. By verifying the obtained results of modeling with the experimental data, discrepancies in the values of natural frequencies were obtained at a level of up to 28%.

In [26], the authors present a solution consisting of replacing a cast iron vertical milling center column with a steel-reinforced polymer concrete structure. The purpose of the presented research was to increase the damping capacity of the structure while ensuring the required rigidity. The authors, using FEM, conducted an analysis of the structure deformations for seven body variants differing in their arrangement and reinforcement method. A similar approach is presented in [27], replacing the cast iron base of the milling center with a steel-reinforced epoxy granite structure.

In [28], the authors present an enhancement of the static and dynamic characteristics of a micro-lathe bed by the use of alternate form designs and composite material (epoxy granite and nettle polyester). Using the finite element method, the authors analyze six design variants differing in rib configurations. Based on the performed analysis, the best solution in terms of bending and torsional stiffness was selected. However, the authors of the papers [26–28] do not provide the results of the experimental verification of the developed finite element models of the hybrid machine tool body elements.

The prognostic value of the results of the calculations made in the design phase will depend directly on the accuracy of the mapping of the static and dynamic properties of the designed body by the calculation model. Since calculations are usually performed for many considered structural solution variants—i.e., they are repeated many times—it is advisable to adopt a design model that is as simple as possible and as complex as necessary. This can be achieved by the rigid finite element model [29], which, compared to the currently most widely used finite element model, has a much smaller number of degrees of freedom, which can significantly reduce the calculation time [30,31].

According to the latest contribution [32], a strongly heterogenous material, such as the polymer concrete analyzed, can be modeled using a linear-elastic isotropic model of material, if present in a steel-polymer concrete beam. Moreover, it is possible to develop a reliable analytical model of a composite steel-polymer concrete beam using a homogeneous beam model with an equivalent bending stiffness and equivalent mass per unit length of the beam. The presented study takes this a step further and proposes a modeling methodology that deals with the problem of modeling steel beams and complex, spatial, steel-welded frames filled with strongly heterogenous material, while maintaining a low dimensionality of the model. The main novelty of the presented study is establishing the model of the composite, steel-polymer concrete beam using the rigid finite element method. The model is built by aggregating matrices describing the inertia and stiffness of the steel outer and polymer concrete filling. The proposed approach differs from the one presented so far in the literature, where mainly the finite element method is used to model the dynamic properties of complex composite structures; in order to reduce the dimensions of the established models, model order reduction methods are used. The presented paper contains the results of comparative studies to determine whether the rigid finite element method can be effectively used in modeling and calculating the dynamic properties—i.e., the natural frequencies, mode shapes, and frequency response functions—of steel structures filled with a highly heterogenous polymer concrete.

The structure of the article is as follows: in Section 2, the steel-polymer concrete beam and frame concepts are described. Next, a static test to determine the properties of the polymer concrete analyzed and a dynamic test to determine the natural frequencies, mode shapes, and frequency response functions are presented. Then, finite element models of both the structural component (beam) and main frame of the vertical lathe are presented. Finally, in Section 3 rigid finite element models are established and compared with the finite element model and experimental results. Section 4 contains the discussion of the presented results. In Section 5, the main conclusions are presented.

2. Materials and Methods

2.1. Steel-Polymer Concrete Frame

The frame considered in this paper was made of commercial steel profiles with a square hollow section. The individual profiles (basic structural components) were welded, creating a spatial structure which provides the construction with the required stiffness. After the steel frame was welded, the insides of the profiles were filled with polymer concrete. The use of polymer concrete was intended to increase the damping properties of the welded steel frame. Moreover, filling a steel profile with polymer concrete causes a local increase in the stiffness and inertia of the frame. Therefore, the proper, intentional placement of polymer concrete enables the shaping of the dynamic properties of the machine frame. More precisely, the proper distribution of the polymer concrete causes a change in the structure and parameters of the mass-spring-damping system, which is a model reflection of the machine tool, thus changing its dynamic properties. The basic structural component (beam) and frame analyzed are shown in Figure 1.

The basic structural component under consideration consisted of a steel profile with section dimensions of 50 mm × 50 mm, a wall thickness of 2 mm, and a length of 1000 mm filled with polymer concrete. The frame analyzed consisted of basic structural components and steel plates connected by welded joints.

The applied polymer concrete consisted of epoxy resin and a mineral filling of various granulations. Considering the grain size as the dividing criteria, the mineral fillings can be grouped into the following categories: (i) ash, (ii) a fine fraction (mainly sand) of grain size 0.25–2 mm, (iii) a medium fraction of grain size 2–10 mm, and (iv) a coarse fraction of grain size 8–16 mm. The coarse (iv) and medium (iii) fractions mainly consisted of irregularly shaped gravel. The mass percentage of the individual components of the polymer concrete applied is shown in Table 1 [32].

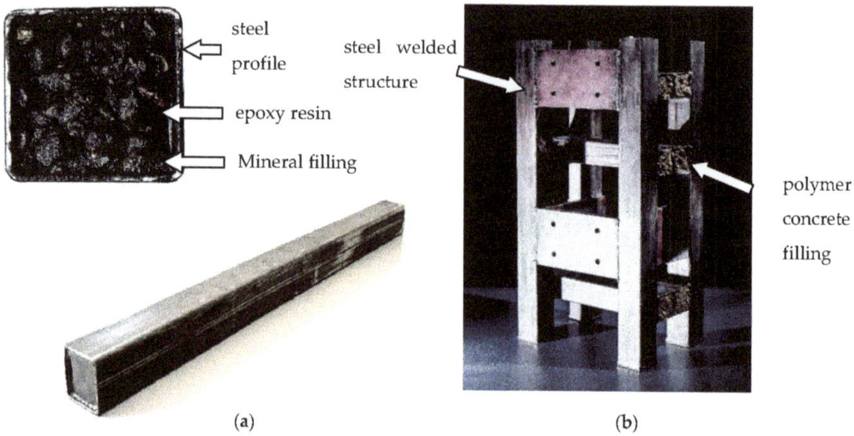

Figure 1. Steel-polymer concrete structure: (**a**) beam, (**b**) steel-polymer concrete frame of vertical lathe.

Table 1. Composition of the applied polymer concrete filling (mass percentage).

Component	Epoxy Resin	Ash	Fine Fraction (0.25–2 mm)	Medium Fraction (2–10 mm)	Coarse Fraction (8–16 mm)
Percentage share of component weight	15%	1%	19%	15%	50%

2.2. Static Tests of Polymer Concrete

To determine the material properties required for the RigFEM modeling—i.e., the Young's modulus and Poisson ratio—static tests were performed using Instron 8850 (Instron, Norwood, MA, USA). The samples (cuboidal samples of steel—50 mm × 50 mm × 110 mm; polymer concrete—45 mm × 45 mm × 110 mm) were conditioned for 72 h in an air-conditioned laboratory at a temperature of 23 °C and relative humidity of 50%. The detailed procedure was presented in [32]. Table 2 contains the material data determined from the static test and the loss factor determined by the impulse test.

Table 2. Material properties.

Parameter	Steel	Polymer Concrete
Young's modulus E	210 ± 5 GPa	16.8 ± 0.2 GPa
Poisson's ratio ν	0.28 ± 0.03	0.20 ± 0.05
Density ρ	7487 ± 35 kg/m^3	2118 ± 6 kg/m^3

2.3. Dynamic Experimental Tests of the Basic Structural Component and Frame

Dynamic tests were performed in the form of an impact test using Siemens TestLab—Desktop Advance 2019.1 (Siemens, Munich, Germany) software and Scadas Mobile Vibco (Siemens, Munich, Germany) hardware. Excitation was carried out using a PCB 086C01 (PCB Piezotronics, Depew, New York, NY, USA) modal hammer in two perpendicular axes in the case of the beam analyzed and in three perpendicular axes in the case of the frame analyzed. The response measurement was carried out using three-axis PCB 356A01 ICP (PCB Piezotronics, Depew, New York, NY, USA) accelerometers at 56 points in the case of the beam and 260 points in the case of the frame. Double impacts or the overload of any channel result in the automatic rejection of measurement. The measurement points' arrangement is shown in Figure 2. The tested beam was suspended on steel cables to approximate free-free boundary conditions, while the frame was founded according to the real operating conditions.

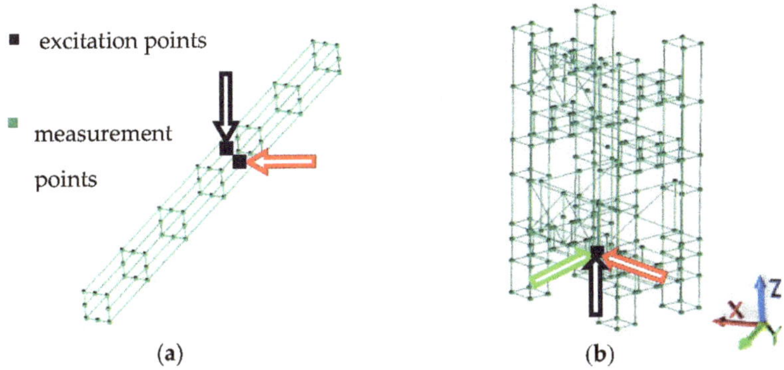

Figure 2. Measurement points' arrangement for (**a**) beam and (**b**) frame.

The selected signal acquisition parameters are presented in Table 3.

Table 3. Parameters of signal acquisition.

Parameter	Value
Sampling rate	4096 Hz
Frequency resolution	0.5 Hz
Signal acquisition time	2 s
Frequency response function estimator	H1
Number of averages	10
Scaling of the frequency response function	global

As a result of the experiment, 112 frequency response functions for the beam and 780 for the frame were determined using the H1 type of FRF estimator. The modal model was estimated using the Polymax algorithm [33] with default values of tolerance on frequency, 1%; damping, 2%; and mode shapes vector, 5%. Before the final interpretation, the models were validated using an modal assurance criterion (MAC) indicator. The detailed procedure for the beam was presented in [32] and for the frame in [34].

Then, using a half-power method (Figure 3c), the value of the loss factors for steel and steel-polymer concrete beams were determined. First, the frequency response functions between the points indicated in Figure 3a were determined. Those points were selected on the basis of the modal analysis in such a way that they were not in the nodes of the experimental mode shapes. Next, the values of the loss factor for steel and the equivalent loss factor for the composite steel-polymer concrete beam were determined as a weighted arithmetic mean, giving weight to the individual coefficients η corresponding to the resonance amplitude for which they were determined. The idea of the described method is presented in Figure 3b; the values of the loss factors determined are presented in Table 4.

Table 4. Loss factors for steel and steel-polymer concrete structure.

Parameter	Steel	Polymer Concrete
Loss factor η	0.00220 ± 0.00005	-
Equivalent loss factor	0.00480 ± 0.00024	

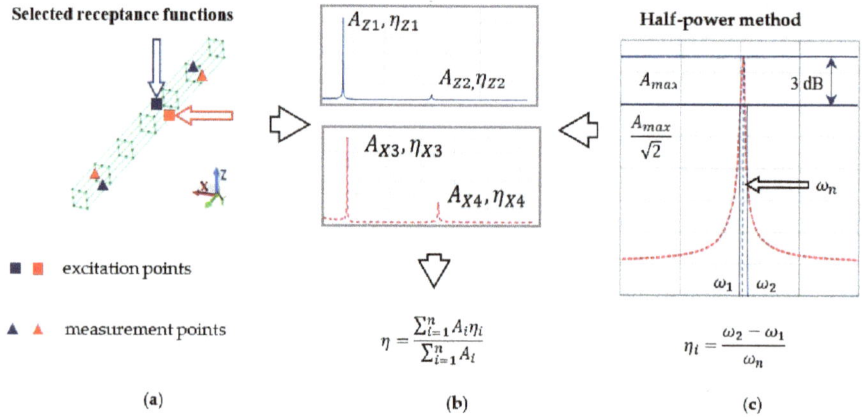

Figure 3. Method of determining the loss factor based on (**a**) selected frequency response functions, (**b**) weighted arithmetic mean, and (**c**) half-power method.

2.4. Finite Element Modeling

As indicated in Section 1, in engineering practice, the finite element method (FEM) is most commonly used to assess the dynamic properties of composite structures. Therefore, finite element models of the steel-polymer concrete beam and frame under consideration were built and the corresponding calculations were made.

The beam was modeled according to the method presented in [34], and the steel profile and the polymer concrete filling were meshed using cubic, isoparametric, eight-noded solid elements (CHEXA) and isoparametric, wedge, and solid elements with six nodes (CPENTA). A structural mesh was applied to increase the efficiency of the FEM. The polymer concrete was modeled as a linear-elastic isotropic material (MAT1) [35]. In total, the model of a single beam was composed of 19,600 elements and 21,985 degrees of freedom.

Alternatively, the steel-polymer concrete beam was modeled using one-dimensional beam finite elements (CBAR), formulated on the basis of classical beam theory [35]. The model was built by aggregating mass and stiffness matrixes representing the steel outer and polymer concrete filling. In total, the one-dimensional steel-polymer concrete beam model was composed of 40 elements and 126 degrees of freedom.

In the case of the analyzed frame, the model was built according to [34] using the CHEXA and CPENTA elements. In total, the model of the machine tool frame was composed of 60,584 elements and 225,384 degrees of freedom. The support was modeled using a beam element (CBAR) with an equivalent stiffness and damping. The discretization process was performed using Midas NFX 2018 R1 (Midas Information Technology Co. Ltd., Seongnam, Korea) [35,36]. The discretized models are shown in Figure 4.

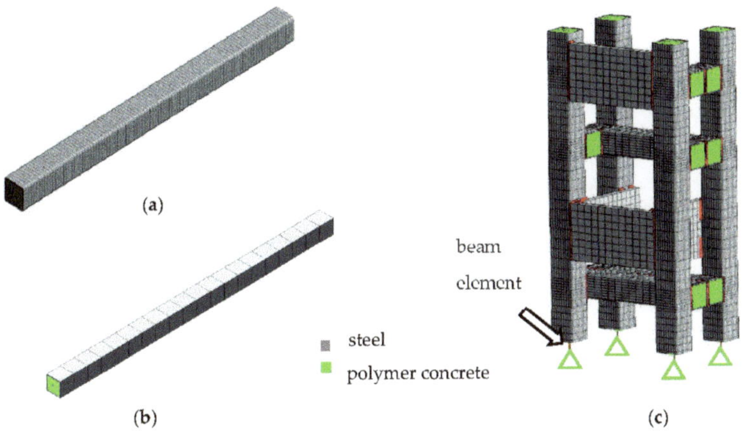

Figure 4. Finite element model (FEM) of a beam discretized using 3D elements (**a**) and 1D elements (**b**) and the model of a frame (**c**).

3. Rigid Finite Element Modeling

3.1. Rigid Finite Element Model of the Basic Structural Component

In previous research [32], it was proved analytically and experimentally confirmed that it is possible to build a reliable model of a composite steel-polymer concrete beam using an equivalent bending stiffness and an equivalent mass per unit length. Adopting the Hamiltonian developed in [32] and the model assumptions that the contact of the steel profile with the polymer concrete filling occurs on the entire inner surface of the steel profile and the adhesion forces prevent tangential movements within the material contact area, the steel-composite beam was modeled using rigid finite elements characterized by mass and moments of inertia, which were connected to one another using spring-damping elements [37,38].

According to the adopted assumptions, a steel-polymer concrete beam can be modeled as an assembly of steel and polymer concrete parts, determining for them the substitute values of inertia, stiffness, and damping coefficients. Therefore, the matrices describing the inertia, stiffness, and damping of the model are defined as follows Equations (1)–(3):

$$\boldsymbol{M} = \boldsymbol{M}^{st} + \boldsymbol{M}^{pc} \tag{1}$$

$$\boldsymbol{K} = \boldsymbol{K}^{st} + \boldsymbol{K}^{pc} \tag{2}$$

$$\boldsymbol{C} = \boldsymbol{C}^{st} + \boldsymbol{C}^{pc} \tag{3}$$

where $\boldsymbol{M}$ is the inertia matrix of the rigid finite element, $\boldsymbol{M}^{st}$ is the inertia matrix of the steel part, $\boldsymbol{M}^{pc}$ is the inertia matrix of the polymer concrete part, $\boldsymbol{K}$ is the stiffness coefficient matrix describing the spring-damping element, $\boldsymbol{K}^{st}$ is the stiffness matrix describing the steel part, $\boldsymbol{K}^{pc}$ is the stiffness matrix describing the polymer concrete part, $\boldsymbol{C}$ is the damping coefficient matrix describing the spring-damping element, $\boldsymbol{C}^{st}$ is the damping coefficient matrix for the steel part, and C^{pc} is the damping coefficient matrix for the polymer concrete part.

The inertia matrix of a rigid finite element for the beam under consideration is a diagonal matrix whose form is as follows Equations (4)–(6):

$$\boldsymbol{M} = diag\left(M_1^{st}, M_2^{st}, \ldots, M_6^{st}\right) + diag\left(M_1^{pc}, M_2^{pc}, \ldots, M_6^{pc}\right) \tag{4}$$

whereby:

$$M_1^{st} = M_2^{st} = M_3^{st} = m^{st},\ M_4^{st} = J_0^{st},\ M_5^{st} = J_2^{st},\ M_6^{st} = M_5^{st} \tag{5}$$

$$M_1^{pc} = M_2^{pc} = M_3^{pc} = m^{pc},\ M_4^{pc} = M_0^{pc},\ M_5^{pc} = J_2^{pc},\ M_6^{pc} = M_5^{pc} \tag{6}$$

where m is the mass of the rigid finite element for the polymer concrete filling (pc) and the steel coating (st), J_0^{st} is the main central moment of inertia for the steel profile, J_0^{pc} is the main central moment of inertia for the polymer concrete filling, J_2^{st} is the inertia for the steel profile relative to the X axis, and J_2^{pc} is the moment of inertia for the polymer concrete filling relative to the X axis—Figure 3.

The stiffness $\mathbf{K}$ matrix of the spring-damping element is defined as follows Equations (7)–(9):

$$\mathbf{K} = diag\left(K_1^{st}, K_2^{st}, \ldots, K_6^{st}\right) + diag\left(K_1^{pc}, K_2^{pc}, \ldots, K_6^{pc}\right) \tag{7}$$

whereby:

$$K_1^{st} = \frac{E^{st}A^{st}}{\Delta l},\ K_2^{st} = \frac{G^{st}A^{st}}{\Delta l},\ K_3^{st} = K_2^{st},\ K_4^{st} = \frac{G^{st}J_0^{st}}{\Delta l},\ K_5^{st} = \frac{E^{st}J_2^{st}}{\Delta l},\ K_6^{st} = K_5^{st} \tag{8}$$

$$K_1^{pc} = \frac{E^{pc}A^{pc}}{\Delta l},\ K_2^{pc} = \frac{G^{pc}A^{pc}}{\Delta l},\ K_3^{pc} = K_2^{pc},\ K_4^{st} = \frac{G^{pc}J_0^{pc}}{\Delta l},\ K_5^{pc} = \frac{E^{pc}J_2^{pc}}{\Delta l},\ K_6^{pc} = K_5^{pc} \tag{9}$$

where E^{st} is the Young's module for steel, E^{pc} is the Young's module for polymer concrete, A^{st} is the steel profile sectional area, A^{pc} is the polymer concrete filling sectional area, G^{st} is the Kirchhoff's module for steel, G^{pc} is the Kirchhoff's module for polymer concrete, and Δl is the length of the rigid element.

Meanwhile, the damping matrix $\mathbf{C}$ of the spring-damping element according to the adopted complex stiffness model [39,40] takes the following form Equation (10):

$$\mathbf{C} = i\eta\left(diag\left(K_1^{st}, K_2^{st}, \ldots K_6^{st}\right) + diag\left(K_1^{pc}, K_2^{pc}, \ldots, K_6^{pc}\right)\right) \tag{10}$$

where i—imaginary unit; η—equivalent loss factor for a steel-polymer concrete beam.

The beam model using the predefined rigid finite elements was built in MSC Adams 2019 (MSC Software, Newport Beach, California, USA) [41]. The division of the beam under consideration into rigid finite elements and spring-damping elements is shown in Figure 5. The model consisted of 20 rigid finite elements of equal length and 19 spring-damping elements (modeled in MSC Adams environment using 6 × 6 field elements). The model had 120 degrees of freedom.

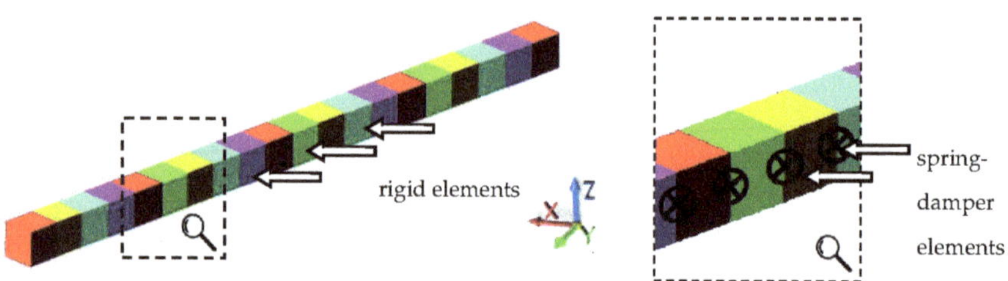

Figure 5. Division of the steel-polymer concrete beam into rigid finite elements.

3.2. *Rigid Finite Element Model of the Steel-Polymer Concrete Body*

Next, a rigid finite elements model for the steel-polymer concrete frame was built—Figure 6. The model consisted of 91 rigid finite elements; 76 spring-damping elements describing the stiffness and damping of the beam elements, defined analogously to the beam model; and 73 spring-damping elements, describing the stiffness and damping of welded joints between the individual beams. As a result, the model had 546 degrees of freedom.

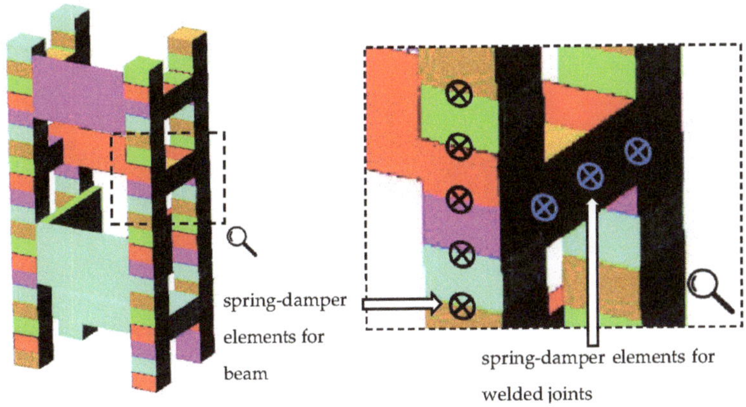

Figure 6. Rigid finite element model of the steel-polymer concrete frame.

3.3. *Comparison of the Results of Model Calculations of the Basic Structural Component with the Results of the Experimental Tests*

The results for the RigFEM and FEM models of the beam were compared to the experimental results. Table 5 compares the values of natural frequencies supplemented by the value of relative error defined as follows Equation (11):

$$\delta = \left| \frac{f_{exp} - f_{model}}{f_{exp}} \right| \times 100\% \quad (11)$$

where f_{exp}—experimentally determined natural frequency; f_{model}—natural frequency determined for the rigid finite element model (RigFEM) and the finite element model (FEM).

Table 5. Comparison of the natural frequencies between the rigid finite element model (RigFEM), 3D FEM, and 1D FEM and the experimental results for the beam analyzed.

Mode Number	Experimental Results	RigFEM Results	Relative Error δ_{RigFEM}	3D FEM Results	Relative Error $\delta_{3D\text{-}FEM}$	1D FEM Results	Relative Error $\delta_{1D\text{-}FEM}$
1	247 Hz	244 Hz	~1%	248 Hz	<1%	243 Hz	<2%
2	251 Hz	245 Hz	~2%	248 Hz	~1%	243 Hz	<2%
3	676 Hz	664 Hz	~2%	685 Hz	~1%	650 Hz	~4%
4	678 Hz	665 Hz	~2%	685Hz	~1%	650 Hz	<4%
5	1276 Hz	1224 Hz	~4%	1241 Hz	<3%	1184 Hz	~7%
6	1282 Hz	1275 Hz	<1%	1326 Hz	~3%	1223 Hz	<5
7	1282 Hz	1275 Hz	<1%	1326 Hz	~3%	1223 Hz	<5

A comparison of the selected mode shapes determined by calculation and experimental studies is presented in Figure 7. Analyzing the presented comparison in Figure 7, one can see that a mode shape agreement in the analyzed frequency range was achieved, therefore it can be stated that a full structural compliance of the model was achieved.

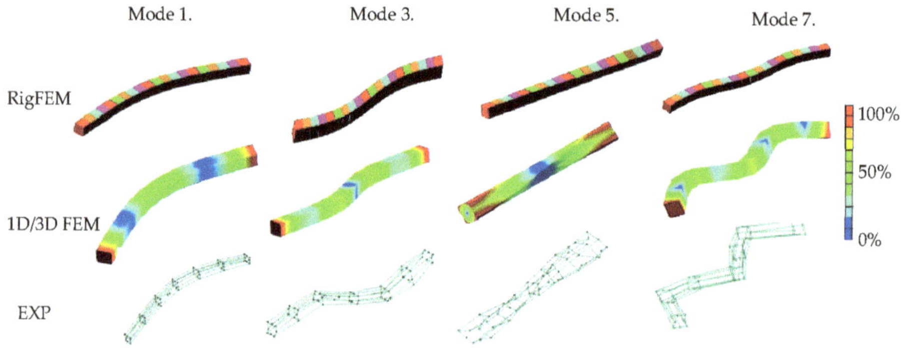

Figure 7. Selected mode shapes comparison for beam.

Figure 8 contains the frequency response functions comparison for the beam.

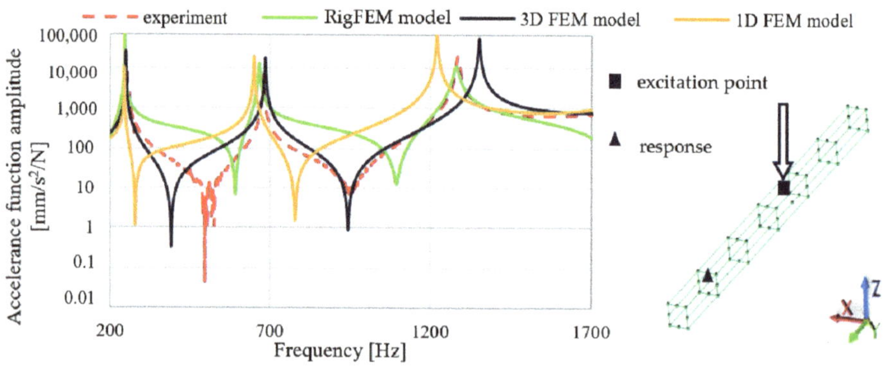

Figure 8. Frequency response function comparison for the 50 × 50 beam.

Analyzing the comparison of accelerance functions presented in Figure 8, it can be seen that a good agreement was achieved for all the models for the first two resonant frequencies both in terms of amplitude and the resonant frequencies values. For the third resonant frequency, the best fit is achieved by the RigFEM model; the 1D and 3D FEM models have a poorer accuracy in terms of the resonant frequency value and the corresponding accelerance amplitude.

Table 6 and Figure 9 compare the results of the calculations made for the RigFEM and FEM models of a steel-polymer concrete frame with the results of the experimental studies.

Table 6. Comparison of the natural frequencies between the RigFEM and FEM models and the experimental results of a machine tool body composed of 50 × 50 mm steel beams filled with polymer concrete.

Mode Number	Experimental Results	RigFEM Results	Relative Error δ_{RigFEM}	FEM Results	Relative Error δ_{FEM}
1	431 Hz	472 Hz	~9%	437 Hz	<1%
2	791 Hz	825 Hz	~4%	858 Hz	<9%
3	928 Hz	926 Hz	<1%	936 Hz	<1%
4	969 Hz	934 HZ	~4%	957 Hz	~1%
5	1041 Hz	1070 Hz	<3%	1032 Hz	<1%
6	1151 Hz	1230 Hz	<7%	1141 Hz	<1%
7	1232 Hz	1372 Hz	~11%	1195 Hz	~3%

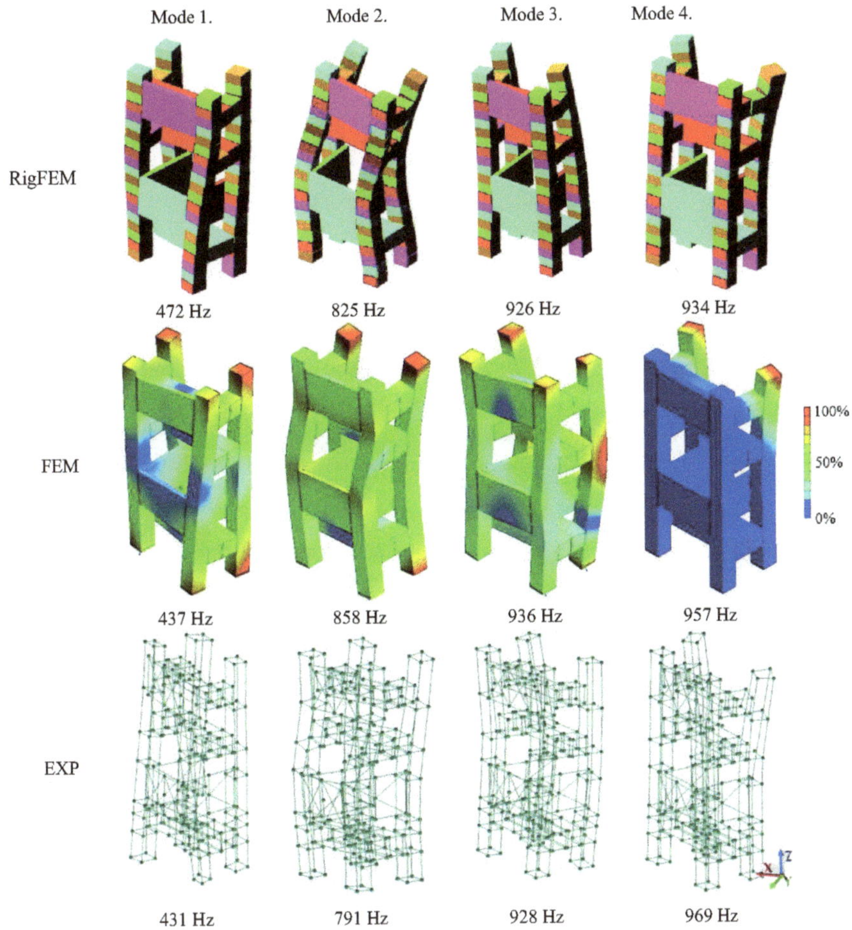

Figure 9. Comparison of the selected mode shapes between the RigFEM, FEM model, and the experimental results for the machine tool body.

Analyzing the comparison presented in Figure 9, it can be seen that, similar to the beam, a mode shape agreement for the frame was achieved, therefore it can be stated that a full structural compliance of the model was achieved.

Figure 10 presents the FRF comparisons at selected points where the spindle is mounted.

Analyzing the comparison presented in Figure 10, it can be seen that the RigFEM model gives less accurate results compared with FEM model, in particular in the lower frequency range. For the RigFEM model, the amplitude mapping of the first resonance is inaccurate; a similar phenomenon can be seen for resonances close to the frequency of 1200 Hz.

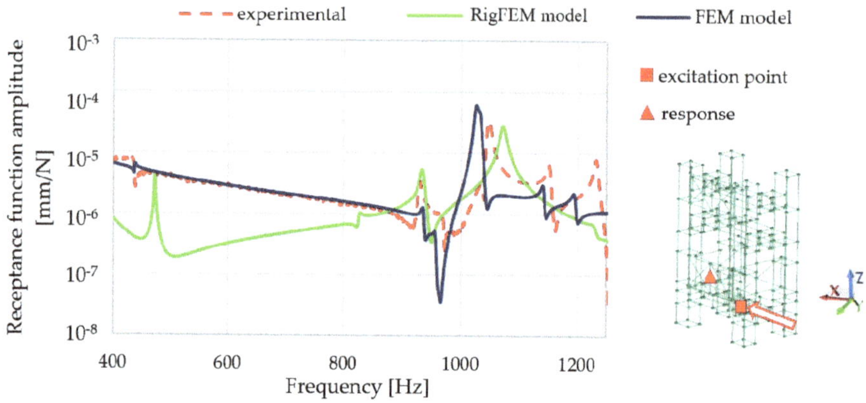

Figure 10. Selected receptance function comparison for the RigFEM, FEM model, and experimental data between the excitation point and the point where spindle is mounted in the X direction.

4. Discussion

To summarize, it can be concluded that for the RigFEM model of the beam analyzed, the full structural compliance of the model—i.e., the compliance of the mode shapes—has been achieved in the frequency range under consideration. The relative error for the natural frequency values for the RigFEM model did not exceed 4%; on average it was 1.8%. Referring these results to those presented in [34,36], it can be seen that for the competitive FEM model, the relative error did not exceed 3%—on average, it was 1.8%. The 3D FEM did not exceed 4%—on average, it was 4.1% (1D FEM). For the continuous model based on Timoshenko beam theory, the relative error did not exceed 4%—on average, it was 3.1%. Additionally, for the Euler–Bernoulli beam model of a similar steel-polymer concrete beam (differing in a steel profile cross-section dimensions—70 × 70 mm) presented in [42], the relative error for the bending mode shapes did not exceed 13.2%—on average, it was 7.6%. Comparing the natural frequencies obtained on the basis of the RigFEM beam model with the models described in the cited works, it can be seen that th eRigFEM model has a slightly lower accuracy than the 3D FEM model and has a better accuracy than the 1D FEM model and the continuous models, especially the Euler–Bernoulli model. It should be also noted that for the RigFEM model, an acceptable compatibility of the frequency response functions for resonant frequencies has been achieved.

Comparing the RigFEM and FEM models, it can be seen that the model developed according to RigFEM is characterized by a 95% reduction in the number of degrees of freedom in the case of the 3D model, and 50% in the case of the 1D model. Reducing the number of degrees of freedom directly shortens the time needed for the calculation, without the need for the model order reduction presented in [43].

The differences between the RigFEM beam model and the experimental tests can result from the adopted simplifications—i.e., the use of a small number of rigid finite elements and spring-damper elements can adversely affect the accuracy of both natural frequency values and frequency response functions [29]. The adoption of an isotropic, homogenous, linear elastic model of material to represent heterogenous polymer concrete may also influence the accuracy of the model due to the possible uneven distribution of the stiffness and mass. However, it should be noted that these assumptions were made deliberately and with full awareness of their consequences and were intended primarily to reduce the dimensionality of the model. One way to improve the accuracy of the model may be to use a model updating algorithm.

It can be also concluded that, similarly to the beam, for the frame, a full structural compliance of the models was obtained, which was manifested by the compliance of the mode shapes. Comparing the natural frequency values for the RigFEM model, it can be seen that the relative error did not exceed 11%—on average, it was 5.5%. For the FEM model built according to methodology presented in [34], the relative error did not exceed 9%—on average, it was 2.4%. In addition, a satisfactory compliance of the frequency response functions was achieved for both the models.

Comparing the models developed in accordance with RigFEM and FEM, it can be seen that the model developed according to RigFEM is characterized by a 99.8% reduction in the number of degrees of freedom, directly influencing the time needed for the calculation.

The differences between the RigFEM frame model and the experimental data may result from similar reasons to the beam model. An additional source of inaccuracy may be the simplified method of modeling the welded joints, which was implemented using spring-damper elements connecting the centers of mass of the rigid end elements.

Therefore, it can be concluded that the RigFEM models developed for both a steel-polymer beam and a frame composed of such beams provide a reliable description of their dynamic properties. In addition, their small size makes them competitive in relation to the models developed in accordance with FEM.

5. Conclusions

The literature sources indicate that the finite element method is the most popular in engineering practice to anticipate the dynamic properties of steel-polymer concrete machine tool frames. However, using FEM models at an early design stage, when several structural variants need to be analyzed, can be time-consuming due to their high dimensionality.

Therefore, the presented rigid finite element methodology of modeling composite steel-polymer concrete structures can be a useful tool to assist engineers in a design decision-making process, owing to the possibility of developing reliable low-dimensional models.

The obtained results confirmed the initially formulated thesis that the rigid finite element method can be effectively used in modeling and calculating the dynamic properties (natural frequencies, mode shapes, and frequency response functions) of both simple steel-polymer beams and complex spatial concrete frames. The accuracy of predicting the dynamic properties of the steel-polymer concrete structures for the RigFEM model is only slightly inferior to the FEM model and is fully acceptable. On the other hand, using RigFEM models with a reduced dimensionality leads to the reduction of time necessary to perform the calculations, which may be beneficial in the case of multiple repetitions of the calculations for many different structural variants (when searching for the optimal solution, favor the RigFEM model over the FEM model).

The main limitation of the presented models is the frequency response function accuracy for the non-resonant frequencies. However, according to the fact that displacements for non-resonant frequencies are small in general, the accuracy of the presented method can be considered as satisfactory.

To summarize, the study presents a method of modeling steel-polymer concrete machine tool frames based on the rigid finite element method. Comparing the RigFEM models with the analytical and finite element ones, it can be stated that the developed method is an interesting alternative to them.

Author Contributions: Conceptualization, P.D. and K.M.; methodology, P.D. and B.M.; software, P.D. and B.M.; validation, P.D.; formal analysis, P.D., S.B., and K.M.; writing—original draft preparation, P.D. and K.M.; writing—review and editing, S.B. and K.M.; visualization, P.D.; supervision, S.B. and K.M. All authors have read and agreed to the published version of the manuscript.

Funding: The research was financed by the Smart Growth Operational Program in the project POIR.04.01.02-00-0078/16 "Vertical lathe of light construction". The research was carried out on research apparatus purchased as part of the project No. RPZP.01.03.00-32-0004/17. The project was co-financed by the European Union from the European Regional Development Fund under the Regional Operational Program of the West Pomeranian Voivodeship 2014–2020. The project was co-financed by the Ministry of Science and Higher Education.

Conflicts of Interest: The authors declare no conflict of interest. The funders had no role in the design of the study; in the collection, analysis, or interpretation of data; in the writing of the manuscript; or in the decision to publish the results.

References

1. Chybowska, D.; Chybowski, L. A Review of TRIZ Tools for Forecasting the Evolution of Technical Systems. *Manag. Syst. Prod. Eng.* **2019**, *27*, 174–182. [CrossRef]
2. Jasiewicz, M.; Miądlicki, K. Implementation of an algorithm to prevent chatter vibration in a CNC system. *Materials* **2019**, *12*, 3193. [CrossRef] [PubMed]
3. Powałka, B.; Okulik, T. Influence of application of special casting compound on dynamic characteristics of the guideway system. *J. Mach. Eng.* **2010**, *10*, 71–81.
4. Kim, M.; Kim, J.; Lee, M.; Lee, S.-K. Surface finish improvement using a damping-alloy sleeve-insert tool holder in the end milling process. *Int. J. Adv. Manuf. Technol.* **2020**, *106*, 2433–2449. [CrossRef]
5. Chodnicki, M.; Kaliński, K.J.; Galewski, M.A. Vibration surveillance during milling flexible details with the use of active optimal control. *J. Low Freq. Noise Vib. Act. Control* **2013**, *32*, 145–155. [CrossRef]
6. Kaliński, K.J.; Galewski, M.A. A modified method of vibration surveillance by using the optimal control at energy performance index. *Mech. Syst. Signal Process.* **2015**, *58*, 41–52. [CrossRef]
7. Hernandez-Vazquez, J.-M.; Garitaonandia, I.; Fernandes, M.H.; Muñoa, J.; de Lacalle, L.N.L. A consistent procedure using response surface methodology to identify stiffness properties of connections in machine tools. *Materials* **2018**, *11*, 1220. [CrossRef]
8. Mahendrakumar, N.; Thyla, P.R.; Mohanram, P.V.; Raja Kumaran, C.; Jayachandresh, J. Study on static and dynamic characteristics of nettle–polyester composite micro lathe bed. *Proc. Inst. Mech. Eng. Part L J. Mater. Des. Appl.* **2019**, *233*, 141–155. [CrossRef]
9. Möhring, H.-C. Composites in production machines. *Procedia CIRP* **2017**, *66*, 2–9. [CrossRef]
10. Möhring, H.-C.; Brecher, C.; Abele, E.; Fleischer, J.; Bleicher, F. Materials in machine tool structures. *CIRP Ann.* **2015**, *64*, 725–748.
11. Murugan, S.; Thyla, P.R. Mechanical and dynamic properties of alternate materials for machine tool structures: A review. *J. Reinf. Plast. Compos.* **2018**, *37*, 1456–1467. [CrossRef]
12. Rangasamy, S.; Loganathan, K.; Natesan, A. Experimental investigation and numerical analysis of the dynamic characteristics of a laminated hybrid composite bed. *Polym. Compos.* **2017**, *38*, 20–26. [CrossRef]
13. Cortés, F.; Castillo, G. Comparison between the dynamical properties of polymer concrete and grey cast iron for machine tool applications. *Mater. Des.* **2007**, *28*, 1461–1466. [CrossRef]
14. Aggogeri, F.; Borboni, A.; Merlo, A.; Pellegrini, N.; Ricatto, R. Vibration damping analysis of lightweight structures in machine tools. *Materials* **2017**, *10*, 297. [CrossRef] [PubMed]
15. Do Suh, J. Thermal characteristics of composite sandwich structures for machine tool moving body applications. *Compos. Struct.* **2004**, *66*, 429–438. [CrossRef]
16. Do Suh, J.; Kim, H.S.; Kim, J.M. Design and manufacture of composite high speed machine tool structures. *Compos. Sci. Technol.* **2004**, *64*, 1523–1530.
17. Hwang, H.Y.; Lee, H.G. Clamping effects on the dynamic characteristics of composite machine tool structures. *Compos. Struct.* **2004**, *66*, 399–407. [CrossRef]
18. Kim, D.I.; Jung, S.C.; Lee, J.E.; Chang, S.H. Parametric study on design of composite–foam–resin concrete sandwich structures for precision machine tool structures. *Compos. Struct.* **2006**, *75*, 408–414. [CrossRef]
19. Kroll, L.; Blau, P.; Wabner, M.; Frieß, U.; Eulitz, J.; Klärner, M. Lightweight components for energy-efficient machine tools. *CIRP J. Manuf. Sci. Technol.* **2011**, *4*, 148–160. [CrossRef]
20. Kulisek, V.; Janota, M.; Ruzicka, M.; Vrba, P. Application of fibre composites in a spindle ram design. *J. Mach. Eng.* **2013**, *13*, 7–23.
21. Suh, J.D.; Lee, D.G.; Kegg, R. Composite machine tool structures for high speed milling machines. *CIRP Ann.* **2002**, *51*, 285–288. [CrossRef]
22. Gao, X.; Qin, Z.; Guo, Y.; Wang, M.; Zan, T. Adaptive Method to Reduce Thermal Deformation of Ball Screws Based on Carbon Fiber Reinforced Plastics. *Materials* **2019**, *12*, 3113. [CrossRef] [PubMed]
23. Suh, J.D.; Lee, D.G. Design and manufacture of hybrid polymer concrete bed for high-speed CNC milling machine. *Int. J. Mech. Mater. Des.* **2008**, *4*, 113–121. [CrossRef]
24. Sonawane, H.; Subramanian, T. Improved Dynamic Characteristics for Machine Tools Structure Using Filler Materials. *Procedia CIRP* **2017**, *58*, 399–404. [CrossRef]
25. Cho, S.-K.; Kim, H.-J.; Chang, S.-H. The application of polymer composites to the table-top machine tool components for higher stiffness and reduced weight. *Compos. Struct.* **2011**, *93*, 492–501. [CrossRef]

26. Venugopal, P.R.; Kalayarasan, M.; Thyla, P.R.; Mohanram, P.V.; Nataraj, M.; Mohanraj, S.; Sonawane, H. Structural investigation of steel-reinforced epoxy granite machine tool column by finite element analysis. *Proc. Inst. Mech. Eng. Part L J. Mater. Des. Appl.* **2019**, *233*, 2267–2279. [CrossRef]
27. Venugopal, P.R.; Dhanabal, P.; Thyla, P.R.; Mohanraj, S.; Nataraj, M.; Ramu, M.; Sonawane, H. Design and analysis of epoxy granite vertical machining centre base for improved static and dynamic characteristics. *Proc. Inst. Mech. Eng. Part L J. Mater. Des. Appl.* **2020**, *234*, 481–495. [CrossRef]
28. Mahendrakumar, N.; Thyla, P.R.; Mohanram, P.V.; Ramu, M.; Prabhu Raja, V.; Raja Kumaran, C.; Manojkumar, K.N.; Siddarth, A. Enhancement of Static and Dynamic Characteristics on Micro-lathe Bed by the Use of Alternate Form Design and Composite Materials. In Proceedings of the Advances in Simulation, Product Design and Development, Chennai, India, 13–15 December 2018; Shunmugam, M.S., Kanthababu, M., Eds.; Springer: Singapore, 2020; pp. 611–621.
29. Wittbrodt, E.; Adamiec-Wójcik, I.; Wojciech, S. *Dynamics of Flexible Multibody Systems: Rigid Finite Element Method*; Springer Science & Business Media: Berlin, Germany, 2007; ISBN 3-540-32352-X.
30. Kono, D.; Lorenzer, T.; Weikert, S.; Wegener, K. Evaluation of modelling approaches for machine tool design. *Precis. Eng.* **2010**, *34*, 399–407. [CrossRef]
31. Zaeh, M.; Siedl, D. A new method for simulation of machining performance by integrating finite element and multi-body simulation for machine tools. *CIRP Ann.* **2007**, *56*, 383–386. [CrossRef]
32. Dunaj, P.; Berczyński, S.; Chodźko, M.; Niesterowicz, B. Finite Element Modeling of the Dynamic Properties of Composite Steel–Polymer Concrete Beams. *Materials* **2020**, *13*, 1630. [CrossRef]
33. Peeters, B.; Van der Auweraer, H.; Guillaume, P.; Leuridan, J. The PolyMAX frequency-domain method: A new standard for modal parameter estimation? *Shock Vib.* **2004**, *11*, 395–409. [CrossRef]
34. Dunaj, P.; Berczyński, S.; Chodźko, M. Method of modeling steel-polymer concrete frames for machine tools. *Compos. Struct.* **2020**, *242*, 112197. [CrossRef]
35. Midas, I.T. *User's Manual of Midas NFX*; MIDAS IT: Seongnam, Korea, 2011.
36. Nozdrzykowski, K.; Chybowski, L.; Dorobczyński, L. Model-based estimation of the reaction forces in an elastic system supporting large-size crankshafts during measurements of their geometric quantities. *Measurement* **2020**, *155*, 107543. [CrossRef]
37. Adamiec-Wójcik, I.; Drąg, Ł.; Metelski, M.; Nadratowski, K.; Wojciech, S. A 3D model for static and dynamic analysis of an offshore knuckle boom crane. *Appl. Math. Model.* **2019**, *66*, 256–274. [CrossRef]
38. Adamiec-Wójcik, I.; Drąg, Ł.; Wojciech, S. A new approach to the rigid finite element method in modeling spatial slender systems. *Int. J. Struct. Stab. Dyn.* **2018**, *18*, 1850017. [CrossRef]
39. Neumark, S. *Concept of Complex Stiffness Applied to Problems of Oscillations with Viscous and Hysteretic Damping*; HM Stationery Office: London, UK, 1962.
40. Holanda, S.A.; Silva, A.A.; de Araújo, C.J.; de Aquino, A.S. Study of the complex stiffness of a vibratory mechanical system with shape memory alloy coil spring actuator. *Shock Vib.* **2014**, *2014*, 162781. [CrossRef]
41. Adams, M. *ADAMS User's Manual*; MacNeal-Schwendler Inc.: Oakdale, IA, USA, 2001.
42. Niesterowicz, B.; Dunaj, P.; Berczyński, S. Timoshenko beam model for vibration analysis of composite steel-polymer concrete box beams. *J. Theor. Appl. Mech.* **2020**, *58*, 799–810. [CrossRef]
43. Dunaj, P.; Dolata, M.; Berczyński, S. Model Order Reduction Adapted to Steel Beams Filled with a Composite Material. In Proceedings of the International Conference on Information Systems Architecture and Technology, Nysa, Poland, 16–18 September 2018; Springer: Cham, Switzerland, 2018; pp. 3–13.

materials

MDPI

Article

An Orthotropic Elastic-Plastic Constitutive Model for Masonry Walls

Piotr Bilko *,† and Leszek Małyszko †

Faculty of Geoengineering, University of Warmia and Mazury in Olsztyn, Oczapowskiego 2, 10-719 Olsztyn, Poland; leszek.malyszko@uwm.edu.pl

* Correspondence: piotr.bilko@uwm.edu.pl

† These authors contributed equally to this work.

Received: 7 August 2020; Accepted: 11 September 2020; Published: 13 September 2020

Abstract: The use of a continuum structural model for the analysis of masonry structures in the plane stress state is discussed in this paper. Attention is paid to orthotropic masonry at the material level and validation of the model after its implementation in a proprietary finite element method (FEM) system via user-supplied subroutine. The constitutive relations are established in the framework of the mathematical elastoplasticity theory of small displacements and deformations. Based on the orthotropic failure criterion that was originally proposed by Hoffman in the spatial stress state, the model includes a generalization of the criterion in the plane stress. As it is the case for isotropic quasi-brittle materials, different yield surfaces are considered for tension and compression, which are both of Hoffman type.

Keywords: orthotropic failure criteria; implementation; finite element; plasticity; masonry

1. Introduction

The mechanical behavior of masonry is subjected to various influencing factors, mostly resulting from its complicated mesoscopic and microscopic structure and two basic materials used. Therefore, different modeling approaches are available for the numerical simulation of the mechanical behavior of masonry structures. Extensive research has been conducted on the development of advanced numerical modeling and the analysis of historical masonry structures for several decades [1–4]. However, a macroscopic approach is required for the viable analysis and the prediction of global structural behavior. The macroscopic composite behavior of masonry can be described assuming a homogeneous continuum and an anisotropic material with directional properties. This alternative of the constitutive modeling proved to be promising in two-dimensional problems, especially for models with closed form of the failure, yield, or limit surface. Some constitutive macro-models that are relied on for the finite element method (FEM) are widely used in the last decades with a different degree of complexity and idealizations—from initial models with simply isotropic, linear elastic behavior to advanced models with non-linear orthotropic behavior that are recently developed in the framework of modern concepts of continuum damage mechanics.

More considerable interest in the biaxially loaded masonry began more than forty years ago, in the late seventies of the last century, both by experiments and theory. In general, experiments were mainly carried out at that time concerning the failure of shear walls. They were usually related to the proposed failure criterion for the masonry in the plane state of stresses in the representative form of the particular tests that had to be involved [5,6]. The representative biaxial test for walls built with any kind of masonry elements was conducted very seldom because of the technical difficulties. The biaxial test data of Page [7,8] were interpolated in [9] in the form of a failure criterion. The criterion was written using three failure surfaces in the form of elliptic cones expressed by a second-order polynomial of stress components in the reference axes coaxial with masonry layers and without any reference to the observed distinct failure modes of masonry panels. On the other hand, an important composite failure criterion which was derived based on the four different failure mechanisms was proposed in [10]. This criterion was next used in the FEM-based constitutive model in [11].

There have been few attempts to use a single failure surface in constitutive models of masonry panels because of the non-acceptable fit of experimental values. The general orthotropic failure criterion of Tsai-Wu [12] was already available for composite materials since 1971. The use of this criterion for masonry was attempted in [13] in the form of a polynomial function of the second degree in stresses. A criterion in the form of a double pyramid with a rectangular base and the slope angle equal to the internal friction angle of the material was assumed in [14]. The non-acceptable fit of Page's experimental values resulting from the Hoffman [15] single surface criterion was discussed in [16], although the criterion itself as a single limit condition is quite flexible and attractive to use. The phenomenological single-surface constitutive model cannot also distinguish between different failure modes. In the framework of computational plasticity, the use of a composite yield criterion containing several failure criteria seems to overcome this drawback. The choice of two failure surfaces, one for the tension regime and the second one for the compression regime, provides better agreement with Page's experimental results. To describe the orthotropic behavior of masonry, two orthotropic yield criteria were used in [17], where a hydrostatic pressure insensitive material of a Hill-type as the yield criterion for compression together with a Rankine-type yield criterion for tension were assumed. The composite criterion containing three failure surfaces, each of them being for tension, compression, and shear regimes, allows distinguishing between different failure modes [18]. Contours consisting of a few surfaces are characterized good compatibility with the experimental results—see, e.g., [19,20] and [18,21]. However, this approach seems to be too demanding a computational task of plasticity.

Concerning masonry inelastic behavior, the closed-form macro-models are more efficient and suitable for complex structural computations. Some constitutive models that are recently developed in the framework of modern concepts of continuum damage mechanics are based on the assumption that the masonry axes of the bed and head joints are also damage principal axes. Usually, scalar damage parameters are assumed in each direction of the fixed axis, see for instance [14], where two independent scalar parameters in each direction of the material axis were used and their evolution is described by the energy-based approach. A similar approach in the framework of computational plasticity was used in [17], where the principal directions of damage are fixed and aligned with the initial orthotropy axes and softening/hardening relationships were adopted for the stress–strain diagrams in tension and compression, with different fracture energies along the axes of each material. The energies were coupled by a single scalar internal parameter used in the plasticity algorithm to measure simultaneously the amount of softening/hardening in two material axes. The main drawback of the closed-form orthotropic macro-models is the identification of the material parameters. To estimate macro-scale properties from mortar and brick parameters and their bonding, homogenization techniques can be used, both in elastic as well as plastic behavior, e.g., [22–24]. An alternative option is to transfer the identification problem to the level of masonry constituents by using multi-scale methods, e.g., [18,20,21], or to use methods developed recently for modeling of anisotropic quasi-brittle fracture, e.g., the so-called phase-field fracture model in the diffusive damage mechanics [25–27].

A continuum damage model in which the orthotropic behavior is simulated using a mapping relationship between the orthotropic behavior and an auxiliary model has been recently published in [28]. By using the concept of the mapped stress tensor the problem can be more efficiently solved in the mapped space and the results can be transported to the real field. Two distinct isotropic failure criteria are assumed in the mapped space and two stress transformation tensors are adopted. In the paper, the computational representation of complex failure loci obtained by experiments on orthotropic masonry is also presented.

This paper discusses an extension of the approach presented in [17]. Although the extension can be done both in a plane stress state as well as in a spatial stress state, the constitutive model with the generalization of the Hoffman criterion in a plane stress state is discussed here. Two orthotropic failure criteria are used, which are formulated in the framework of the representation theory of orthotropic tensor functions based on the Hoffman criterion [15]. The use of two Hoffman-type failure criteria as the yield criteria in the plasticity model seems to be particularly attractive and may give the better fit of the experimental values [29,30]. The composite masonry is treated as a homogenized orthotropic continuum. Since the failure criteria are scalar-valued functions of the stress tensor, the invariant representations of these criteria are dependent only on orthotropic invariants of the stress tensors. It is also the purpose of the paper to show the possibility of formulating robust numerical algorithms of the model implementation into a commercial finite element code at the integration point level using user-defined subroutines. Some tests of the proposed numerical algorithm for an anisotropic continuum are presented in the paper, both at the single element level and at the structural level and in the plane stress state.

2. The Orthotropic Hoffman-Type Failure Criteria in an Invariant Form

To model such effects as a marked difference observed between strengths in tension and compression, Hoffman [15] proposed a fracture criterion for brittle orthotropic materials as an extension to the Hill yield criterion. The criterion was proposed in the spatial stress state and in the $\{\mathbf{m}_i\}$ frame of reference that coincides with the axes of orthotropy. The criterion was originally described by the function with nine material constants C_i that are dependent on the three uniaxial tensile strengths Y_{ti} and the three uniaxial compressive strengths Y_{ci}, along with the orthotropy directions i and also on the three shearing strengths k_{ij} on the planes of material orthotropy. In the case of the plane stress state, when the normal vector to the stress plane is coincided with the axis of orthotropy $\mathbf{n} = \mathbf{m}_3 = \mathbf{b}_3$ (Figure 1), stress components $\sigma_{33} = \sigma_{31} = \sigma_{23} = 0$ and the Hoffman criterion:

$$\begin{gathered} C_1(\sigma_{22}-\sigma_{33})^2 + C_2(\sigma_{33}-\sigma_{11})^2 + C_3(\sigma_{11}-\sigma_{22})^2 + \\ C_4\,\sigma_{11} + C_5\,\sigma_{22} + C_6\,\sigma_{33} + C_7\,\sigma_{23}^2 + C_8\,\sigma_{13}^2 + C_9\,\sigma_{12}^2 - 1 = 0\,, \end{gathered} \tag{1}$$

where:

$$\begin{gathered} C_1 = \frac{1}{2}\left(\frac{1}{Y_{t2}Y_{c2}} + \frac{1}{Y_{t3}Y_{c3}} - \frac{1}{Y_{t1}Y_{c1}}\right), \quad C_2 = \frac{1}{2}\left(\frac{1}{Y_{t3}Y_{c3}} + \frac{1}{Y_{t1}Y_{c1}} - \frac{1}{Y_{t2}Y_{c2}}\right), \\ C_3 = \frac{1}{2}\left(\frac{1}{Y_{t1}Y_{c1}} + \frac{1}{Y_{t2}Y_{c2}} - \frac{1}{Y_{t3}Y_{c3}}\right), \quad C_4 = \frac{1}{Y_{t1}} - \frac{1}{Y_{c1}}, \quad C_5 = \frac{1}{Y_{t2}} - \frac{1}{Y_{c2}}, \\ C_6 = \frac{1}{Y_{t3}} - \frac{1}{Y_{c3}}, \quad C_7 = \frac{1}{k_{23}^2}, \quad C_8 = \frac{1}{k_{31}^2}, \quad C_9 = \frac{1}{k_{12}^2}, \end{gathered} \tag{2}$$

takes the following form with the six material constants:

$$C_1(\sigma_{22})^2 + C_2\,(\sigma_{11})^2 + C_3(\sigma_{11}-\sigma_{22})^2 + C_4\,\sigma_{11} + C_5\,\sigma_{22} + C_9\,\sigma_{12}^2 - 1 = 0\,, \tag{3}$$

where the constants $C_1 \div C_3$ are dependent on uniaxial strengths Y_{t3} and Y_{c3} in the direction perpendicular to the stress plane.

The theory of tensor functions together with the theorems on their representations has been recognized to be an efficient mathematical tool for the formulation of constitutive relationships with both the desirable analytical clarity and required generality.

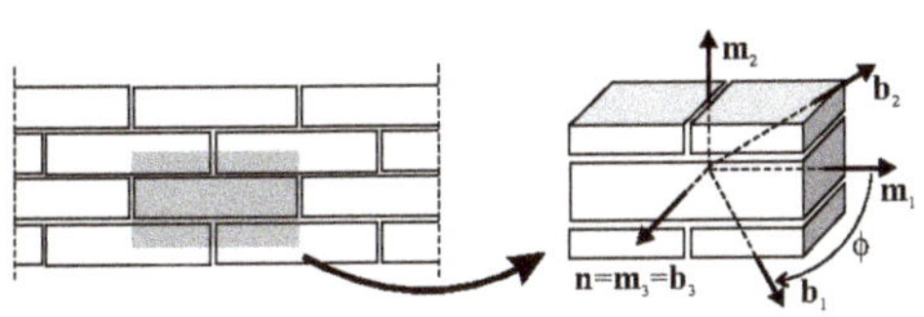

Figure 1. Axes of orthotropy $\{\mathbf{m}_i\}$ and a Cartesian system $\{\mathbf{b}_i\}$.

For some other recent applications of tensor functions see, e.g., [31,32]. It also allows accounting straightforwardly the invariance requirements of the principle of the space isotropy and the material symmetries so that the orientation of the material in space does not affect on its constitutive relation. Using this theory with Boehler's results [33], we can assume that the orthotropic criterion (3) is a particular case of the more general scalar-valued orthotropic function of three invariants $tr\left(\mathbf{M}_1\boldsymbol{\sigma}\right)$, $tr\left(\mathbf{M}_2\boldsymbol{\sigma}\right)$, $tr\boldsymbol{\sigma}^2$ of the following form:

$$f\left(tr\left(\mathbf{M}_1\boldsymbol{\sigma}\right), tr\left(\mathbf{M}_2\boldsymbol{\sigma}\right), tr\boldsymbol{\sigma}^2\right) - 1 = 0, \tag{4}$$

where $\boldsymbol{\sigma}$ is the symmetric plane stress tensor ($\sigma \in T_2^s, dim T_2^s = 3$) and $\mathbf{M}_\alpha$ are the parametric (structural) tensors defined as:

$$\mathbf{M}_1 = \mathbf{m}_1 \otimes \mathbf{m}_1, \quad \mathbf{M}_2 = \mathbf{m}_2 \otimes \mathbf{m}_2. \tag{5}$$

The unit vectors $\mathbf{m}_\alpha$ are the privileged directions of the orthotropic material, so they have to be perpendicular to each other. The invariants are very useful for the interpretation of the failure surface in any coordinate systems of the plane stress tensor that are different from the principal axes of orthotropy. Following the paper [34] or [35], we can choose another set of invariants K_p in the form:

$$K_1 = tr\left(\mathbf{M}_1\boldsymbol{\sigma}\right), \; K_2 = tr\left(\mathbf{M}_2\boldsymbol{\sigma}\right), \; K_3 = tr\boldsymbol{\sigma}^2 - \left(tr\left(\mathbf{M}_1\boldsymbol{\sigma}\right)\right)^2 - \left(tr\left(\mathbf{M}_2\boldsymbol{\sigma}\right)\right)^2, \tag{6}$$

where the symbol "*tr*" denotes the trace of a second order tensor ($tr(\mathbf{AB}) = A_{ij}B_{ij}$). The form (6) is very convenient because in the $\{\mathbf{m}_\alpha\}$ axes the invariants are:

$$K_1 = \sigma_{11}, \quad K_2 = \sigma_{22}, \quad K_3 = \left(\sqrt{2}\sigma_{12}\right)^2. \tag{7}$$

Using the invariants (7), the criterion (3) can be treated as a particular case of the criterion proposed in [34] and may be written in the following invariant form:

$$f(K_i) - 1 = a_\alpha K_\alpha + b_{\alpha\beta}K_\alpha K_\beta + cK_3 - 1 = 0, \tag{8}$$

where $\alpha, \beta = 1, 2$, $i = 1, 2, 3$ and the material constants are defined as:

$$\begin{aligned} a_\alpha &= \frac{1}{Y_{t\alpha}} - \frac{1}{Y_{c\alpha}}, \; b_{11} = \frac{1}{Y_{t1}Y_{c1}}, \; b_{22} = \frac{1}{Y_{t2}Y_{c2}}, \\ b_{12} &= \frac{1}{2}\left(\frac{1}{Y_{t3}Y_{c3}} - \frac{1}{Y_{t1}Y_{c1}} - \frac{1}{Y_{t2}Y_{c2}}\right), \; c = \frac{1}{2k_{12}^2}. \end{aligned} \tag{9}$$

Note that in the constant b_{12} there are again uniaxial strengths Y_{t3} and Y_{c3} in the direction perpendicular to the stress plane.

The criterion (8) can be also written in the following invariant form:

$$\frac{1}{2}\boldsymbol{\sigma} \cdot \mathbf{P} \cdot \boldsymbol{\sigma} + \mathbf{p} \cdot \boldsymbol{\sigma} - 1 = 0, \tag{10}$$

where a dot means a double contraction of two tensors, $\mathbf{p}$ is the symmetric tensor function of the second-order:

$$\mathbf{p} = a_1\mathbf{M}_1 + a_2\mathbf{M}_2, \tag{11}$$

$\mathbf{P}$ is the double symmetric tensor function of the fourth-order:

$$\mathbf{P} = 2b_{11}\mathbf{M}_1 \otimes \mathbf{M}_1 + 2b_{22}\mathbf{M}_2 \otimes \mathbf{M}_2 + 2b_{12}\left(\mathbf{m}_1 \otimes \mathbf{m}_2 + \mathbf{m}_2 \otimes \mathbf{m}_1\right) + c\mathbb{M} \tag{12}$$

with the fourth-order tensor $\mathbb{M}$:

$$\mathbb{M} = 4\,\mathbf{N} \otimes \mathbf{N}\,, \quad \mathbf{N} = \frac{1}{2}\left(\mathbf{m}_1 \otimes \mathbf{m}_2 + \mathbf{m}_2 \otimes \mathbf{m}_1\right). \tag{13}$$

Several criteria proposed in the literature for orthotropic materials are special cases of the quadratic limit surface (8), including an elliptic failure surface according to Tsai and Wu [12] and criteria discussed recently in [30]. However, a phenomenological single-surface model may give an insufficient description of the mechanical behavior. It does not permit easy identification of failure modes and thus renders the description of different post-failure mechanisms very difficult. At least two failure criteria should be taken into consideration, the one for the compression regime and the second for tension regime. Each of them may be of the form (8) as proposed in [34] where the failure criterion for orthotropic materials in the spatial stress state is represented by two quadratic functions of the six invariants of the stress tensor and parametric tensors. It may more accurately describe the failure data distribution than classical limit surfaces, although it may include fifteen independent material parameters for the description of failure surfaces. On the other hand, the concept of a smooth single-surface description seems to be attractive from a numerical point of view and also allows for modeling of different inelastic behavior by changing size, shape, and location of a quadratic state function according to the form (8) in orthotropic stress space.

It is possible to propose a generalization of the Hoffman criterion for the orthotropic material in the plane state of stresses. The determination of the six material parameters in the criterion (3) requires six strength tests. The five standard tests are uniaxial loading along the axes of orthotropy (two tests for tension and two for compression) and the shearing test in the plane of stresses. If the test of the uniform biaxial compression is used for determining the strength Y_{cc} as the sixth test in addition to conventional tests, we will get the criterion in the invariant form (8) or (10), in which the material parameter b_{12} is changed to the form:

$$b_{12}^{(1)} = \frac{1}{2Y_{cc}^2} - \frac{1}{2Y_{t1}Y_{c1}} - \frac{1}{2Y_{t2}Y_{c2}} + \frac{1}{2Y_{cc}}\left(\frac{1}{Y_{t1}} + \frac{1}{Y_{t2}} - \frac{1}{Y_{c1}} - \frac{1}{Y_{c2}}\right). \tag{14}$$

If as the additional test we will use the test of the uniform biaxial tension for determining the strength Y_{tt}, we will get the criterion in the invariant form with the following material parameter b_{12}:

$$b_{12}^{(2)} = \frac{1}{2Y_{tt}^2} - \frac{1}{2Y_{t1}Y_{c1}} - \frac{1}{2Y_{t2}Y_{c2}} - \frac{1}{2Y_{tt}}\left(\frac{1}{Y_{t1}} + \frac{1}{Y_{t2}} - \frac{1}{Y_{c1}} - \frac{1}{Y_{c2}}\right). \tag{15}$$

Note that now there are not the uniaxial strengths Y_{t3} and Y_{c3} in the material constant b_{12}. Finally, using two functions of the form (8) or (10), one with the parameter (14) for compression regime $f_1(K_i) - 1 = 0$ and the second with the parameter (15) for tension regime $f_2(K_i) - 1 = 0$, we can construct the composite failure criterion in such a way that the following set:

$$B = B_1 \cap B_2 \wedge B_\alpha \equiv \{\boldsymbol{\sigma} \in T_2^s \,|\, f_\alpha(K_i) - 1 < 0\}\,, \tag{16}$$

is convex and

$$f_\alpha(K_i) - 1 = a_\alpha^{(\alpha)} K_\alpha + b_{\alpha\beta}^{(\alpha)} K_\alpha K_\beta + c^{(\alpha)} K_3 - 1 = 0, \tag{17}$$

We can also assume that $c^{(1)} = c^{(2)} \equiv c$ and $a_\alpha^{(\alpha)} = a_\alpha$, $b_{\alpha\alpha}^{(\alpha)} = b_{\alpha\alpha}$ for simplification and we then have the following seven material parameters: the two uniaxial tensile strengths ($Y_{t\alpha} > 0$), the two uniaxial compressive strengths ($Y_{c\alpha} > 0$), the pure shear strength ($k_{12} > 0$), the biaxial uniform compressive strength $Y_{cc} > 0$ and the biaxial uniform tensile strength $Y_{tt} > 0$. This procedure is proposed in the paper [34], although the number of material parameters may be increased to 12 even though some of them can be used just to adjust failure surfaces to experimental data.

Figure 2 shows the criterion in the orthotropy axes with the following parameters for material: simple shear strength $k = 0.4$ [MPa], vertical compression strength $Y_{c2} = 6.34$ [MPa] and vertical tensile strength $Y_{t1} = 0.31$ [MPa] perpendicular to the horizontal joint, compression strength $Y_{c1} = 4.95$ [MPa] and tensile strength $Y_{t2} = 0.14$ [MPa] parallel to the horizontal joint, biaxial uniform compression strength $Y_{cc} = 9.0$ [MPa] and biaxial uniform tension $Y_{tt} = 0.14$ [MPa]. Figure 2a shows general view and Figure 2b cross section of $\sigma_{12} = 0$ (solid line) with contours every 0.2 [MPa] (dashed lines). Figure 2c shows cross section of $\sigma_{22} = -6.34$ [MPa] (solid line) with parallel contours every 0.81 [MPa] (dashed lines). Figure 2d shows cross section of $\sigma_{11} = -4.95$ [MPa] (solid line) with parallel contours every 0.81 [MPa] (dashed lines).

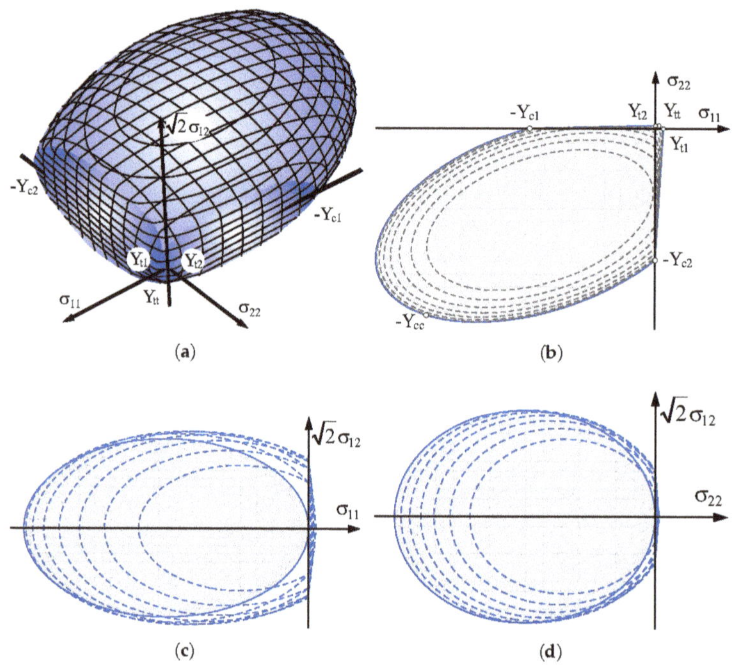

Figure 2. Criterion in orthotropy axes: (**a**) 3D view, (**b**–**d**) contours (described in text).

An alternative connection of two surfaces is proposed, different from that used in [34], which is shown schematically in Figure 3. It allows for shifting a common edge. Figure 3 shows the method for constructing a boundary surface of the criterion in the normal stress components, in the orthotropy axes (at zero shear stresses). The surface in the tension range conventionally adopts compressive strength values $Y^t_{c\alpha}$ greater than twice those for the area of the compression range $Y^c_{c\alpha}$, that is $Y^t_{c\alpha} = 2\,Y^c_{c\alpha}$. On the other hand, the surface in the compression range conventionally adopts tensile strength $Y^c_{t\alpha}$, by making them the value of compressive strength $Y^c_{c\alpha}$. It is assumed that they will be two-fold smaller, that is $Y^c_{t\alpha} = 0.5\,Y^c_{c\alpha}$.

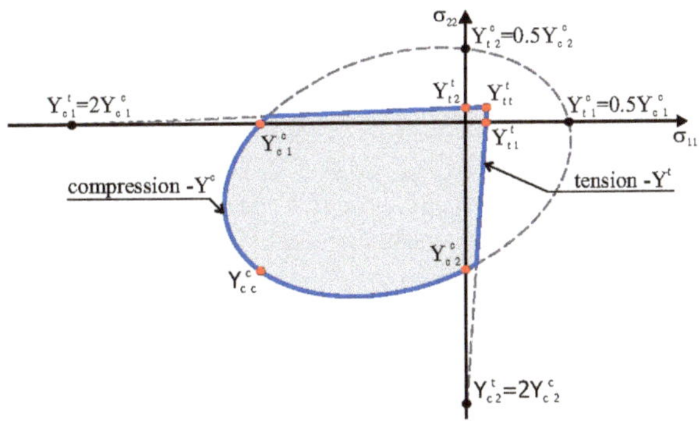

Figure 3. The proposed criterion and material parameters determining the surfaces.

3. Comparison with the Experimental Results

The composite failure criterion is shown in Figure 4 in the orthotropy axes and in comparison with one of the most complete sets of experimental data of biaxially loaded masonry that was given by Page [7,8], who tested 102 panels of half-scale solid clay brick masonry. Tested elements were made on a scale of 1:2 with dimensions 360 × 360 × 50 [mm]. Tests were differentiated due to the rotation of the principal stress terms of the horizontal joint (axis of the material). The ratio of the principal stresses was changed so that the wall was considered in any possible state of stress. The following values of the parameters are adopted: axial tensile strength along the horizontal joint $Y_{t1} = 0.43$ [MPa] and perpendicular to it $Y_{t2} = 0.32$ [MPa], the axial compression strength along the horizontal joint $Y_{c1} = 8.74$ [MPa] and perpendicular to it $Y_{c2} = 8.03$ [MPa], the pure shear strength $k = 0.33$ [MPa], uniform biaxial tensile strength $Y_{tt} = 0.32$ [MPa] and compression strength $Y_{cc} = 8.38$ [MPa]. Good agreement was found in the shape of the failure surface for principal stresses, which may mean that the criterion would appear to be sufficiently well validated for further investigations. Finally, the Rankine–Hill failure criterion [16] is shown in Figure 4 by dashed lines as a comparison to the proposed criterion.

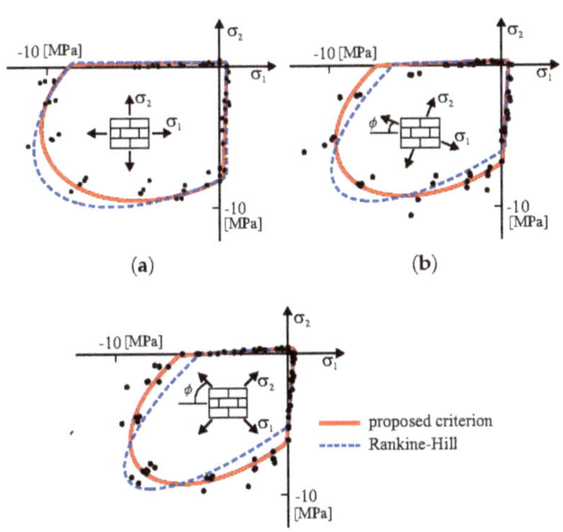

Figure 4. Comparison of the proposed criterion with the experimental results of Page [7,8] depending on the inclination axes of orthotropy relative to the principal axis: (**a**) $\phi = 0.0$, (**b**) $\phi = 22.5$, (**c**) $\phi = 45.0$.

Another discussed example is the comparison of the proposed failure criterion with the results of the experimental research from work [36]. The research program was carried out at the ETH Polytechnic in Zurich. Ganz and Thürliman's team investigated biaxially loaded wall panels (designated K1–K12) with dimensions of 1200 × 1200 × 150 [mm^3] and with different orientation of material axes to the principal stress directions. The results of experimental tests [36] are presented in Table 1 excluding two tests (K5 and K9), which concerned samples of reinforced walls. The second column of the Table 1 contains the proportion of stresses, while the third column gives the angle by which the system of material axes has been rotated to the load directions. The ratio of principal stresses allows the direction of the load path to be determined. The intersection point of the load path direction with the criterion surface determines the stress state, the components of which are placed in columns 7–9 of the Table 1. The criterion surfaces are determined by the following parameters [MPa]: for a tension regime — $Y_{t1} = 0.28$, $Y_{t2} = 0.01$, $Y_{c1} = 3.74$, $Y_{c2} = 15.72$, $k_t = 0.048$, $Y_{tt} = 0.01$ and for a compression regime — $Y_{t1} = 0.94$, $Y_{t2} = 3.81$, $Y_{c1} = 1.87$, $Y_{c2} = 7.61$, $k_c = 2.868$, $Y_{cc} = 2.06$. The anisotropy ratio of the compressive strength is $Y_{c2}/Y_{c1} = 4.07$ and is related to the arrangement of cores in ceramic masonry elements. The length of the load vector should be determined as $\sqrt{\sigma_{11}^2 + \sigma_{22}^2 + 2\sigma_{12}^2}$. The last column of the Table 1 presents the values of the ratios of the length of the vectors resulting from the experiment and from the initial surfaces of the proposed criterion. The results show good compliance of the criterion with the experimental results.

Table 1. Comparison of the proposed criterion with the experimental results of [36].

Panel	σ_1/σ_2	ϕ	Experimental [MPa]			Numerical [MPa]			Exp/Num
		[°]	σ_{11}	σ_{22}	τ_{12}	σ_{11}	σ_{22}	τ_{12}	[-]
K1	−0.09	22.5	−0.08	−0.92	0.42	−0.09	−1.04	0.47	0.88
K2	−0.05	22.5	−0.17	−1.42	0.62	−0.17	−1.35	0.59	1.05
K3	0.0	0.0	0.00	−7.63	0.00	0.00	−7.61	0.00	1.01
K4	0.0	90.0	−1.83	0.00	0.00	−1.87	0.00	0.00	0.98
K6	0.0	45.0	−0.32	−0.32	0.32	−0.43	−0.43	0.43	0.74
K7	0.0	22.5	−0.39	−2.25	0.93	−0.37	−2.15	0.89	1.05
K8	0.0	67.5	−0.22	−0.04	0.09	−0.22	−0.04	0.08	1.01
K10	0.33	0.0	−2.11	−6.44	0.00	−2.02	−6.15	0.00	1.05
K11	0.31	22.5	−2.04	−4.49	1.23	−2.02	−4.46	1.22	1.01
K12	0.30	45.0	−2.03	−2.03	1.08	−2.01	−2.01	1.07	1.01

Figure 5a shows the initial surface of the criterion with points marked on the surface, indicating the stress limits of individual experimental tests conducted by Ganz and Thürliman. Analyzing the contour lines, it can be seen that most of the safe area is limited by the tensile regime. This is more clearly seen in Figure 5b,c, in which the plane cross-sections of the criterion are shown. One can see cross-sections with assigned to them, the values of the ultimate stress obtained from the individual experimental tests which are marked with diamonds.

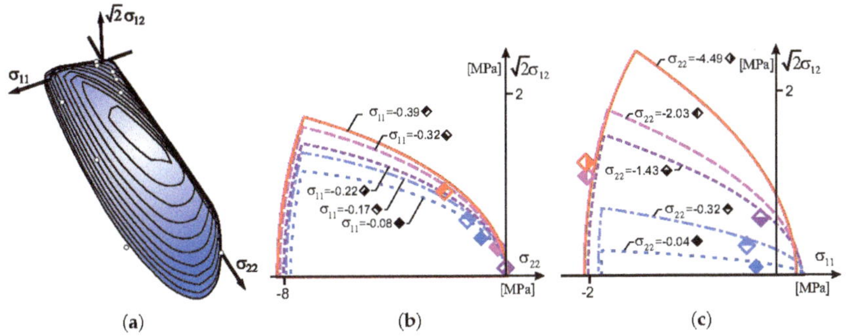

Figure 5. Comparison of the proposed base contour of the failure surface with the experimental results of Ganz and Thürliman [36]: (**a**) 3D view, (**b**) by cross sections of the body with $\sigma_{22} = const$, (**c**) by cross sections of the body with $\sigma_{11} = const$.

At the ETH Polytechnic in Zurich, a test program for walls made of concrete, hollow masonry elements ZSW1-ZSW12 was also carried out [37]. The results of the experimental tests are given in Table 2, and in Figure 6. Figure 6a shows failure surface of the criterion with the experimental results of [37]. Figure 6b,c shows cross-sections with planes of the constant normal stress. A position of the yield stress from experimental tests are marked with diamonds.

Table 2. Comparison of the proposed criterion with the experimental results of [37].

Panel	σ_1/σ_2	ϕ	Experimental [MPa]			Calculated [MPa]			Exp/Num
		[°]	σ_{11}	σ_{22}	τ_{12}	σ_{11}	σ_{22}	τ_{12}	[-]
ZSW1	0.0	0.0	0.00	−9.12	0.00	0.00	−9.21	0.00	0.99
ZSW2	0.14	0.0	−6.12	−0.83	0.00	−6.05	−0.82	0.00	1.01
ZSW4	1.53	0.0	−5.98	−9.13	0.00	−5.58	−8.52	0.00	1.07
ZSW5	0.0	45.0	−3.06	−3.06	3.06	−3.06	−3.06	3.06	1.00
ZSW6	0.22	45.0	−4.60	−4.60	2.93	−4.69	−4.69	2.98	0.98
ZSW7	1.0	45.0	−6.12	−6.12	0.00	−6.36	−6.36	0.00	0.96
ZSW8	0.0	67.5	−2.34	−0.40	0.97	−2.34	−0.40	0.97	1.00
ZSW9	0.0	22.5	−0.97	−5.66	2.35	−0.97	−5.66	2.35	1.00

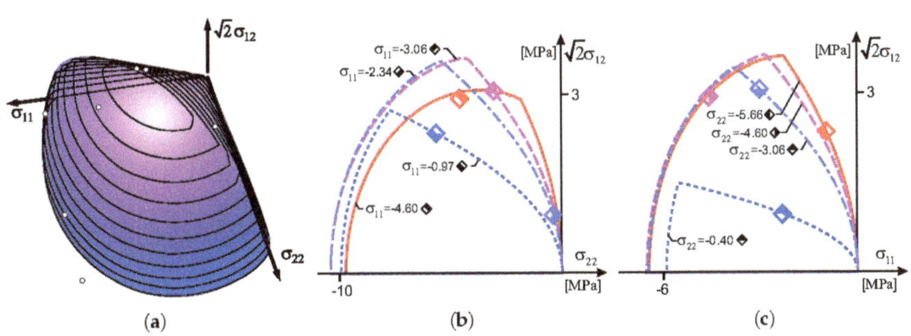

Figure 6. Comparison of the proposed criterion with the experimental results of of Lurati [37]: (**a**) 3D view, (**b**) by cross-sections of the body with $\sigma_{22} = const$, (**c**) by cross-sections of the body with $\sigma_{11} = const$.

The criterion surfaces are determined by the following parameters [MPa]: for a tension regime — $Y_{t1} = 0.01$, $Y_{t2} = 0.01$, $Y_{c1} = 11.52$, $Y_{c2} = 18.42$, $k_t = 0.01$, $Y_{tt} = 0.01$ and for a compression regime— $Y_{t1} = 2.88$, $Y_{t2} = 4.61$, $Y_{c1} = 5.76$, $Y_{c2} = 9.21$, $k_c = 3.98$, $Y_{cc} = 6.36$. The anisotropy ratio of the compressive strength is $Y_{c2}/Y_{c1} = 1.60$. The material has almost zero tensile strength. As before, the load vector length was determined for each limit point as $\sqrt{\sigma_{11}^2 + \sigma_{22}^2 + 2\sigma_{12}^2}$. The maximum difference of vectors resulting from the experiment data and those of the initial surface compared criterion does not exceed 7 percentage points, which shows a very good agreement of the proposed model with the experiment results.

4. The Constitutive Relations and Implementation of the Model

The elastic-plastic orthotropic material is considered assuming an additive decomposition of the strain tensor into the elastic part $\boldsymbol{\varepsilon}_e$ and the plastic part $\boldsymbol{\varepsilon}_p$. The elastic part is defined by orthotropic Hooke's law. The plastic part of the strain tensor is defined by a flow rule associated with the yield function given by the plasticity (failure) criterion written based on the forms (10) and (17) as:

$$f_\alpha\left(\boldsymbol{\sigma}, z_\alpha\right) = \frac{1}{2}\boldsymbol{\sigma} \cdot \mathbf{P}_\alpha \cdot \boldsymbol{\sigma} + \mathbf{p}_\alpha \cdot \boldsymbol{\sigma} - K_\alpha\left(z_\alpha\right) = 0, \tag{18}$$

where the $K_\alpha(z_\alpha)$ are given functions with the real functional value from a closed interval [0, 1] that describes the type of hardening/softening (Figure 7), z_α are internal scalar hardening variables. The softening behavior is modeled with a smeared approach, where the localized damage is represented by the scalar, which is related by an equivalent length h to the released energy per unit cracked area, G_f. The length h should correspond to a dimension of the finite element mesh. As one can see in Figure 7, different fracture energies are introduced in the model, as additional parameters—the tensile fracture energy G_{ft} and the compressive energy G_{fc}.

In the frame of reference coinciding with the orthotropy axes, we have the following matrix representations of tensors $\mathbf{P}_\alpha$ and $\mathbf{p}_\alpha$ in the Voigt notations for the plane stress $\boldsymbol{\sigma} \Rightarrow \begin{bmatrix}\sigma_{11} & \sigma_{22} & \sqrt{2}\sigma_{12}\end{bmatrix}^{\mathrm{T}}$, ($\alpha = t$ for tension surface, $\alpha = c$ for compression surface):

$$\mathbf{P}_\alpha \Rightarrow \boldsymbol{P}_\alpha = \begin{bmatrix} 2b_{11}^\alpha & 2b_{12}^\alpha & 0 \\ 2b_{21}^\alpha & 2b_{22}^\alpha & 0 \\ 0 & 0 & 4c^\alpha \end{bmatrix}, \qquad \mathbf{p}_\alpha \Rightarrow \boldsymbol{p}_\alpha = \begin{bmatrix} a_1^\alpha & a_2^\alpha & 0 \end{bmatrix}. \tag{19}$$

For composite criteria, the subscript α also denotes the number of the active surface. Let us first assume that only one surface is active, which will allow this marking to be omitted. In the framework of the mathematical theory of the elastic-plastic material a permissible stress state is any state of stresses for which $f \leq 0$. A stress state is called the elastic stress state if $f < 0$. A plastic state refers to a stress state at the boundaries of the current elastic region for which $f = 0$. The plastic part of the strain tensor is defined by a flow rule associated with the yield function given by the Equation (18). The flow rule defines the sign (direction) of plastic-strain increment in the following form:

$$\dot{\boldsymbol{\varepsilon}}_p = \gamma \left.\frac{\partial f\left(\boldsymbol{\sigma}, z\right)}{\partial \boldsymbol{\sigma}}\right|_{\boldsymbol{\sigma}=\boldsymbol{\sigma}^T} = \gamma\left(\mathbf{P} \cdot \boldsymbol{\sigma} + \mathbf{p}\right) \equiv \gamma\,\mathbf{r}, \tag{20}$$

where $\gamma > 0$ is a plastic multiplier. After applying differentiation to orthotropic Hooke's elastic law with the respect to the time and after substituting Equation (20) we obtain:

$$\dot{\boldsymbol{\sigma}} = \mathbf{C} \cdot \left(\dot{\boldsymbol{\varepsilon}} - \gamma\,\mathbf{r}\right) \equiv \mathbf{C}_{ep} \cdot \dot{\boldsymbol{\varepsilon}}, \tag{21}$$

where the elasto-plastic tangent operator $\mathbf{C}_{ep}$ can be calculated after the parameter γ is known. Assuming that

$$\dot{z} = \gamma(\mathbf{r} \cdot \mathbf{r}) \equiv \gamma\|\mathbf{r}\|^2. \tag{22}$$

one can compute from the consistency condition

$$\gamma \dot{f}(\boldsymbol{\sigma}, z) = 0, \quad \gamma > 0, \tag{23}$$

the plastic multiplier

$$\gamma = \frac{\langle \mathbf{r} \cdot \mathbf{C} \cdot \dot{\boldsymbol{\varepsilon}} \rangle}{\mathbf{r} \cdot \mathbf{C} \cdot \mathbf{r} + \partial K / \partial z \, \|\mathbf{r}\|^2}, \tag{24}$$

and the operator $\mathbf{C}_{ep}$ in the following form:

$$\mathbf{C}_{ep} = \mathbf{C} - \frac{(\mathbf{C} \cdot \mathbf{r}) \otimes (\mathbf{r} \cdot \mathbf{C})}{\mathbf{r} \cdot \mathbf{C} \cdot \mathbf{r} + \partial K / \partial z \, \|\mathbf{r}\|^2}. \tag{25}$$

The double symmetric fourth-order tensor of elastic material constants $\mathbf{C}$ in the orthotropic Hooke's law can be conveniently defined by the compliance tensor $\mathbf{S} \equiv \mathbf{C}^{-1}$, which in the frame of reference aligned with the orthotropic axes can be written in Voigt notation for the plane stress as

$$\mathbf{S} \rightarrow \begin{bmatrix} \frac{1}{E_1} & -\frac{\nu_{21}}{E_2} & 0 \\ -\frac{\nu_{12}}{E_1} & \frac{1}{E_2} & 0 \\ 0 & 0 & \frac{1}{G_{12}} \end{bmatrix}. \tag{26}$$

where we have five technical in-plane moduli: E_1, E_2 are Young's moduli, G_{12} is the shear modulus and ν_{12}, ν_{21} are Poisson's ratios ($\nu_{12}/E_1 = \nu_{21}/E_2$). It should be noted that two yield criteria are combined in the model into a composite yield surface and the intersection of different yield surfaces defines corners that require special attention in a numerical algorithm according to Koiter's generalization.

4.1. Implementation into Finite Elements under the Plane Stress Condition

In this subsection, according to the convention adopted in many finite element programs the components of a symmetric second-order tensor are presented as a single column array, whereas fourth-order tensors are presented as two-dimensional arrays. The matrix representations of the tensors are shown in terms of the Cartesian components in the frame coinciding with the materialaxes of orthotropy.

The constitutive relationship in (21) is in the form of the "highly non-linear" differential equation which can be solved by the modified Euler method (usually the implicit Euler backward algorithm). Therefore, it is replaced by the incremental equation of the form:

$$\Delta\sigma = \boldsymbol{C}(\Delta\varepsilon - \gamma\tilde{\boldsymbol{r}}) \equiv \tilde{\boldsymbol{C}}_{ep}\,\Delta\varepsilon, \tag{27}$$

where $\tilde{\boldsymbol{C}}_{ep}$ is called the operator consistent with the integration algorithm of constitutive relations. We assume that for each $t_n \in [0, T]$ the strain increment $\Delta\varepsilon = \varepsilon_{n+1} - \varepsilon_n$ is known, thus the problem is strain driven, and we want to compute the stress state σ_{n+1} for t_{n+1}. We assume:

$$\sigma_{n+1} = \sigma_{n+1}^{trl} - \Delta\gamma\,\boldsymbol{C}\boldsymbol{r}^{trl}, \tag{28}$$

where $\sigma_{n+1}^{trl} = \boldsymbol{C}\,\varepsilon_{n+1}$ is called the trial stress state and $\boldsymbol{r}^{trl}$ is the gradient of $f\left(\sigma_{n+1}^{trl}, z_n\right)$. The calculation of the multiplier $\Delta\gamma > 0$ and the tensor function $\tilde{\boldsymbol{r}}$ (Equation (27)) is significantly dependent on the realization of σ_{n+1}^{trl}. A detailed description of the numerical implementation into the FEM is given in [38] for the elastic-plastic material with the yield criteria of Huber–Mises–Hencky (isotropy) and Hill (orthotropy). The most important step is to calculate the multiplier using the quadratic equation of the variable $\Delta\gamma$. This step is significantly different from the case of the Hill yield criterion.

Based on the consistency condition (23) in the algorithmic form:

$$f\left(\sigma_{n+1}^{T}, z_{n+1}\right) = \frac{1}{2}\sigma_{n+1}^{T} \boldsymbol{P}\, \sigma_{n+1} + \boldsymbol{p}^{T} \sigma_{n+1} - K\left(z_{n+1}\right) = 0, \tag{29}$$

we obtain after the substitution of (28) into (29) a following quadratic equation (indices $n+1$ suppressed):

$$A\,\Delta\gamma^{2} + B\,\Delta\gamma + C = 0, \tag{30}$$

where:

$$\begin{aligned} A &= \tfrac{1}{2}\left(\boldsymbol{C}\,\boldsymbol{r}_{n+1}^{trl}\right)^{T} \boldsymbol{P}\,\boldsymbol{C}\,\boldsymbol{r}_{n+1}^{trl} - \tfrac{\partial^{2} K}{\partial z^{2}}\left[\left(\boldsymbol{r}_{n+1}^{trl}\right)^{T} \boldsymbol{r}_{n+1}^{trl}\right]^{2}, \\ B &= -\left[\tfrac{1}{2}\sigma_{n+1}^{trl} \boldsymbol{P}\,\boldsymbol{C}\boldsymbol{r}_{n+1}^{trl} + \tfrac{1}{2}\left(\boldsymbol{C}\,\boldsymbol{r}_{n+1}^{trl}\right)^{T} \boldsymbol{P}\,\sigma_{n+1}^{trl} + \boldsymbol{p}^{T}\boldsymbol{C}\,\boldsymbol{r}_{n+1}^{trl} + \tfrac{\partial K}{\partial z}\left(\boldsymbol{r}_{n+1}^{trl}\right)^{T} \boldsymbol{r}_{n+1}^{trl}\right], \\ C &= f^{trl}\left(\sigma_{n+1}^{trl}, z_{n}\right). \end{aligned} \tag{31}$$

The solution of the Equation (30) is particularly simple if the hardening/softening function $K(z)$ is the second degree polynomial of the form as shown in the Figure 7. The last part of (30) may become then equal to the yield function of the trial state f^{trl}. It can be seen that the linearization of the Equation (30) does not lead to significant errors. Determining the plastic multiplier from the linear part of Equation (30) in the form $\Delta\gamma = -B/C$ does not lead to large errors, although it depends on the assumed strain increment step length.

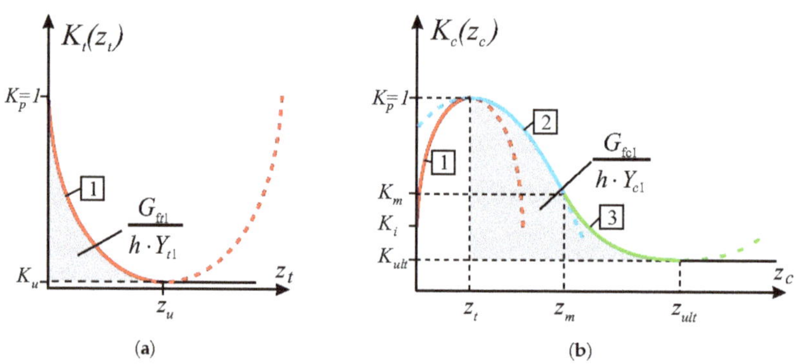

Figure 7. Degradation of material strength parameters curves adopted in the model: (**a**) tension regime, (**b**) compression regime.

A model based on two criteria complicates the procedure algorithm. If only one of the criteria is exceeded, e.g., the stresses are reduced to the exceeded surface according to the algorithm (28)–(30), as for a model based on only one criterion. It remains, to establish which criterion is actually exceeded.

A separate case is exceeding both conditions at the same time:

$$f_{1}^{trl} > 0 \quad \wedge \quad f_{2}^{trl} > 0.$$

This is the case where the point is outside the edge joining both surfaces and requires a different procedure. A linear combination of gradients has been used here (compare Koiter's law). The component of the plastic multiplier is calculated separately for the tensile criterion and for the compression criterion. Plastic strains and stress state components at time t_{n+1} can be determined on the basis of the formulas:

$$
\begin{gathered}
\varepsilon_{n+1}^{p} = \varepsilon_{n}^{p} + \Delta\gamma_1 \, r_{n+1}^{1,trl} + \Delta\gamma_2 \, r_{n+1}^{2,trl}, \\
\sigma_{n+1} = \sigma_{n+1}^{trl} - \Delta\gamma_1 \, C \, r_{n+1}^{1,trl} - \Delta\gamma_2 \, C \, r_{n+1}^{2,trl}, \\
C_{ep} = C - \frac{C \, r_{n+1}^{1,trl} \left(r_{n+1}^{1,trl}\right)^T C^T}{\left(r_{n+1}^{1,trl}\right)^T C \, r_{n+1}^{1,trl} + \partial K_1/\partial z_1 \left\| r_{n+1}^{1,trl} \right\|^2} - \frac{C \, r_{n+1}^{2,trl} \left(r_{n+1}^{2,trl}\right)^T C^T}{\left(r_{n+1}^{2,trl}\right)^T C \, r_{n+1}^{2,trl} + \partial K_2/\partial z_2 \left\| r_{n+1}^{2,trl} \right\|^2}.
\end{gathered} \tag{32}
$$

After similar transformations as before, the following system of quadratic equations is obtained.

$$
\begin{cases}
A_1(\Delta\gamma_1)^2 + B_1\Delta\gamma_1 + C_1 + D_1(\Delta\gamma_2)^2 + E_1\Delta\gamma_2 + F_1\Delta\gamma_1\Delta\gamma_2 = 0 \\
A_2(\Delta\gamma_2)^2 + B_2\Delta\gamma_2 + C_2 + D_2(\Delta\gamma_1)^2 + E_2\Delta\gamma_1 + F_2\Delta\gamma_1\Delta\gamma_2 = 0
\end{cases}, \tag{33}
$$

where some of the coefficients are analogous to the problem with one active surface and are marked with letters $A_\alpha, B_\alpha, C_\alpha$ $(\alpha = 1, 2)$:

$$
\begin{aligned}
A_\alpha &= \frac{1}{2}\left(C r_{n+1}^{\alpha,trl}\right)^T P_\alpha C r_{n+1}^{\alpha,trl} - K''_{\alpha,n}\left[\left(r_{n+1}^{\alpha,trl}\right)^T r_{n+1}^{\alpha,trl}\right]^2, \\
B_\alpha &= -\frac{1}{2}\sigma_{n+1}^{trl} P_\alpha C r_{n+1}^{\alpha,trl} - \frac{1}{2}\left(C r_{n+1}^{\alpha,trl}\right)^T P_\alpha \sigma_{n+1}^{trl} - (p_\alpha)^T C r_{n+1}^{\alpha,trl} - K'_{\alpha,n}\left(r_{n+1}^{\alpha,trl}\right)^T r_{n+1}^{\alpha,trl}, \\
C_\alpha &= f_\alpha^{trl}\left(\sigma_{n+1}^{trl}, z_n^\alpha\right).
\end{aligned} \tag{34}
$$

The remaining coefficients depend on the data related to both boundary surfaces and the formulas can be expressed as follows:

$$
\begin{aligned}
D_1 &= \frac{1}{2}\left(C r_{n+1}^{2,trl}\right)^T P_1 C r_{n+1}^{2,trl}, \qquad D_2 = \frac{1}{2}\left(C r_{n+1}^{1,trl}\right)^T P_2 C r_{n+1}^{1,trl}, \\
E_1 &= -\frac{1}{2}\sigma_{n+1}^{trl} P_1 C r_{n+1}^{2,trl} - \frac{1}{2}\left(C r_{n+1}^{2,trl}\right)^T P_1 \sigma_{n+1}^{trl} - p_1^T C r_{n+1}^{2,trl}, \\
E_2 &= -\frac{1}{2}\sigma_{n+1}^{trl} P_2 C r_{n+1}^{1,trl} - \frac{1}{2}\left(C r_{n+1}^{1,trl}\right)^T P_2 \sigma_{n+1}^{trl} - p_2^T C r_{n+1}^{1,trl}, \\
F_1 &= \left(C r_{n+1}^{1,trl}\right)^T P_1 C r_{n+1}^{2,trl}, \qquad F_2 = \left(C r_{n+1}^{2,trl}\right)^T P_2 C r_{n+1}^{1,trl}.
\end{aligned} \tag{35}
$$

According to the above numerical algorithm, the several models with the failure criterion of the type (18) has been coded in the programming language FORTRAN and next implemented into a commercial finite element code DIANA [39]. There are standard Newton–Raphson and Riks algorithms for the solving nonlinear equilibrium equations in the program. Because of this, implementation of the model into a finite element code is done at the integration point level by means of user-defined subroutine USRMAT. The subroutine lets the user specify a general nonlinear material behavior by updating the state variables over the equilibrium step $n \rightarrow n+1$ within a framework of an incremental–iterative algorithm of finite element method. Both the return-mapping algorithm allowing the stresses to be returned to the yield surface and a consistent tangent stiffness operator have been coded. The implementation of the model is presented both at the single element tests (see Figures 8–12) and at the structural level test (next section).

4.2. Results of the Single-Element Tests

Tests in the plane stress were performed in the homogeneous stress state and with one isoparametric continuum finite element (four-node, linear interpolation and Gaussian integration). The displacement-controlled load diagram is shown in Figure 8. The following material data are adopted: the moduli of elasticity in two directions of orthotropy: $E_1 = E_2 = 8$ [GPa], the shear modulus: $G_{12} = 3.478$ [GPa], the Poissonś coefficient: $\nu_{12} = 0.15$, that is, as for the isotropic material. The strength parameters associated with the initial tensile failure surface of the criterion are: $Y_{c1} = 17$, $Y_c2 = 17$, $Y_{t1} = 0.35$, $Y_{t2} = 0.25$, $Y_{tt} = 0.22$, $k_{12} = 0.296$ [MPa]. The strength parameters associated with the compressive surface are: $Y_{c1} = 8.5$, $Y_{c2} = 8.5$, $Y_{t1} = 8.5$, $Y_{t2} = 8.5$, $Y_{tt} = 8.5$, $k_{12} = 4.9$ [MPa]. The material parameters are typical for unreinforced walls in terms of the values and the proportion between them. The strength degradation curves are adopted with two internal scalar parameters: z_t for the tensile softening and z_c for the softening during compression (Figure 7). The curves are matched to set the fracture energies along the first orthotropic material axis as $G_{f1} = 54.0$ [J/m^2] during tension and as $G_{fc1} = 20.0$ [kJ/m^2] during compression.

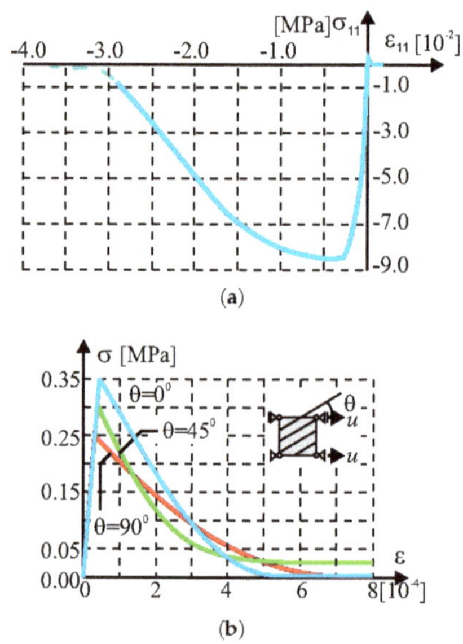

Figure 8. Single-element test results. Relationship $\sigma - \varepsilon$: **(a)** compression along 1(x) axis, **(b)** tensile softening in the direction rotated by an angle θ to the first orthotropy axis.

In the test with the homogeneous deformation field, all eight degrees of freedom in the four-node finite element were fixed to force the desired linear deformation. To compare the behavior of our model in the tests, the model of the Rankine–Hill criterion was used, which is standard in the DIANA system and dedicated to the analysis of masonry structures. The following parameters of the Rankine–Hill model are adopted: $Y_{c1} = 8.5Y$, $Y_{c2} = 8.5Y$, $\beta = -1.0$, $\gamma = 3.0$, $Y_{t1} = 0.35Y$, $Y_{t2} = 0.25Y$, $\alpha = 1.0$, where $Y = 1$ [MPa] and the tensile fracture energies $G_{fX} = 54.0$ [J/m^2], $G_{fY} = 18.0$ [J/m^2] and the compressive fracture energies $G_{fcX} = 20.0$ [kJ/m^2] and $G_{fcY} = 15.0$ [kJ/m^2]. The algorithm of the own model was also programmed in MATHEMATICA. As a result, it was possible to control the correctness of the algorithm and its implementation in both computing environments. Results of the tests are presented in Figures 9–12.

		Own Criterion MATHEMATICA	Own Criterion FEM DIANA	Criterion R-H FEM DIANA
$\varepsilon_{11} = 3.85 \times 10^{-5}$	σ_{11} [MPa]:	0.315090	0.315090	0.315090
	σ_{22} [MPa]:	0.0472634	0.0472634	0.0472634
$\varepsilon_{11} = 3.5 \times 10^{-3}$	σ_{11} [MPa]:	0.283159	0.283125	0.130216
	σ_{22} [MPa]:	−1.05447	−1.05355	−1.65648

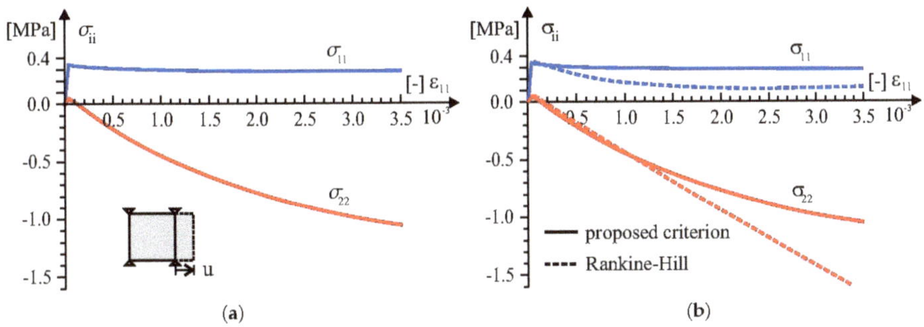

Figure 9. A homogeneous strain field $\varepsilon_{11} > 0$ and $\varepsilon_{22} = 0$ single element test results: (**a**) symbolic computation by MATHEMATICA [40], (**b**) finite element method (FEM) analysis in TNO DIANA [39].

		Own Criterion MATHEMATICA	Own Criterion FEM DIANA	Criterion R-H FEM DIANA
$\varepsilon_{11} = -3.2 \times 10^{-3}$	σ_{11} [MPa]:	−9.79701	−9.79701	−9.78427
	σ_{22} [MPa]:	−4.38814	−4.38809	−4.39655
$\varepsilon_{11} = -5.0 \times 10^{-2}$	σ_{11} [MPa]:	−0.967462	−0.967462	−0.92013
	σ_{22} [MPa]:	−0.626954	−0.626954	−0.491048

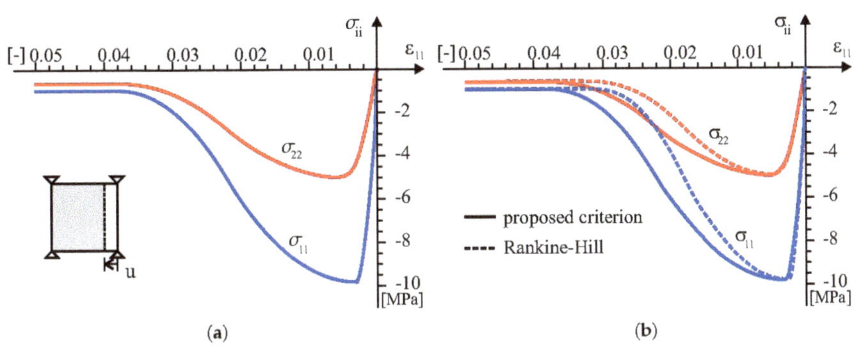

Figure 10. A homogeneous strain field $\varepsilon_{11} < 0$ and $\varepsilon_{22} = 0$ single element test results: (**a**) symbolic computation by MATHEMATICA [40], (**b**) FEM analysis in TNO DIANA [39].

		Own Criterion MATHEMATICA	Own Criterion FEM DIANA	Criterion R-H FEM DIANA
$\varepsilon_{11}^{Own} = 5.25 \times 10^{-5}$	σ_{11} [MPa] :	0.276645	0.276645	0.344186
$\varepsilon_{11}^{R-H} = 4.2 \times 10^{-5}$	σ_{22} [MPa] :	0.171389	0.171389	0.241186
$\varepsilon_{11} = 3.5 \times 10^{-3}$	σ_{11} [MPa] :	0.0592608	0.0592608	0.0144707
	σ_{22} [MPa] :	−0.047245	−0.047245	0.0002711

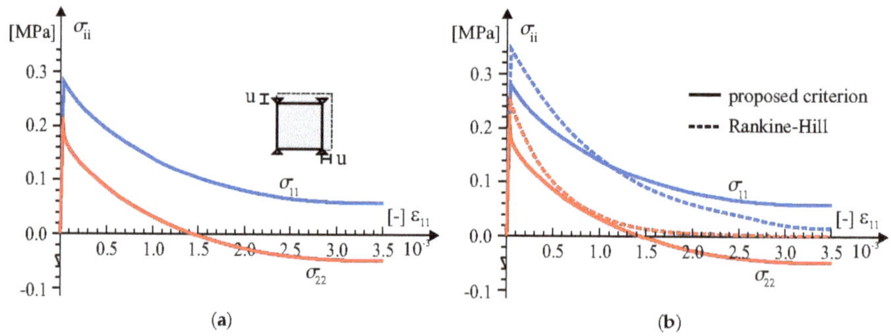

Figure 11. A strain field $\varepsilon_{11} > 0$ and $\varepsilon_{22} > 0$ single element test results: (**a**) symbolic computation by MATHEMATICA [40], (**b**) FEM analysis in TNO DIANA [39].

		Own Criterion MATHEMATICA	Own Criterion FEM DIANA	Criterion R-H FEM DIANA
$\varepsilon_{11}^{Own} = -2.4 \times 10^{-3}$	σ_{11} [MPa]:	−8.49975	−8.49975	−8.49992
$\varepsilon_{11}^{R-H} = -1.5 \times 10^{-3}$	σ_{22} [MPa]:	−8.49975	−8.49975	−8.49992
$\varepsilon_{11} = -5.0 \times 10^{-2}$	σ_{11} [MPa]:	−0.850000	−0.850000	−0.850075
	σ_{22} [MPa]:	−0.850000	−0.850000	−0.849990

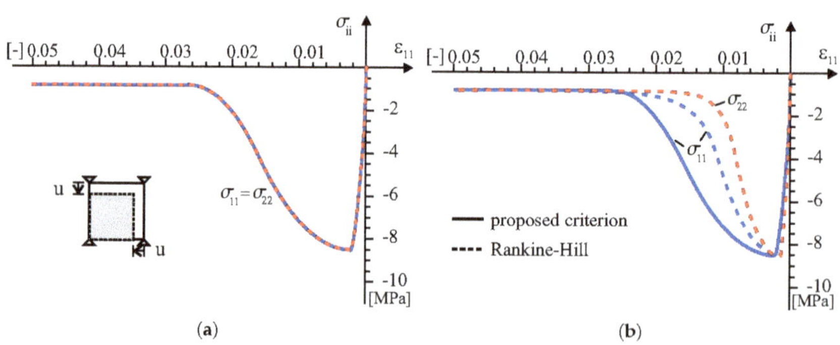

Figure 12. A strain field $\varepsilon_{11} < 0$ and $\varepsilon_{22} < 0$ single element test results: (**a**) symbolic computation by MATHEMATICA [40], (**b**) FEM analysis in TNO DIANA [39].

5. Tests at the Structural Level

Two examples of the proposed constitutive model are briefly presented here as the tests at the structural level. More information on the tests can be found in the works [41,42].

The first example is restricted to the numerical simulation of the load capacity tests of masonry structures in the plane stress state that were conducted experimentally in [43]. In laboratory tests, two similar wall fragments marked as J2G and J3G were tested, Figure 13. The masonry shear walls with an opening and the thickness of 10 mm were built from 18 courses of masonry cement-sand units 210 × 52 × 100 [mm]. The top and bottom courses were fully clamped in steel beams. The wall is shearing with the horizontal force F under displacement control. The top edge can move with a horizontal displacement (Figure 13). The wall was first vertically loaded through a steel beam to the value p = 0.3 kN/m^2 that remains constant through the subsequent loading steps of the horizontal force up to the maximum horizontal displacement of Δ = 16 mm. The beam movement was limited to the horizontal direction, i.e., the lower and upper edges of the wall were kept parallel. Figure 14 shows the forms of cracks obtained in the tests.

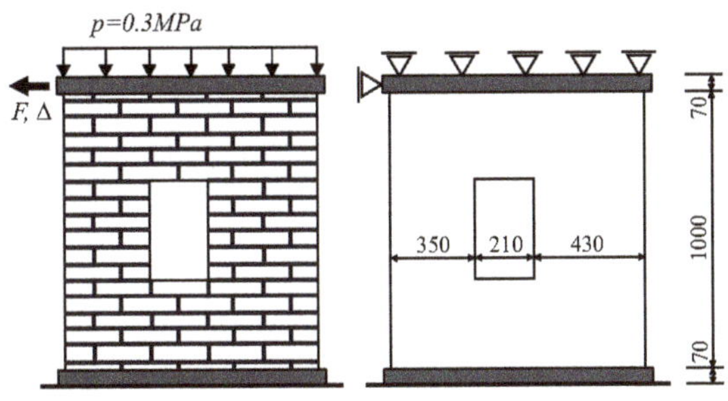

Figure 13. Walls tested by Raijmakekers and Vermeltfoort [43]: (**a**) load scheme for shear walls, (**b**) geometry and static sheme for numerical analysis.

Cracks appeared diagonally between the opening and the lower and upper corners of the wall. In addition, tensile cracks appeared on the outer vertical edges of the wall, on both sides of the opening at the top of the left pillar and at the bottom of the right pillar, and their development ran from the outside to the center of the wall. The failure mechanism resulting from the wall tests is shown schematically in the Figure 14. As can be seen, the kinematics of the system focuses on the movement of four blocks connected by hinges. Due to the development of the material crushing zone, also marked in the drawing, the mechanism will activate the compression criterion.

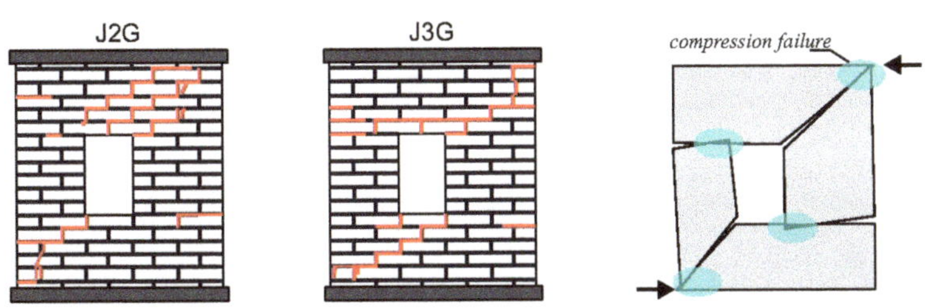

Figure 14. Experimental crack pattern at failure load [43] (**a**,**b**). (**c**) Mechanism of destruction.

For the numerical simulation of the wall failure using the model based on the own criterion, a 1989 mesh of four-node flat finite elements with an average side length of 23 [mm] was built. The values of material parameters used in the numerical simulation are shown in the Table 3. The values given in the table were adopted on the basis of work [44], where the data were adopted based on the results of experimental tests [43], and the supplementary parameters result from the homogenization procedures and were taken from work [16].

Table 3. Masonry shear wall. Material properties of the model.

Elastic Moduli						
	E_{11} [MPa]	E_{22} [MPa]	ν_{12} –	G_{12} [MPa]		
	7520	3960	0.09	1460		
Uniaxial, Biaxial and Shear Strengths in the Orthotropic Axes [MPa]						
	Y^{α}_{c1}	Y^{α}_{c2}	Y^{α}_{t1}	Y^{α}_{t2}	$Y_{\alpha\alpha}$	k_{12}
Compression $\alpha = c$	5.25	3.75	2.625	1.825	3.0	0.45
Tension $\alpha = t$	11.0	7.5	0.35	0.25	0.30	0.30
Fracture energies in $\mathrm{J/m^2}$		$G_{fc1} = 2350$		$G_{ft1} = 43.3$		

The comparison between numerical and experimental load–displacement diagrams, for wall J2G and J3G, is given in Figure 15. Apart from own calculation, results obtained by Pelá in [44] are also shown. Good agreement is found in the elastic range and satisfactory agreement in the inelastic range, although slightly worse in terms of the load capacity than the results obtained in [44]. The maximum divergence is around 17%. It is possible to better match the results of the experiment, however it requires additional calculations and time-consuming calibration of strength parameters and fracture energies. The behavior of the wall at the horizontal displacement $\Delta = 12$ mm is depicted in Figures 16 and 17 in terms of the maximum and minimum principal plastic strains, respectively. Both tensile and compressive failure zones are captured by the model. This indicates that the wall deformability and the general mechanism of its destruction are correctly reflected.

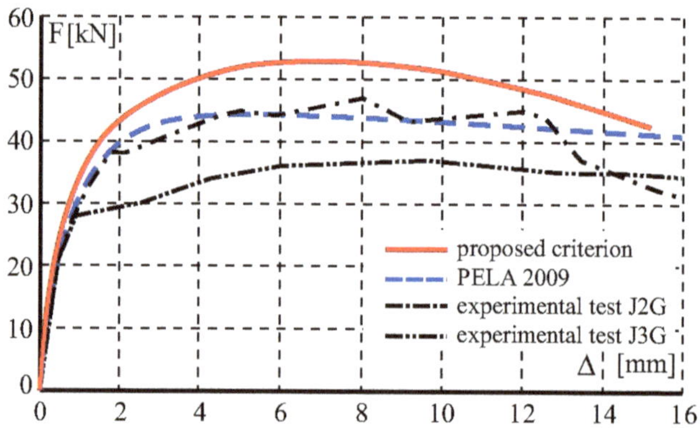

Figure 15. Horizontal load–horizontal displacement diagrams.

Figure 16 shows the principal plastic strains ε_1^p obtained from own simulations. They can be interpreted as the distribution of material failure zones as a result of exceeding the criterion in tension. These are also zones of development of cracks in the structure. Figure 16a shows the strain results in the form of maps plotted on a deformed finite element mesh. For a better effect, the deformation has been scaled 10 times.

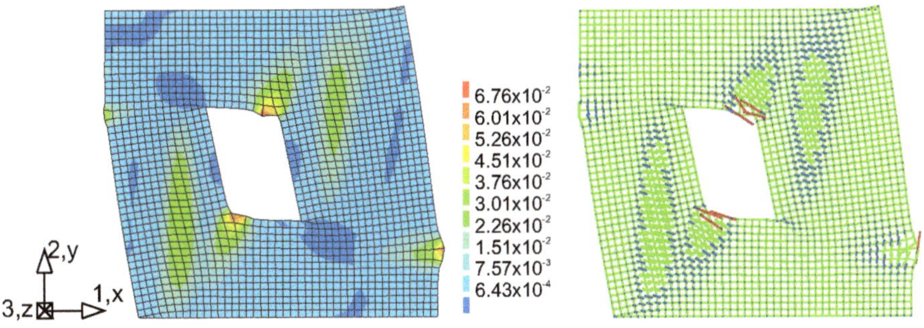

Figure 16. Maps of maximum principal plastic strain at $\Delta = 12$ mm: (**a**) crack (tensile) zones, (**b**) directions of the tensile strain.

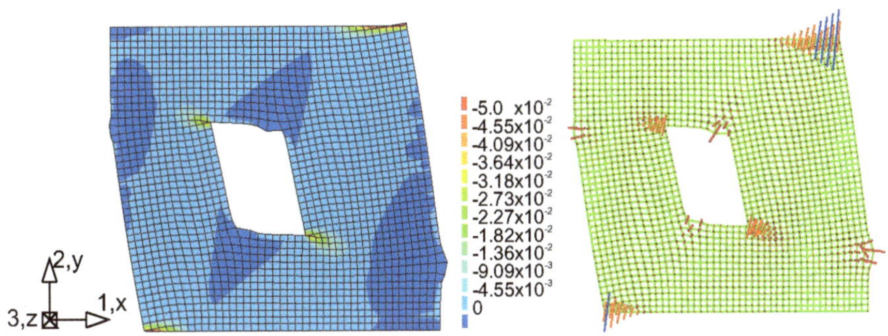

Figure 17. Maps of minimum principal plastic strain at $\Delta = 12$ mm: (**a**) compressive failure zones, (**b**) directions of the compressive strain.

In Figure 16b, the results in the vector form were superimposed on the mesh, wich allows to compare the length of the strain and the direction. Figure 17a shows the principal plastic deformations ε_3^p, which in Figure 17b are shown in vector form. Areas with large ε_3^p strains can be interpreted as zones of material failure due to possible material crushing.

In the second example, the behavior of the masonry infill wall that is built within a reinforced concrete frame is numerically simulated (Figure 18). The wall was experimentally tested in [45]. The frame and the wall were first subjected to the vertical loads $P_2 = 97.8$ kN and $P_3 = 48.9$ kN that remain constant through the subsequent loading steps of the horizontal force P_1 up to the failure.

For the numerical analyses 4-noded quadrilaterals, linear plane stress, and continuum elements are utilized. The subsequent loading steps of the horizontal force up to the maximum horizontal displacement $\Delta = 20$ mm were analyzed under a displacement control. The material properties of the model are given in Table 4. The comparison between numerical and experimental load–displacement diagrams is given in Figure 19. The low initial vertical load combined with the confinement provided by the reinforced concrete frame yields extremely ductile behavior.

In Figure 20, a crack zone is plotted as the map of the tensile principal plastic strain. A good agreement is found with respect to the calculated collapse load value. The model gives the response that is close to the experimental equilibrium path. At the ultimate stage, a well-defined failure mechanism is formed with a final diagonal shear band going from one corner of the specimen to the other.

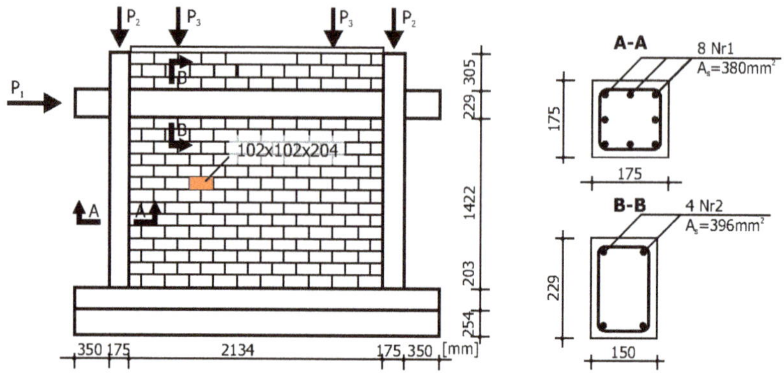

Figure 18. Geometry and loads for masonry infill wall tested in [45].

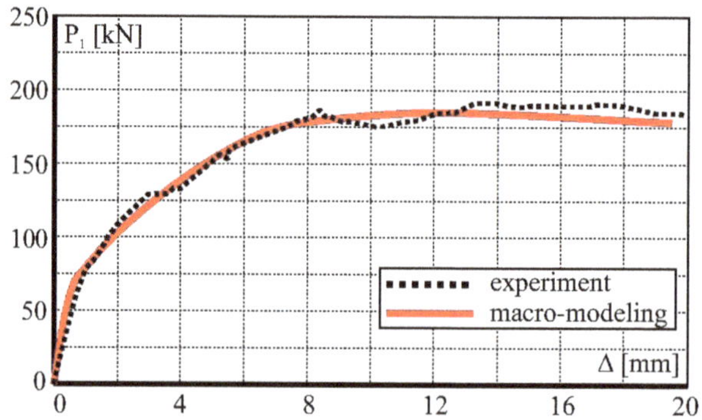

Figure 19. Comparison of load–displacement diagram.

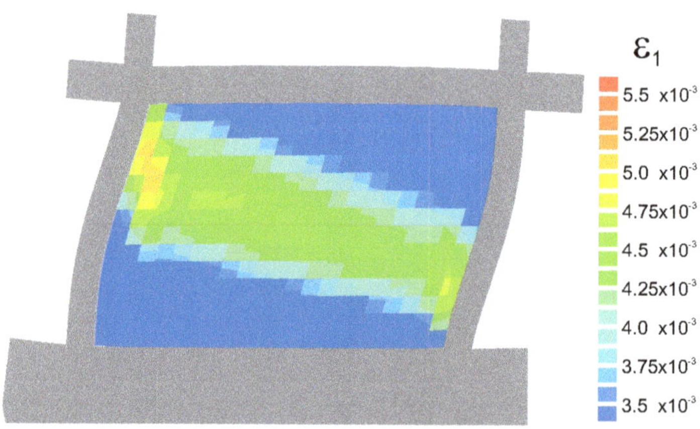

Figure 20. Failure mechanism obtained by numerical simulation as map of maximum principal plastic strain at $\Delta = 10$ mm.

Table 4. Characteristics of materials for orthotropic model of masonry.

Elastic Moduli						
	E_{11}, MPa	E_{22}, MPa	$\nu_{12} = \nu_{21}$	G_{12}		
	[MPa]	[MPa]	[-]	[MPa]		
	13,790	13,790	0.16	3480		
Characteristics of the Strength Surface in MPa						
	Y^{α}_{c1}	Y^{α}_{c2}	Y^{α}_{t1}	Y^{α}_{t2}	$Y_{\alpha\alpha}$	k^{α}_{12}
Compression ($\alpha = c$)	20.7	20.7	20.7	20.7	23.00	5.0
Tension ($\alpha = c$)	40.0	40.0	1.38	1.38	1.2	0.5
Fracture energies in $\mathrm{J/m^2}$		$G_{fc1} = 350$		$G_{ft1} = 16.7$		

The precise localization of cracks can also be achieved using models based on a smeared crack approach [46,47], fracture-based models with advanced crack tracking techniques [44] or by using micro-modeling [16]. To the evaluation of historic buildings for dynamic actions methods based on ultimate load-bearing capacity are used. Especially in combination with the monitoring based on operational modal analysis (OMA) techniques, it is becoming a popular practice [48]. An alternative may also be the relatively young topic of multi-scale modeling [49] or to describe diffuse cracks a gradient-enhanced damage model based on nonlocal displacements and the extended finite element method (XFEM) for sharp cracks [50,51]. However, if we compare the numerical result presented here with the failure mechanisms shown in Figures 16 and 20, we find a good agreement and confirm the ability of our own model to correctly reflect the behavior of the structure.

6. Conclusions

The mechanical behavior of the continuum material model can be described using constitutive relations based on the theory of tensor functions. This theory, together with the theorems on the representation of tensor functions, constitute an effective mathematical tool for the formulation of constitutive relations of the orthotropic material. The invariance requirements of the isotropy of space and the orthotropic symmetry of materials are easily considered, so that the orientation of the material in space does not affect its constitutive relation. It is possible to obtain the analytical transparency as well as to maintain the required universalism of constitutive equations, see, e.g., [30]. The composite orthotropic failure criterion of the proposed model is constructed from two square scalar functions that are dependent on three orthotropic invariants of the plane stress tensor. In general, the criterion needs to have the twelve material parameters to define the failure surface. However, in practice, only seven of them are used that are obtained from the appropriate uniaxial, biaxial and shear strength tests. The criterion is an example of the orthotropic failure criteria, which can be treated as a generalization of the well-known Hoffman failure criterion that is often used for brittle orthotropic materials in which compressive strengths and tensile strengths are significantly different.

The numerical tests confirm correctness of the implementation and the ability of the models to reproduce failure modes in the structural tests in certain situations. They also show that it is possible to incorporate strain softening into the proposed class of models with a single yield surface. At present, the implementation of the other models within the framework of multisurface plasticity is being tested. Although for the multisurface plasticity the intersection of the different yield surfaces defines corners that require special attention in a numerical algorithm, it has the advantage of engaging different hardening laws for each surface, which might be more physically realistic due to the distinction of tension and compression regimes. The choice of a hardening parameter is not crucial because the available experimental data are scarce.

The commercial version of Diana with the so-called user-supplied subroutine mechanism can be used as the development environment for computational materials research. We have used the user-supplied subroutine USRMAT to implement new material models for continuum spatial and plane stress elements. The new models can be applied to a variety of structural problems, to single element tests but also to simulate physical experiments, using different elements types and using standard features of the program such as advanced solution procedures, for instance indirect displacement control with full Newton–Raphson. Also, the use of the post-processing capabilities of the program is an advantage, although the post-processing of the user-defined status variables might be more user-friendly.

Author Contributions: Conceptualization, P.B. and L.M.; methodology, L.M.; software, P.B.; validation, L.M.; visualization, P.B.; supervision, L.M.; writing—original draft preparation, P.B.; writing—review and editing, L.M.; All authors have read and agreed to the published version of the manuscript.

Funding: This research received no external funding.

Conflicts of Interest: The authors declare no conflict of interest.

References

1. Lourenço, P. Recent advances in Masonry modelling: Micromodelling and homogenisation. *Multiscale Model. Solid Mech. Comput. Approaches* **2009**. [CrossRef]
2. Giamundo, V.; Sarhosis, V.; Lignola, G.; Sheng, Y.; Manfredi, G. Evaluation of different computational modelling strategies for the analysis of low strength masonry structures. *Eng. Struct.* **2014**, *73*, 160–169. [CrossRef]
3. Lemos, J. Discrete Element Modeling of the Seismic Behavior of Masonry Construction. *Buildings* **2019**, *9*, 43. [CrossRef]
4. Asteris, P.; Moropoulou, A.; Skentou, A.; Apostolopoulou, M.; Mohebkhah, A.; Cavaleri, L.; Rodrigues, H.; Varum, H. Stochastic Vulnerability Assessment of Masonry Structures: Concepts, Modeling and Restoration Aspects. *Appl. Sci.* **2019**, *9*, 243. [CrossRef]
5. Yokel, F.; Fattal, S. Failure hypothesis for masonry shear walls. *J. Struct. Div.* **1976**, *102*, 515–532.
6. Mann, W.; Muller, H. Failure of shear-stressed masonry—An enlarged theory, tests and application to shear walls. *Proc. Br. Ceram. Soc.* **1982**, *30*, 223–235.
7. Page, A. The biaxial compressive strength of brick masonry. *Proc. Inst. Civ. Eng.* **1981**, *71*, 893–906. [CrossRef]
8. Page, A. The strength of brick masonry under biaxial tension-compression. *Int. J. Mason. Constr.* **1983**, *3*, 26–31.
9. Dhanasekar, M.; Page, A.; Kleeman, P. The failure of brick masonry under biaxial stresses. *Proc. Inst. Civ. Eng.* **1985**, *79*, 295–313. [CrossRef]
10. Ganz, H.; Thürlimann, B. Strength of brick walls under normal force and shear. In Proceedings of the 8th International Symposium on Load Bearing Brickwork, London, UK, November 1983.
11. Seim, W. Nümerische Modellierung Des Anisotropen Versagens Zweiachsig Beanspruchter Mauerwerksscheiben. Ph.D. Thesis, Universität Karlsruhe, Karlsruhe, Germany, 1995.
12. Tsai, S.; Wu, E. A general theory of strength for anisotropic materials. *J. Compos. Mater.* **1971**, *5*, 58–80. [CrossRef]
13. Syrmakezis, C.; Asteris, P. Masonry failure criterion under biaxial stress state. *J. Mater. Civ. Eng.* **2001**, *13*, 58–64. [CrossRef]
14. Berto, L.; Saetta, A.; Scotta, R.; Vitaliani, R. An orthotropic damage model for masonry structures. *Int. J. Numer. Methods Eng.* **2002**, *55*, 127–157. [CrossRef]
15. Hoffman, O. The brittle strength of orthotropic materials. *J. Compos. Mater.* **1967**, *1*, 200–206. [CrossRef]
16. Lourenço, P. Computational Strategies for Masonry Structures. Ph.D. Thesis, Delf University of Technology, Delft, The Netherlands, 1996.
17. Lourenço, P.; Borst, R.; Rots, J. Plane stress softening plasticity model for orthotropic materials. *Int. J. Numer. Methods Eng.* **1997**, *40*, 4033–4057. [CrossRef]
18. Małyszko, L. *Modelowanie Zniszczenia w Konstrukcjach Murowych z Uwzględnieniem Anizotropii (Failure Modeling in Masonry Structures Taking into Account Anisotropy)*; Wydawnictwo UWM: Olsztyn, Poland, 2005.
19. Geniev, G.A.; Kurbatov, A.; Samedov, F. *Voprosy Procznosti I Plasticznosti Anizotropnyh Materialov*; Interbuk: Moskva, Russia, 1993.

20. Lourenço, P. *An Anisotropic Macro-Model for Masonry Plates and Shells: Implementation and Validation;* Delft University of Technology: Delft, The Netherlands, 1997.
21. Jasiński, R. Research and Modeling of Masonry Shear Walls. Ph.D. Thesis, Silesian University of Technology, Gliwice, Poland, 2017.
22. Gambarotta, L.; Lagomarsino, S. Damage models for the seismic response of brick masonry shear walls. Part II: The continuum model and its applications. *Earthq. Eng. Struct.* **1997**, *D 26*, 441–462. [CrossRef]
23. Massart, T.; Peerlings, R.; Geers, M. Mesoscopic modeling of failure and damage-induced anisotropy in brick masonry. *Eur. J. Mech. Solids* **2004**, *23*, 719–735. [CrossRef]
24. Milani, G.; Lourenço, P.; Tralli, A. Homogenised limit analysis of masonry walls. Part I: Failure Surfaces. *Comput. Struct.* **2006**, *84*, 166–180. [CrossRef]
25. Noii, N.; Aldakheel, F.; Wick, T.; Wriggers, P. An adaptive global-local approach for phase-field modeling of anisotropic brittle fracture. *Comput. Methods Appl. Mech. Eng.* **2020**, *361*, 112744. [CrossRef]
26. Noii, N.; Khodadadian, A.; Wick, T. Bayesian Inversion for Anisotropic Hydraulic Phase-Field Fracture. *arXiv* **2020**, arXiv:2007.16038.
27. Khodadadian, A.; Noii, N.; Parvizi, M.; Abbaszadeh, M.; Wick, T.; Heitzinger, C. A Bayesian estimation method for variational phase-field fracture problems. *Comput. Mech.* **2020**. [CrossRef]
28. Pelá, L.; Cervera, M.; Roca, P. Continuum damage model for orthotropic materials: Application to masonry. *Comput. Methods Appl. Mech. Eng.* **2011**, *200*, 917–930. [CrossRef]
29. Małyszko, L.; Jemioło, S.; Bilko, P. On the use of the Hoffman criterion in the continuum structural model for masonry panels. In Proceedings of the 7th International Conference AMCM 2011, Kraków, Poland, 13–15 June 2011.
30. Małyszko, L.; Jemioło, S.; Gajewski, M.; Bilko, P. FEM and Constitutive Modelling in Failure Analyses of Masonry Structures. Orthotropic Failure Criteria. *WTA-Schriftenreihe* **2009**, *33*, 371–394.
31. Wagner, W.; Klinkel, S.; Gruttmann, F. Elastic and plastic analysis of thin-walled structures using improved hexahedral elements. *Comput. Struct.* **2002**, *80*, 857–869. [CrossRef]
32. Rolfes, R.; Vogler, M.; Czichon, S.; Ernst, G. Exploiting the structural reserve of textile composite structures by progressive failure analysis using a new orthotropic failure criterion. *Comput. Struct.* **2011**, *89*, 1214–1223. [CrossRef]
33. Boehler, J. *Applications of Tensor Functions in Solid Mechanics*; CISM International Centre for Mechanical Sciences, No. 292; Springer: Berlin/Heidelberg, Germany, 1987.
34. Jemioło, S.; Małyszko, L. New failure criteria for orthotropic materials. In *Monograph Computer Systems Aided Science and Engineering Work in Transport, Mechanics and Electrical Engineering*; Kazimierz Pułaski Technical University of Radom: Radom, Poland, 2008; pp. 223–228.
35. Jemioło, S.; Gajewski, M.; Małyszko, L.; Bilko, P. Orthotropic elastic-plastic model of masonry for numerical analysis in spatial stress state. In *Lightweight Structures in Civil Engineering, Contemporary Problems, 15th International Seminar of IASS PC*; Micro-Publisher C-P Jan B. Obrebski: Warsaw, Poland, 2009; Volume XV, pp. 64–67.
36. Ganz, H.; Thürlimann, B. *Tests on the Biaxial Strength of Masonry*; Report No. 7502-3; Institute of Structural Engineering, ETH Zurich: Zurich, Switzerland, 1982. (In German)
37. Lurati, F.; Graf, H.; Thürlimann, B. *Experimental Determination of the Strength Parameters of Concrete Masonry*; Report No. 8401-2; Institute of Structural Engineering, ETH Zurich: Zurich, Switzerland, 1990. (In German)
38. Simo, J.; Hughes, T. *Computational Inelasticity;* Interdisciplinary Applied Mathematics; Springer: Berlin/Heidelberg, Germany, 1998.
39. DIANA. *User's Manual Release 9.3*; TNO DIANA BV: Delft, The Netherlands, 2009.
40. Wolfram. *Mathematica*; Version 8.0; Wolfram Research Inc.: Champaign, IL, USA, 2010.
41. Małyszko, L.; Bilko, P. Symulacje numeryczne wyczerpania nośności ramy żelbetowej wypełnionej murem (Numerical simulations of bearnig capacity of reinforced concrete frame with infill masonry). *Inżynieria I Bud.* **2016**, *6*, 316–320.
42. Klovanych, S.; Małyszko, L. *Plasticity in Structural Engineering;* Wydawnictwo Uniwersytetu Warmińsko-Mazurskiego w Olsztynie: Olsztynie, Poland, 2019.
43. Raijmakers, T.M.J. *Deformation Controlled Tests in Masonry Shear Walls: Report B-92-1156;* Technical Report; TNO-Bouw: Delft, The Netherlands, 1992.

44. Pelá, L. Continuum Damage Model for Nonlinear Analysis of Masonry Structures. Ph.D. Thesis, Universitat degli studi di Ferrara, Ferrara, Italy, 2009.
45. Mehrabi, A.B.; Shing, P.B.; Schuller, M.P.; Noland, J.L. *Performance of Masonry-Infilled R/C Frames under in-Plane Lateral Loads. CU/SR-94/6*; Technical Report; Department of Civil, Environmental and Architectural Engineerlng, Colorado University: Boulder, CO, USA, 1994.
46. Gattulli, V.; Lofrano, E.; Paolone, A.; Pirolli, G. Performances of FRP reinforcements on masonry buildings evaluated by fragility curves. *Comput. Struct.* **2017**, *190*, 150–161. [CrossRef]
47. Ungermann, J.; Adam, V. Fictitious Rough Crack Model (FRCM): A Smeared Crack Modelling Approach to Account for Aggregate Interlock and Mixed Mode Fracture of Plain Concrete. *Materials* **2020**, *13*, 2774. [CrossRef]
48. Milani, G.; Shehu, R.; Valente, M. *Simple Limit Analysis Approach for the Optimal Strengthening of Existing Masonry Towers*; AIP Publishing LLC: College Park, MD, USA, 2018; Volume 1978, p. 450007. [CrossRef]
49. Qian, Z.; Schlangen, E.; Ye, G.; Breugel, K. Modeling Framework for Fracture in Multiscale Cement-Based Material Structures. *Materials* **2017**, *10*, 587. [CrossRef]
50. Meschke, G.; Dumstorff, P. Energy-based modeling of cohesive and cohesionless cracks via X-FEM. *Comput. Methods Appl. Mech. Eng.* **2007**, *196*, 2338–2357. [CrossRef]
51. Tamayo-Mas, E.; Feliu-Fabá, J.; Casado-Antolin, M.; Rodríguez-Ferran, A. A continuous-discontinuous model for crack branching. *Int. J. Numer. Methods Eng.* **2019**, *120*, 86–104. [CrossRef]

materials

MDPI

Article

Numerical Analysis of Bowing Phenomenon Due to Thermal Stresses in Marble Slabs

William Hideki Ito [1,2,*], Anna Maria Ferrero [1] and Paulo Ivo Braga de Queiroz [2]

1 Department of Earth Sciences, University of Turin, 35 Valperga Caluso Street, 10125 Turin, Italy; anna.ferrero@unito.it

2 Division of Civil Engineering, Aeronautics Institute of Technology; 50 Marechal Eduardo Gomes Square, 12228-900 São José dos Campos, Brazil; pi@ita.br

* Correspondence: williamhideki.ito@unito.it

Received: 26 August 2020; Accepted: 27 September 2020; Published: 30 September 2020

Abstract: Bowing is a pathology known by the deformation experienced in some external covering systems in ornamental stones, especially in marble, and thermal action is one of the key factors that lead to this degradation. Previous studies presented remarkable contributions about the mechanical behavior of bowing but they were based on classical beam's theory and improper assumptions might mislead the evaluation of internal stresses. This study proposes to evaluate internal stresses in bowing due to thermal loading considering the true deformed shape in continuum media. Finite displacement concepts are proposed to calculate stress-strain relationship and comparison with linear elastic theory is also addressed. Internal stresses not predictable in the Euler-Bernoulli beam were found in parametric analyses. Moreover, the numerical analysis accomplished in this paper indicates that transient heat flux should induce higher stresses than just considering higher gradients of temperature in steady flux which could explain the larger decohesion through width in bowing tests.

Keywords: geometric nonlinearity; FEM; thermoelasticity; bowing; transient heat flux

1. Introduction

Nonlinear mechanics theory has once been strictly applied in special problems in engineering, such as aeronautical and mechanical fields, due to higher computational effort and accuracy required in those analyses. Other areas have used small strain approach because it could well represent the great majority of problems adopting simplified hypotheses. However, the advances in technology have allowed widespread uses of nonlinear models in some areas where small strain was predominant, such as geotechnical and rock engineering, where assumption of small deformations, in deep underground excavation or landslides, might not be respected anymore [1–4].

In accordance with this trend it is proposed in this work to evaluate the effect of geometric nonlinearity and internal stress development in a specific problem in rock mechanics known as bowing. It can be defined as a pathology that affects some kind of marbles and causes permanent deformation due to weathering of these ornamental stones when used in external environment. In ventilated façade the deformation of marble slabs might reach large displacement easily noticed visually (see [5,6]) and similar decay might be responsible to deteriorate some sculptures in marble.

Thermal load due to sunlight exposure was found as an important factor to contribute to this degradation and previous studies of bowing in marble were mainly focused to understand the micromechanical phenomenon trying to simulate the intergranular decohesion due to anisotropic thermal expansion of calcite grains. Royer-Carfagni [7] presented an initial discussion about induced shear stresses by thermal actions in a chessboard like mineralogical arrangement. After that, Weiss et al. [8] improved that analysis using finite element analysis (FEA) to evaluate the stress distribution in microscale using the contours of calcite grains from a thin section image. One of the first attempts to represent the macrophenomenon was proposed by Ferrero et al. [9], improved later by Ferrero et al. [10]. These works presented models to predict the macromechanical behavior through linear elastic fracture mechanics considering the Euller–Bernoulli beam theory.

Beam's theories, such those used in previous studies, presume deformed shape which will fit to equilibrium equations. For instance, in Euler-Bernoulli beam the orthogonality of planes in relation to neutral axis in reference and deformed configuration are assumed, i.e., according to this theory the material is infinitely rigid to distortion by shear stresses. Linear and third order approaches considering shear strain were proposed by Timoshenko and Reddy [11] to consider this effect in thick beams; nonetheless, deformation is still imposed by these hypotheses.

Since the thermal gradient in a ventilated façade is relatively small to generate great internal stresses if were compared to those in thermal machines (turbines, diesel engines, etc.) it is proposed to evaluate the problem considering a more refined solution. A different approach could be solving the differential equations of equilibrium respecting the constitutive relationship and compatibility, which means that the equilibrium in deformed shape will occur according to the displacement field that minimizes the total potential energy, just as predicted in variational theorems. Advantages over previous studies are based on the true deformed shape since no restriction to deformation is previously imposed.

Inexact solution considering linear elasticity could also be reviewed to get more accurate solution; hence it is proposed to evaluate the problem via FEA considering geometric nonlinearity. Different ways could be chosen to approach the problem, such as, Cosserat continuum, true (logarithmic) strain or corotational formulation [12,13]. Herein will be considered the updated Lagrangian formulation using polar decomposition to overcome the approximations assumed in linear elastic theory. Nonlinear concepts are applied to evaluate transient thermal loads in marble slabs and a MATLAB code was implemented to compute the development of internal stresses via finite element method (FEM), including pre and post processing analyses. The formulation herein proposed presents relevant contribution not only to the understanding of bowing in marble slabs but also in other fields, for instance, the influence of nonlinear thermoelasticity in thin membranes [14,15] since the consideration of thermal stresses at micro and nanoscale, respectively, is important to understand the behavior of these materials.

2. Review of Bowing

Despite the fact that degradation of ornamental stones has long been studied in cultural heritage field mainly focused on preservation of monuments, historical heritages and statues [16–20], bowing was identified just few decades ago and has degraded some specific kinds of ornamental stones in short periods. Several types of marbles and limestones from different countries were affected by this degradation process, although it affects the most calcitic marble. White Carrara marble is an example of ornamental stone affected and the one more frequent in literature about this subject. It comes from the Alpi-Apuane complex formed by metamorphic processes in limestone deposits and can be considered a monomineralic rock composed mainly by calcite.

Many efforts were made to understand this complex phenomenon but still remains some open questions about it. The major research projects about this theme were MARA [21] and TEAM [22] which produced good insight into this problem. Some hypotheses proposed by Winkler [23] were verified in laboratory tests, such as the influence of thermal anisotropy and the microstructure, while other aspects are not yet quite understood.

One of the most known cases was verified at the Finland Hall's façade in which bowing affected the external covering system twice. According to Gelk et al. [24], the first time the panels were composed by white Carrara marble fixed at vertical joints by dowels and the largest ones were 140 cm length and 3 cm thickness. Just some years later the deflection of bowed panels reached up to 5 cm inward relative to the façade (concave). In the renewal façade, though, bowing occurred outward (convex). Similar problems were also verified in other buildings spread in Europe and North America [22].

Siegesmund et al. [6] verified correlation between height of the buildings and deflection of bowed slabs, even if it were not possible to apply for general cases.

Thermal anisotropy of calcite grains has been considered as one of the major causes responsible to contribute to this degradation since that mineral is the primary component of limestones and marbles. Calcite belongs to the hexagonal crystal family and the thermal expansion coefficients aligned with these optical axes present antagonist behavior. When a calcite grain is submitted to heating load it tends to expand in C-axis direction, but contracts in A-axis direction, which generates internal stresses during heating-cooling cycles.

Gelk et al. [25] verified that cyclic thermal loading could cause some deformation on marble specimens, nonetheless, it is extremely influenced by humidity. Plastic deformation rate is larger at the first cycles and decreases. In dry environments the residual deformation stabilizes after some heating-cooling cycles. On the other hand, in those marble specimens tested in wet environment, the residual deformation could increase continuously. Similar results were found by Koch and Siegesmund [26], Rodriguez-Gordillo and Saez-Perez [27], Luque et al. [28] and others. Bellopede et al. [29] verified that artificial weathering could be used to represent the natural one in accelerated way. These authors also verified that the decohesion of calcite grains increases from the less to the more weathered faces, which agrees with deformed shapes of weathered marbles.

Correlation between microstructure and bowing were also found in previous studies. Regular structures (granoblastic texture) tend to bow more than irregular (xenoblastic) ones. Influence of locked-in stresses due to metamorphic process, exploitation procedures, external forces applied by supporting anchorage systems can also be cited as contribution factors.

3. Fundamentals

Typical problems in rock engineering are usually dealt with engineering strain because rocks are assumed as brittle materials and small deformations are expected previous failure. Engineering strain, or small strain, assumes some linearities to approach stress-strain relationship, however, cases in which deformation becomes non infinitesimal are not so uncommon. Limitation previously imposed by small strain might lead to inexact evaluation when the hypothesis of infinitesimal deformation no longer occurs, or refined approach is required.

Even considering the small gradient of temperature between internal and external faces in ventilated façades studies have shown the importance of these induced stresses in bowing. For this reason, it is proposed to study the problem using nonlinear concepts, which take into account the continuous deformation of a body and due to corrections at each step even large deformations can be assessed.

3.1. Finite Displacement Concepts

In continuum mechanics the deformation gradient ($\boldsymbol{F}$) is used to express the current configuration in relation to a reference state. There are several ways to represent this tensor and a useful one to understand it could be in terms of principal stretches and principal directions as presented in Equation (1), in which λ represents the stretch in each principal direction. Due to polar decomposition it can be decomposed as the product of a pure rotational ($\boldsymbol{R}$) and a pure stretch tensor ($\boldsymbol{U}$ or $\boldsymbol{V}$). Right polar decomposition considers that stretch of fibers occurs first and then rotation tensor is applied; in the left polar decomposition the order of deformation is changed.

$$\boldsymbol{F} = \sum_{i=1}^{3} \lambda_i \boldsymbol{e}_{i,x} \otimes \boldsymbol{e}_{i,X} \tag{1}$$

Green–Lagrange strain is defined as the quadratic changing of a fiber prior to its rotation and can be expressed in terms of deformation gradient as given by Equation (2). Since the stretches are rotated, it can be used in problems of finite deformation without the approximation assumed in small strain theory [30]. Substituting $\boldsymbol{F}$ by the product of right polar decomposition in Equation (2) one can get to Equation (3) which proves that rotation does not affect this strain measure.

$$2\,\epsilon = \boldsymbol{F}^T\boldsymbol{F} - \boldsymbol{I} \tag{2}$$

$$2\,\epsilon = (\boldsymbol{RU})^T\boldsymbol{RU} - \boldsymbol{I} = \boldsymbol{U}^T\left(\boldsymbol{R}^T\boldsymbol{R}\right)\boldsymbol{U} - \boldsymbol{I} = \boldsymbol{U}^2 - \boldsymbol{I} \tag{3}$$

The energetic conjugate of Green strain is the 2nd Piola–Kirchhoff stress tensor. It does not have direct physical interpretation such as Cauchy stresses because it considers stresses acting on the reference configuration rotated to the current state, but it can be related with real stress as given in Equation (4) [30]:

$$\boldsymbol{S} = |\boldsymbol{F}|\,\boldsymbol{F}^{-1}\,\sigma\,\boldsymbol{F}^{-T} \tag{4}$$

Since these stress-strain measures are energetic conjugate pairs, the product of these tensors produces a scalar quantity, energy per volume, such as the integration using Cauchy stress and real strain tensors, as indicated by Equation (5). It must be pointed out that the left-hand side is integrated over the reference configuration, i.e., known shape, and the other is related with the deformed shape.

$$\int_{\Omega_0} \boldsymbol{S}\;\delta\epsilon d\Omega_0 = \int_{\Omega} \sigma\;\delta\varepsilon d\Omega \tag{5}$$

3.2. Constitutive Model

As described in the previous section, the correlation between deformation due to weathering and the influence of some factors which induce bowing are known, but the real mechanisms are not yet fully understood. Internal stresses developed by thermal gradient have shown as an important factor acting on bowing, thus, it will be the object of this study. Since the purpose of this work remains on the evaluation of induced strains it is proposed a 2-D solid formulation in plane strain state with homogeneous characteristics.

According to theory of elasticity no stress should be induced in a continuous homogeneous isotropic material subject to uniform temperature increasing. Nonetheless, gradient of temperature occurring in the inner part of a body would induce stress due to differential strain [31–33]. In such analysis the total strain could be decomposed in thermal and mechanical strain, in which the last one would be responsible for induce internal stresses. Then, the total strain is given by:

$$\epsilon = \epsilon_\theta + \epsilon_{mec} \tag{6}$$

Assuming St. Venant-Kirchhoff nonlinear material, internal stresses in bowing could be calculated via the constitutive stress-strain relationship written in Equation (7):

$$\boldsymbol{S} = \boldsymbol{C}\,\epsilon_{mec} = \boldsymbol{C}\,(\epsilon - \epsilon_\theta) \tag{7}$$

where

$$\boldsymbol{C} = \mathrm{A}\begin{pmatrix} 1-\nu' & \nu & 0 \\ \nu' & 1-\nu' & 0 \\ 0 & 0 & \frac{1}{2}-\nu' \end{pmatrix},\ \mathrm{A} = \frac{\mathrm{E}'}{(1-2\nu)(1+\nu)},\ \mathrm{E}' = \frac{\mathrm{E}}{1-\nu^2}\ \text{and}\ \nu' = \frac{\nu}{1-\nu}$$

For sake of simplicity, one considers a coarse discretization of a slab laid on the ground subject to sunlight exposure on the external face, as illustrated in Figure 1. In this example the deformation will be limited only to y direction and uniform increase of temperature inside each element is assumed. According to classical elastic theory, for a generic material with positive thermal expansion coefficient, the right-hand side should present larger expansion than the other. However, due to compatibility, the displacements of pairs 3A–3B and 4A–4B must coincide, therefore, shear stresses on this interface should induce compressive stress at the upper element and tensile stresses at the lower. It is clear that the shear stress along the interface must not be constant once it should be zero at the boundary (nodes 3 and 4) and at the symmetry plane. The same concept can be applied to a more refined mesh and to other directions.

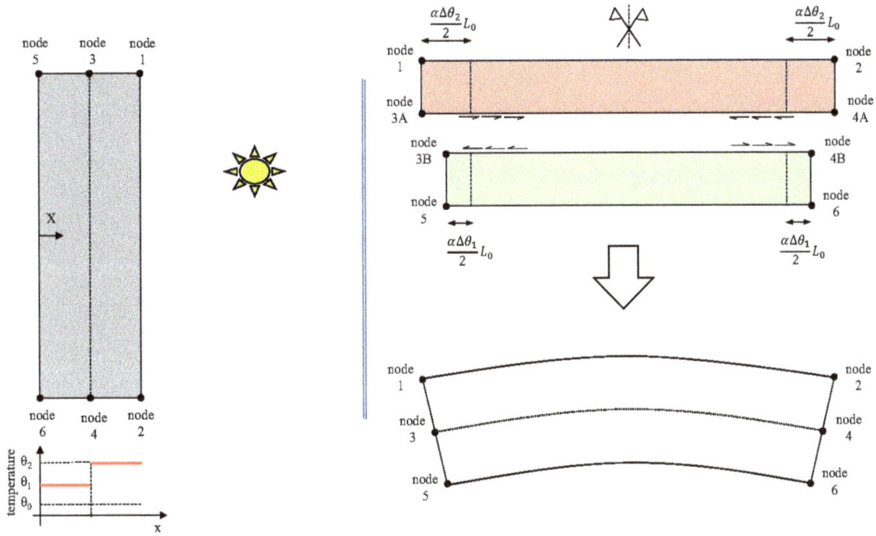

Figure 1. Idealization of the deformation due to thermal loading in continuum media.

In continuum media the stress-strain relationship could be solved for a displacement field which satisfies the equilibrium, compatibility and constitutive relationship by the principle of virtual work. Considering that only mechanical strain should induce stresses in the body, the variation of total potential energy can be described as presented in Equation (8), and the equivalent internal and external works as follow in Equations (9) and (10), respectively.

$$\delta\Pi = \int_{\Omega_0} \delta\epsilon_v{}^T C\,(\epsilon - \epsilon_\theta) d\Omega_0 - \int_{\Omega_0} \delta u_v^T f^B d\Omega_0 - \int_{\Gamma_0} \delta u_v^T f^S d\Gamma_0 \quad (8)$$

$$\delta W_{int} = \int_{\Omega_0} \delta\epsilon_v{}^T \boldsymbol{C}\, \epsilon d\Omega_0 \quad (9)$$

$$\delta W_{ext} = \int_{\Omega_0} \delta E_v{}^T \boldsymbol{C}\,(\epsilon_{residual}) d\Omega_0 + \int_{\Omega_0} \delta u_v^T f^B d\Omega_0 + \int_{\Gamma_0} \delta u_v^T f^S d\Gamma_0 \quad (10)$$

4. Numerical Implementation

Using the discretization via FEM, the displacement field in a continuum media is described by interpolation of nodal displacement according to specifics shape functions:

$$\boldsymbol{u} = \sum \mathrm{N}_i(\xi, \eta)\, \widetilde{\boldsymbol{u}} \quad (11)$$

Deformation gradient and the Green strain tensor are given by [34]:

$$F_{IJ} = \delta_{iI} + \frac{\partial u_i}{\partial X_I} \quad (12)$$

$$\epsilon_{IJ} = 0.5\left(\delta_{iI}\frac{\partial u_i}{\partial X_J} + \delta_{iJ}\frac{\partial u_i}{\partial X_I} + \frac{\partial u_i}{\partial X_I}\frac{\partial u_i}{\partial X_J}\right) \tag{13}$$

where δ_{iI} and δ_{iJ} are Kronecker delta

Finite strain equations are solved similarly as in small strain problems but considering the reference configuration instead. Through Newton–Raphson procedure, the variation of internal work is calculated by increments of displacements that minimize the variation of total energy, then, the new value is updated from previous step:

$$\delta\delta W_{int}^{t+\Delta t} = \delta W_{ext}^{t} - \delta W_{int}^{t} \tag{14}$$

$$\delta W_{int}^{t+\Delta t} = \delta W_{int}^{t} + \delta\delta W_{int}^{t+\Delta t} \tag{15}$$

4.1. Discretization of Equivalent Internal Work

Variation of equivalent internal work can be described as:

$$\delta W_{int} = \int_{\Omega_0} S_{IJ}\delta\epsilon_{IJ} d\Omega_0 \tag{16}$$

where the variation of Green strain is given by:

$$\delta\epsilon_{IJ} = 0.5\left(\frac{\partial \delta u_i}{\partial X_I}F_{iJ} + \frac{\partial \delta u_i}{\partial X_J}F_{iI}\right) \tag{17}$$

Increments of internal work is given by linearization of Equation (16) through directional derivatives in displacement direction:

$$\delta\delta W_i = \int \delta\epsilon_{IJ}\delta S_{IJ} d\Omega_0 + \int \delta\delta\epsilon_{IJ} S_{IJ} d\Omega_0 \tag{18}$$

At the first step, initial guess for increment of nodal displacements is set as zero, therefore, nonlinear terms ($\boldsymbol{K}_{II}$ and $\boldsymbol{K}_{III}$) would not change tangent stiffness matrix and both (linear and nonlinear) could be calculated at the same way ($\boldsymbol{K}_I$). Difference between linear and nonlinear approaches appears at the next incremental steps of Newton–Raphson procedure. According to de Borst et al. [35], the geometric term ($\boldsymbol{K}_{III}$) is important because it considers numerical instability due to geometry changes.

$$[\boldsymbol{K}_I + \boldsymbol{K}_{II} + \boldsymbol{K}_{III}]\widetilde{\boldsymbol{u}}^{t+\Delta t} = \delta W_{ext,i}^{t} - \delta W_{int,i}^{t} \tag{19}$$

4.2. Discretization of Equivalent External Work

Since the variation of equivalent external work is function of temperature it was necessary to calculate the temperature field through Fourier's law. It was considered heat flow in 1-D direction with fixed boundary conditions (Equation (20)):

$$\frac{\partial \theta}{\partial t} = \frac{\partial^2 \theta}{\partial x^2} \tag{20}$$

Dimensionless functions were assumed as solution to previous equation, then, the dimensionless Fourier's law could be written as presented in Equation (22):

$$\Theta = \frac{\theta(x,t)}{\theta_0},\ \chi = \frac{x}{X_0} \text{ and } T = \frac{\alpha\, t}{X_0{}^2} \tag{21}$$

$$\frac{\partial \Theta}{\partial T} = \frac{\partial^2 \Theta}{\partial \chi^2} \tag{22}$$

Using the concept of separation of variables and assuming that each function will individually satisfy the boundary condition, it is possible to rewrite a multivariable function as a sum of the product of these functions. Then, the dimensionless temperature function could be rewritten as:

$$\Theta(\mathrm{T},\chi) = \sum c_i \mathrm{T}_i(\mathrm{T})\chi_i(\chi) \tag{23}$$

Substituting Equation (23) into Equation (22) one gets that each individual equation is related as:

$$\dot{\mathrm{T}}_i\chi_i = \mathrm{T}_i\ddot{\chi}_i \tag{24}$$

Previous equality could be represented by a constant value as given in Equation (25) just rearranging Equation (24), since dimensionless time function is not related with dimensionless position function and vice-versa:

$$\frac{\dot{\mathrm{T}}_i}{\mathrm{T}_i} = \frac{\ddot{\chi}_i}{\chi_i} = -\mathrm{M}_i^2 \tag{25}$$

Equations (26) and (27) can be assumed as solutions that satisfy Equation (25):

$$\mathrm{T}_i = \mathrm{C}_i e^{-\mathrm{M}_i^2\mathrm{T}} \tag{26}$$

$$\chi_i = \mathrm{A}_i \cos(\mathrm{M}_i\chi) + \mathrm{B}_i \sin(\mathrm{M}_i\chi) \tag{27}$$

Substituting Equations (26) and (27) into Equation (23) and applying the boundary restraint conditions for a nontrivial solution one can find the first unknown coefficient:

$$\Theta(T,\chi) = \sum e^{-M_i^2T}[\mathrm{A}'_i \cos(M_i\chi) + \mathrm{B}'_i \sin(M_i\chi)] \tag{28}$$

$$\Theta(T=0,\chi=0) = 0 \rightarrow A'_i = 0 \tag{29}$$

$$sin(\mathrm{M}_i\cdot 1) = 0 \rightarrow \mathrm{M}_i = n\cdot\pi \tag{30}$$

The other coefficient can be calculated considering the boundary conditions at initial configuration:

$$\Theta(T=0,\chi) = \sum \mathrm{B}'_i \sin(\mathrm{M}_i\chi) \tag{31}$$

$$\int_0^1 \Theta(T=0,\chi)\sin\left(M_j\chi\right)d\chi = \mathrm{B}\prime_i \int_0^1 \sin(\mathrm{M}_i\chi)\sin\left(\mathrm{M}_j\chi\right)d\chi = \mathrm{B}\prime_i\frac{\mathrm{X}_0}{2} \tag{32}$$

$$\mathrm{B}\prime_i = \frac{2}{\mathrm{X}_0}\int_0^1 \Theta(\mathrm{T}=0,\chi)\sin\left(\mathrm{M}_j\chi\right)d\chi \tag{33}$$

Accordingly, the temperature could be calculated at each integration point just recalling Equation (21). Considering known the temperature distribution at each time t, the deformation gradient due to thermal expansion could be calculated as:

$$\boldsymbol{F}_\theta = \alpha\cdot\Delta\theta\cdot\boldsymbol{I} + \boldsymbol{I} \tag{34}$$

and the equivalent external work can be represented as function only of the thermal actions since isostatic condition is considered in this study:

$$\delta W_e = \int \delta\epsilon_{IJ}S(\theta)_{IJ}d\Omega_0 \tag{35}$$

5. Validation and Discussion

In the first part of this section it is presented the verification of numerical formulation with analytical solutions to verify the accuracy, then comparison with results in literature are proposed. The validation tests were accomplished in two steps, one to evaluate the convergence of the nonlinear model and other to verify the numerical formulation of thermal loading. The flowchart used to code the MATLAB program is presented in Appendix A.

The nonlinear model was verified considering mechanical loads applied on a beam. Since finite displacements are assumed, the external loads used in this example are not constant, as indicated in Equation (8), because they will vary according to geometry changes. A way to solve them could be expressing the follower loads in terms of parametric variables which will be updated according to the current geometry, as indicated in Equation (36). The linearization of these surface forces must also be considered when the tangent stiffness matrix is calculated, as explained by Bonet and Wood [36].

An example with a built-in beam with bending moment applied at the end is now considered. Since the bending moment is constant along the beam, rotated Cauchy stresses (recall Equation (4)) must remain invariant in perpendicular sections along this beam, as indicated in Figure 2.

$$\delta W_{ext} = \int \mathrm{N}_a \overline{p} \left(\frac{\partial x}{\partial \xi} \times e_3 \right) d\xi \tag{36}$$

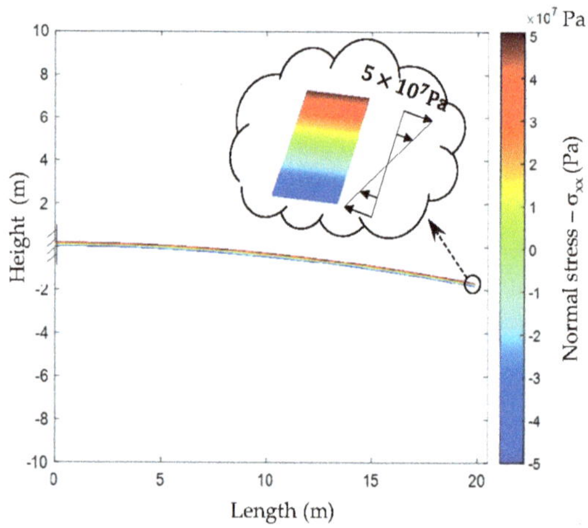

Figure 2. Normal stress induced by bending moment applied at the end of a built-in beam.

Henceforth the numerical formulation of thermal stresses is considered. Thermal stresses in unrestrained slab could be calculated adding the thermal component to the Airy stress function [37,38] which, in rectangular slab in plane strain analysis, could be given by Equation (37). Since the slab is free to expand no stress is expected at the borders; hence the first term of Equation (37) corresponds to suppression of internal stresses generated due to thermal expansion (second term). The third term is related with equilibrium since asymmetric thermal distribution could result in eccentric load.

$$\sigma = -\frac{\mathrm{E}}{1-\nu}\alpha T + \frac{1}{2\mathrm{b}} \int_{-\mathrm{b}}^{\mathrm{b}} \frac{\mathrm{E}}{1-\nu} \alpha T dx + \frac{3x}{2\mathrm{b}^3} \int_{-\mathrm{b}}^{\mathrm{b}} \frac{\mathrm{E}}{1-\nu} \alpha T x dx \tag{37}$$

In these analyses the transient heat flux was considered with fixed boundaries conditions. The temperature on the external face was assumed as 50 °C along the whole surface, for the internal side it was assumed 20 °C. Thermal and mechanical properties of marble were assumed the same used in Ferrero et al., [9], as indicated in Figure 3. Transient heat flow was calculated to four elapsed times (5 s, 10 s, 25 s and 50 s) and good correlation was achieved between linear and analytical solution, as presented in Figure 4.

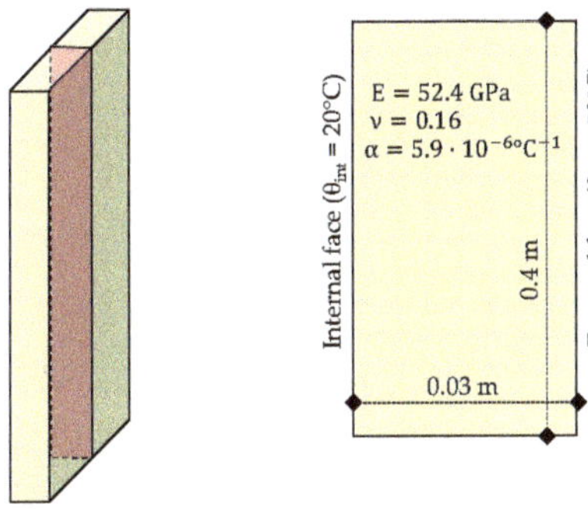

Figure 3. Model used in validation test.

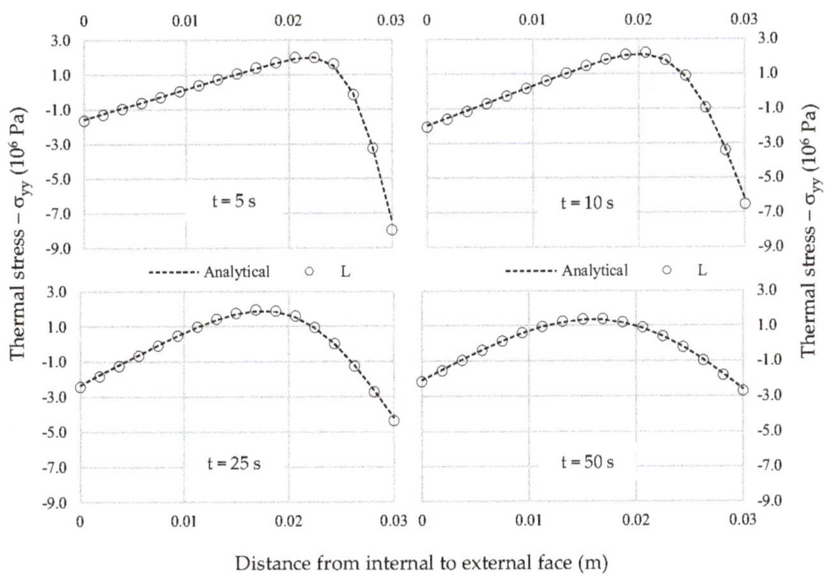

Figure 4. Comparison between analytical and numerical (linear) solution for several elapsed times. (t = 5 s, t = 10 s, t = 25 s, t = 50 s).

Pescara's Building

Ferrero et al. [9] monitored the temperature of the internal and external face of a ventilated façade at the Pescara's building during six days in 2007. An automatic system recorded the temperatures at each 15 min. The highest gradient of temperature was 4.7 °C which induced about 0.04 MPa in tensile stresses according to that model. Similar analyses were performed using the formulation presented in this work for several elapsed times to evaluate the differences between that model and the numerical approach herein proposed.

In Figures 5–7 are presented the thermal stresses for each analysis. Normal stress distributions were considered at the symmetric planes and the shear stresses were represented in two orthogonal planes. The results show that internal stress distribution is different from the one usually found in classical beam theories, in which the traction and compression stresses belong to opposite sides of the neutral axis. Nevertheless, considering the theoretical formulation presented in the previous section it would be expected to find such a distribution because the fibers on the external face, where the thermal expansion is greater, would try to expand but they will be constrained by the adjacent fibers in which the thermal deformations are smaller. Then, the external fibers should be under compression and those on the left side would be under tensile stresses. Symmetric behavior is expected on the left-hand side.

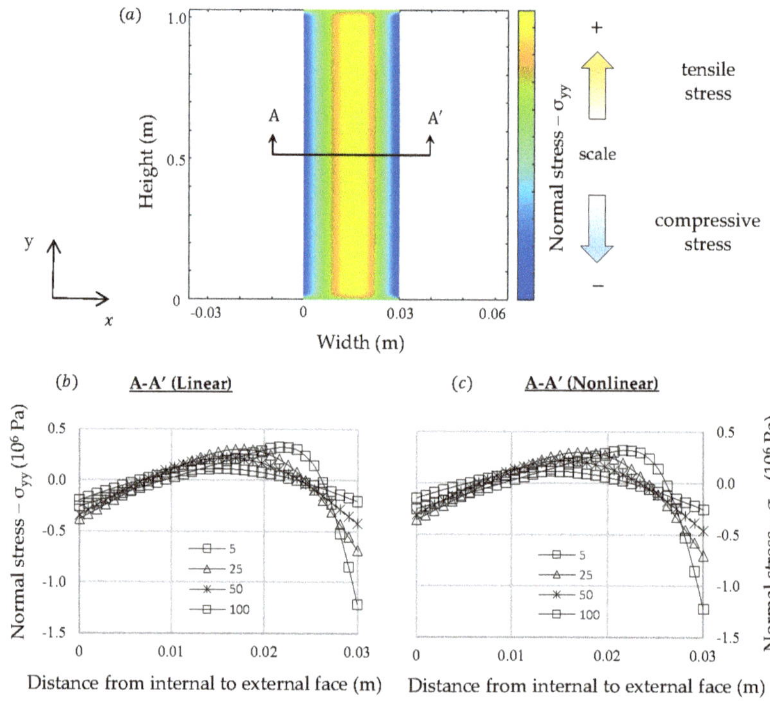

Figure 5. Comparison between linear and nonlinear numerical analysis of δ_{yy}: (**a**) Representation of a generic normal stress distribution (**b**) linear analysis (**c**) nonlinear analysis.

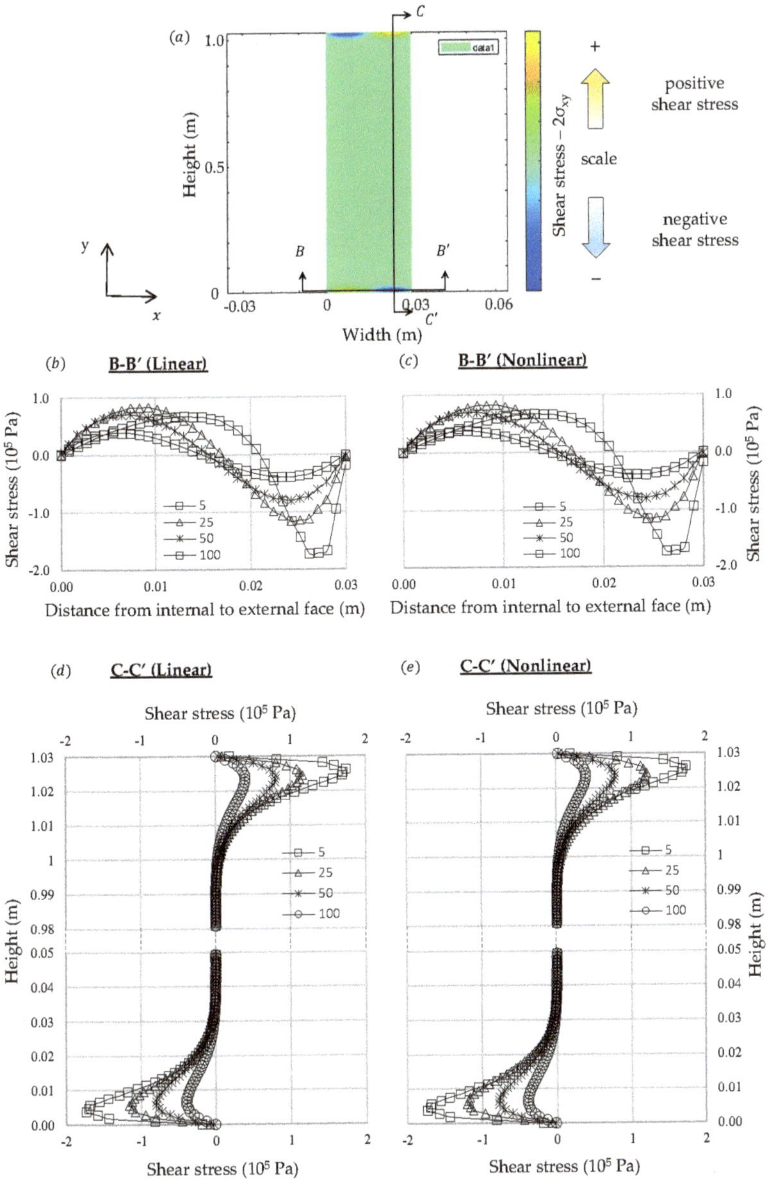

Figure 6. Comparison between linear and nonlinear numerical analysis of $2\delta_{xy}$: (**a**) representation of a generic shear stress distribution, (**b**,**d**): linear analysis, (**c**,**e**): nonlinear analysis.

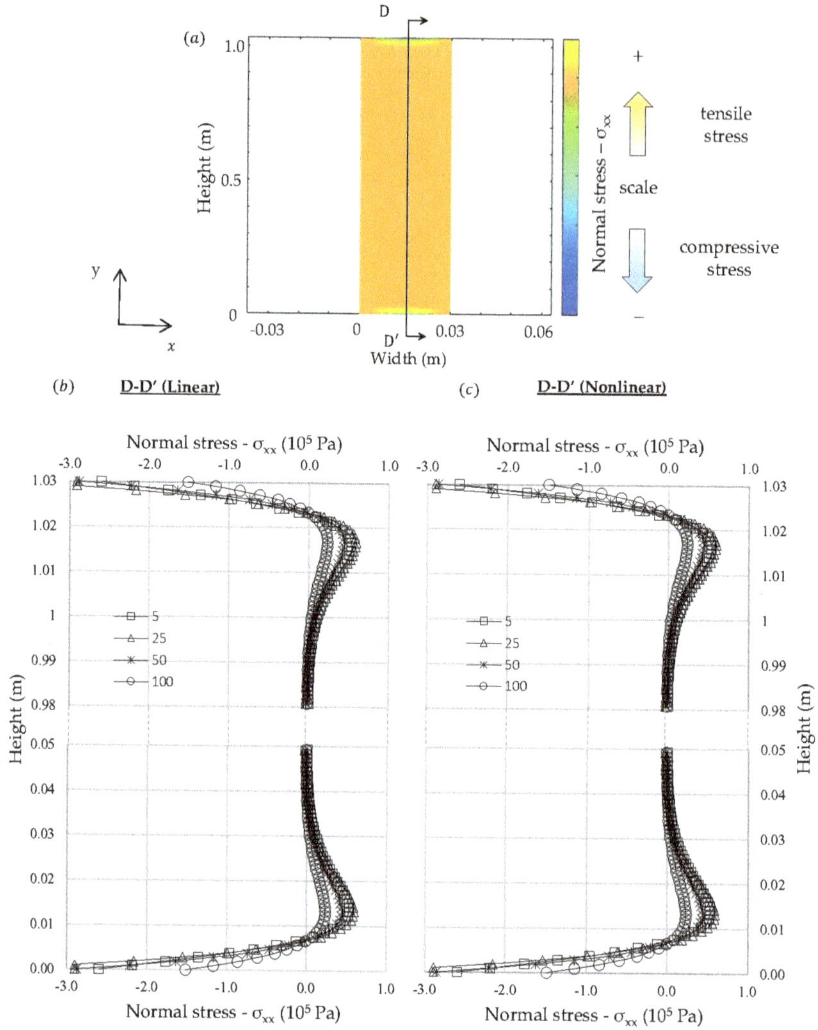

Figure 7. Comparison between linear and nonlinear numerical analysis of δ_{xx}: **(a)** representation of a generic normal stress distribution **(b)** linear analysis **(c)** nonlinear analysis.

It must be noted that internal stresses are greater at the initial instant of the transient regime and they reduce until it becomes constant in stationary condition. The results are consistent with those expected if the shape of the internal temperature curve in transient regime is considered. At the beginning only those fibers close to the right-hand side are deformed by thermal expansion, thus, greater amount of stress will be induced to counterpoise these unbalanced loads. Nonetheless, when the transient gradient approaches to stationary condition these unbalanced loads reduce. The greater stresses developed in transient analysis at the beginning could explain why the intergranular decohesion in thermal treated marble is higher close to the exposed surface, as identified by Bellopede et al. [29]. This phenomenon can also be related with the increase of bowing with the height of buildings related by Siegesmund et al. [6], since the thermal exchanges could increase according to elevation due to wind action.

Shear stresses and normal stresses in x direction were concentrated at the ends of the slab. Similar results were found by Carpinteri and Paggi [39] in nonhomogeneous beams using the classical beam's theory, however, in that formulation shear stresses tend to infinity when they approach the boundaries. In the numerical formulation proposed in this study the shear stresses at the boundaries in simply supported slabs tend to zero.

The shear stress distribution is consistent with the theory of long thin beams, in which most of its length follow a strain profile almost translation-symmetric. At the slab edges, in which the translation-symmetry cited does not hold, there are other bidimensional deformation modes, which dissipates the energy induced by thermal strains in a more efficient way. These edge deformation modes explain why the stress calculated by the finite element method does not fit exactly to the analytical solution.

In transient condition the difference between stress distribution in linear and nonlinear analysis is minimal, just as expected due to the magnitude of the deformation. However, for stationary condition, the internal stresses completely vanish over the entire body in linear analysis just as expected by the analytical solution if one recalls Equation (36) considering linear distribution of temperature though width. Although, in nonlinear analysis, residual stresses still remain, as indicated in Figure 8.

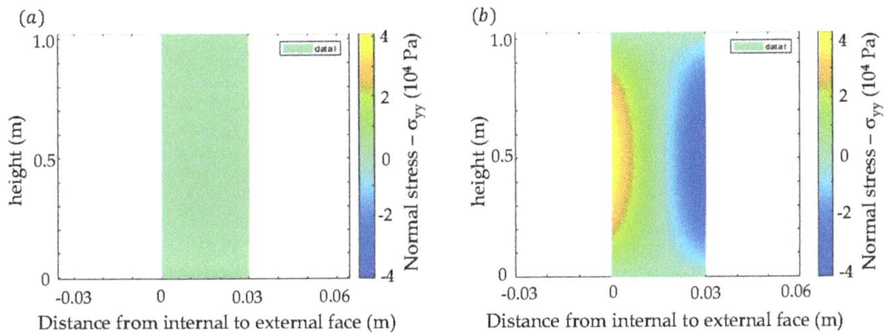

Figure 8. Evaluation of normal stress distribution (σ_{yy}) in steady heat flux ($t \rightarrow \infty$): (**a**) linear (**b**) nonlinear.

6. Conclusions

This paper presents a numerical formulation to evaluate stresses in nonlinear thermoelastic materials. According to our model transient thermal analysis has proven to induce higher levels of stress at the beginning, i.e., at the early stages of the transient heat flux higher stresses were developed to balance the locally differential strains. These results are important to understand bowing phenomenon because transient heat flux continuously occurs in ventilated façades due to the variation of insolation and thermal exchanges between marble slabs and external environment (due to rain, wind, etc.). Therefore, even the small gradients of temperature through width can be harmful in long-term and it help to explain why the intergranular decohesion is more intense at the exposed face. The nonlinear analysis also identified remaining stresses in the slab even in stationary thermal flow. Moreover, shear stresses appear at the ends of the body, hence, for short beams or greater loads, more sophisticated models should be considered.

Author Contributions: Conceptualization, W.H.I.; methodology, W.H.I.; supervision, A.M.F. and P.I.B.d.Q.; writing—original draft, W.H.I.; writing—review and editing, A.M.F. and P.I.B.d.Q. All authors have read and agreed to the published version of the manuscript.

Funding: This research received no external funding.

Conflicts of Interest: The authors declare no potential conflict of interests.

List of Symbols

A	constant
b	slab width
B	constant
C	constitutive elastic matrix
$e_{i,x}$	orthonormal basis in the deformed configuration
$e_{i,X}$	orthonormal basis in the reference configuration
E	Young modulus
F	deformation gradient
I	identity tensor
K	stiffness matrix
M	constant
N	shape function relative to node i
$\overline{p}$	pressure load function
R	rotation tensor
S	second Piola-Kirchhoff stress tensor
t	current time
T	dimensionless time function
U	right Cauchy-Green tensor
V	left Cauchy-Green tensor
u	displacement vector
$\widetilde{u}$	nodal displacement
W_{ext}	external virtual work
W_{int}	internal virtual work
x	coordinates in current configuration
X	Coordinates in reference configuration
α	thermal expansion coefficient
δ	Kronecker delta
δ	directional derivative
ε	engineering strain tensor
ϵ	Green strain tensor
ϵ_θ	strain tensor due to thermal action
ϵ_{mec}	strain tensor due to mechanical action
Π	total potential energy
θ	temperature function
Θ	dimensionless temperature function
λ	eigenvalues of the tensor U
ν	Poisson ratio
σ	Cauchy stress tensor
τ	shear stress
Ω	current configuration
Ω_0	reference configuration
$(\cdot)_\theta$	relative to the thermal actions
χ	dimensionless position function

Appendix A

Workflow

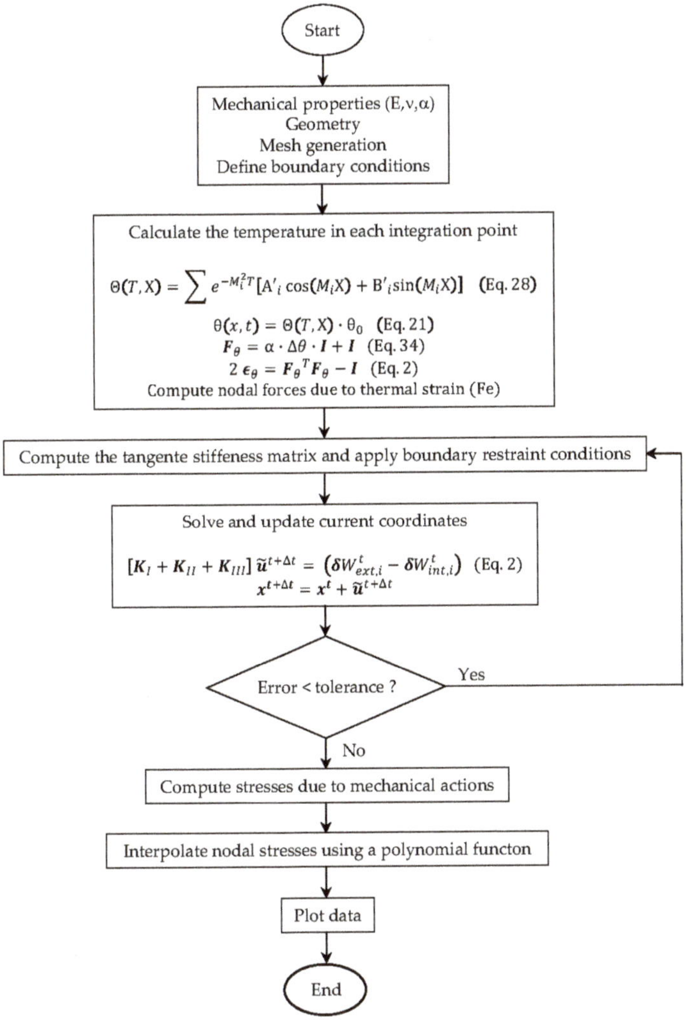

Figure A1. Procedure used to evaluate thermal stresses in marble slabs.

References

1. Chen, X.; Guo, H.; Zhao, P.; Peng, X.; Wang, S. Numerical modeling of large deformation and nonlinear frictional contact of excavation boundary of deep soft rock tunnel. *J. Rock Mech. Geotech. Eng.* **2011**, *3*, 421–428. [CrossRef]
2. Guan, K.; Zhu, W.; Wei, J.; Liu, X.; Niu, L.; Wang, X. A finite strain numerical procedure for a circular tunnel in strain-softening rock mass with large deformation. *Int. J. Rock Mech. Min. Sci.* **2018**, *112*, 266–280. [CrossRef]
3. Lee, K.; Jeong, S. Large deformation FE analysis of a debris flow with entrainment of the soil layer. *Comput. Geotech.* **2018**, *96*, 258–268. [CrossRef]
4. Yuan, J.; Hicks, M.A. Large deformation elastic electro-osmosis consolidation of clays. *Comput. Geotech.* **2013**, *54*, 60–68. [CrossRef]
5. Royer-Carfagni, G.F. Some considerations on the warping of marble façades: The example of Alvar Aalto's Finland Hall in Helsinki. *Constr. Build. Mater.* **1999**, *13*, 449–457. [CrossRef]

6. Siegesmund, S.; Ruedrich, J.; Koch, A. Marble bowing: Comparative studies of three different public building façades. *Environ. Geol.* **2008**, *56*, 473–494. [CrossRef]
7. Royer-Carfagni, G.F. On the thermal degradation of marble. *Int. J. Rock Mech. Min. Sci.* **1999**, *36*, 119–126. [CrossRef]
8. Weiss, T.; Siegesmund, S.; Fuller, E.R., Jr. Thermal degradation of marble: Indication from finite element modelling. *Build. Environ.* **2003**, *38*, 1251–1260. [CrossRef]
9. Ferrero, A.M.; Migliazza, M.; Spagnoli, A. Theoretical modelling of bowing in cracked marble slabs under cyclic thermal loading. *Constr. Build. Mater.* **2009**, *23*, 2151–2159. [CrossRef]
10. Ferrero, A.M.; Migliazza, M.; Spagnoli, A.; Zucali, M. Micromechanics of intergranular cracking due to anisotropic thermal expansion in calcite marbles. *Eng. Fract. Mech.* **2014**, *130*, 42–52. [CrossRef]
11. Reddy, J.N. On locking-free shear deformable beam finite elements. *Comput. Meth. Appl. Mech. Eng.* **1997**, *149*, 113–132. [CrossRef]
12. Burzynski, S.; Chroscielewski, J.; Daszkiewicz, K.; Witlowski, W. Geometrically nonlinear FEM analysis of FGM shells based on neutralphysical surface approach in 6-parameter shell theory. *Compos. Part. B Eng.* **2016**, *107*, 203–213. [CrossRef]
13. Felippa, C.A.; Haugenb, B. A unified formulation of small-strain corotational finite elements: I. Theory. *Comput. Methods Appl. Mech. Eng.* **2005**, *194*, 2285–2334. [CrossRef]
14. Graczykowski, B.; Gomis-Bresco, J.; Alzina, F.; Reparaz, J.S.; Shchepetov, A.; Prunnila, M.; Ahopelto, J.; Sotomayor Torres, C.M. Acoustic phonon propagation in ultra-thin Si membranes under biaxial stress field. *New J. Phys.* **2015**, 16. [CrossRef]
15. Graczykowski, B.; Sledzinska, M.; Placidi, M.; Saleta Reig, D.; Kasprzak, M.; Alzina, F. Elastic Properties of Few Nanometers Thick Polycrystalline MoS2 Membranes: A Nondestructive Study. *Nano Lett.* **2017**, *17*, 7647–7651. [CrossRef] [PubMed]
16. Aydan, O.; Ulusay, R. Geomechanical Evaluation of Derinkuyu Antique Underground City and its Implications in Geoengineering. *Rock Mech. Rock Eng.* **2013**, *42*, 731–754. [CrossRef]
17. Pascalle, G.; Lolli, A. Crack assessment in marble sculptures using ultrasonic measurements: Laboratory rests and application on the statue of David by Michelangelo. *J. Cult. Herit.* **2015**, *16*, 813–821. [CrossRef]
18. Boldini, D.; Guido, G.L.; Margottini, C.; Spizzichino, D. Stability Analysis of a Large-Volume Block in the Historical Rock-Cut City of Vardzia (Georgia). *Rock Rock Eng.* **2018**, *51*, 341–349. [CrossRef]
19. Tiennot, M.; Mertz, J.; Bourges, A. Influence of Clay Minerals Nature on the Hydromechanical and Fracture Behaviour of Stones. *Rock Mech. Rock Eng.* **2018**, *52*, 1599–1611. [CrossRef]
20. Liu, J.; Zhang, Z.; Li, B. Microscopic and macroscopic characterization of Beijing marble as a building material for UNESCO heritage sites: New insights into physico-mechanical property estimation and weathering resistance. *Constr. Build. Mater.* **2019**, *225*, 510–525. [CrossRef]
21. MARA. *Developing Long-Term Durability of Marble Façades*; Final Technical Report; (Raphael program); European Commission: Bruxelles, Belgium, 2001.
22. TEAM. Testing and assessment of marble and limestone. In *Final Technical Report.EC-Project: TEAM—G5RD-CT-2000-00233*; European Committee: Bruxelles, Belgium, 2005.
23. Winkler, E.M. Properties of marble as building veneer. *Int. J. Rock Mech. Min. Sci. Geomech. Abstr.* **1996**, *33*, 215–218. [CrossRef]
24. Grelk, B.; Christiansen, C.; Schouenborg, B.; Malaga, K. Durability of Marble Cladding—A comprehensive literature review. *J. ASTM Int.* **2007**, *4*, 1–19. [CrossRef]
25. Grelk, B.; Goltermann, P.; Koch, A.; Alnaes, L. The Laboratory Testing of Potential Bowing and Expansion of Marble. In Proceedings of the Dimension Stone-New Perspectives for a Traditional Building Material, Prague, Czech Republic, 14–17 June 2004; pp. 253–260.
26. Koch, A.; Siegesmund, S. The combined effect of moisture and temperature on the anomalous expansion behaviour of marble. *Environ. Geol.* **2004**, *46*, 350–363. [CrossRef]
27. Rodriguez-Gordillo, J.; Saez-Perez, M.P. Effects of thermal changes on Macael marble: Experimental282study. *Constr. Build. Mater.* **2006**, *20*, 355–365. [CrossRef]
28. Luque, A.; Cultrone, G.; Mosch, S.; Siegesmund, S.; Sebastian, E.; Leiss, B. Anisotropic behaviour of White Macael marble used in the Alhambra of Granada (Spain): The role of thermohydric expansion in stone durability. *Eng. Geol.* **2010**, *115*, 209–216. [CrossRef]

29. Bellopede, R.; Castelletto, E.; Marini, P. Ten years of natural ageing of calcareous stones. *Eng. Geol.* **2016**, *211*, 19–26. [CrossRef]
30. Bathe, K.J. Finite Element Nonlinear Analysis in Solid and Structural Mechanics. In *Finite Element Procedure*; Peterson, M., Ed.; Prentice-Hall: New York, NJ, USA, 1996; pp. 485–528.
31. Przemieniecki, J.S. Thermal Stresses in Rectangular Plates. *Aero Quart* **1959**, *10*, 65–78. [CrossRef]
32. Keramidas, G.A.; Ting, E.C. A finite element formulation for thermal stress analysis. Part I: Variational formulation. *Nucl. Eng. Des.* **1976**, *39*, 267–275. [CrossRef]
33. Laura, P.A.A.; Gutierrez, R.H.; Sanchez Sarmiento, G.; Basombrio, F.G. Thermal stresses in rectangular plates: Variational and finite element solutions. *Nucl. Eng. Des.* **1978**, *47*, 297–303. [CrossRef]
34. Zienkiewics, O.C.; Taylor, R.L. Geometrically non-linear problems–finite deformation. In *The Finite Element Method for Solid and Structural Mechanics*; Butterworth-Heinemann: New York, NY, USA, 2014; pp. 127–157.
35. De Borst, R.; Crisfield, M.A.; Remmers, J.J.C.; Verhoosel, C.V. Geometrically Non-linear Analysis. In *Non-Linear Finite Element Analysis of Solids and Structures*, 2nd ed.; John Wiley and Sons: West Sussex, UK, 2012; pp. 63–111.
36. Bonet, J.; Wood, R.D. Linearized equilibrium equations. In *Nonlinear Continuum Mechanics for Finite Element Analysis*; Cambridge University Press: Cambridge, UK, 2012; pp. 216–235.
37. Johns, D.J. Two-dimensional formulations and solutions. In *Thermal Stress Analyses*; Pergamon: London, UK, 1965; pp. 21–32.
38. Timoshenko, S.; Goodier, J.N. Thermal stress. In *Theory of Elasticity*; McGraw Hill: New York, NY, USA, 1970; pp. 399–437.
39. Carpinteri, A.; Paggi, M. Thermo-elastic mismatch in nonhomogeneous beams. *J. Eng. Math.* **2008**, *61*, 371–384. [CrossRef]

materials

MDPI

Article

Material Parameter Identification for Acoustic Simulation of Additively Manufactured Structures

Sebastian Rothe [1,*], Christopher Blech [1], Hagen Watschke [2], Thomas Vietor [2] and Sabine C. Langer [1]

1 Institute for Acoustics, Technische Universität Braunschweig, Langer Kamp 19, 38106 Braunschweig, Germany; c.blech@tu-braunschweig.de (C.B.); s.langer@tu-braunschweig.de (S.C.L.)
2 Institute for Engineering Design, Technische Universität Braunschweig, Hermann-Blenk-Straße 42, 38108 Braunschweig, Germany; h.watschke@tu-braunschweig.de (H.W.); t.vietor@tu-braunschweig.de (T.V.)
* Correspondence: sebastian.rothe@tu-braunschweig.de; Tel.: +49-531-391-8774

Abstract: One possibility in order to manufacture products with very few restrictions in design freedom is additive manufacturing. For advanced acoustic design measures like Acoustic Black Holes (ABH), the layer-wise material deposition allows the continuous alignment of the mechanical impedance by different filling patterns and degrees of filling. In order to explore the full design potential, mechanical models are indispensable. In dependency on process parameters, the resulting homogenized material parameters vary. In previous investigations, especially for ABH structures, a dependency of the material parameters on the structure's thickness can be observed. In this contribution, beams of different thicknesses are investigated experimentally and numerically in order to identify the material parameters in dependency on the frequency and the thickness. The focused material is polyactic acid (PLA). A parameter fitting is conducted by use of a 3D finite element model and it's reduced version in a Krylov subspace. The results yield homogenized material parameters for the PLA stack as a function of frequency and thickness. An increasing Young's modulus with increasing frequency and increasing thickness is observed. This observed effect has considerable influence and has not been considered so far. With the received parameters, more reliable results can be obtained.

Keywords: acoustic black holes; acoustic-oriented design; additive manufacturing; finite element method; vibroacoustics; material parameter identification; model order reduction

Citation: Rothe, S.; Blech, C.; Watschke, H.; Vietor, T.; Langer, S.C. Material Parameter Identification for Acoustic Simulation of Additively Manufactured Structures. *Materials* **2021**, *14*, 168. https://doi.org/10.3390/ma14010168

Received: 24 November 2020
Accepted: 29 December 2020
Published: 31 December 2020

1. Introduction

The development of new measures to reduce noise in our environment is an important contribution to comfort and from a health point of view. New advanced design measures are needed to achieve these objectives. This is accompanies the challenge of using unconventional manufacturing processes and providing mechanical models for reliable simulations. This paper aims to contribute to this by new insights on material parameters.

The possibilities of integrating acoustic measures and other properties are often limited due to the given manufacturing constraints of conventional processes. With the help of additive manufacturing (AM), completely new possibilities open up. In this contribution, the additive manufacturing process material extrusion (MEX) is focused upon. If AM is used to produce structures with integrated acoustic measures or functions, design restrictions are significantly reduced. Due to the layer-wise build-up of the parts, AM allows almost every design, even a fully integrated measure. Furthermore, multiple materials can be applied directly during the manufacturing process, e.g., as additional damping material. However, mechanical models or assumptions based on experience with standard materials such as steel and aluminium are no longer sufficient. The step-wise build-up leads to anisotropies and inhomogeneities, which have to be considered depending on the individual task.

In literature, contributions exists in which the use of AM for generating structures that influence airborne and structure-borne sound is described. For example, 3D-printed tailored absorbers. Setaki et al. [1] design absorber duct lengths in such a way that the sound travelling through them receives a phase shift of 180° due to the different lengths and interferes destructively. Ring and Langer [2] link the geometric parameters of lattice structures with the resulting BIOT parameters. This allows the microstructure to be designed to achieve a specific absorber behavior. For the measures influencing structure-borne sound, new possibilities arise, e.g., for the method of Acoustic Black Holes (ABH). Here the production process is one of the biggest challenges [3]. The required thickness reduction of the structure is usually achieved with the help of milling cutters afterwards, which generates a high effort and high costs [3]. In addition, there are many restrictions regarding the placement of the ABHs and their design. Since this method reduces weight and at the same time improves the acoustic properties, it may be possible to overcome the major conflict of objectives between lightweight design and acoustics. The ABH are therefore the focus of this study as an exemplary application.

The ABH effect, first described in 1988 by Mironov [4], can be utilized in thin structures bearing structure-borne sound to focus the radiation critical bending waves in an area where they can be damped very efficiently. For this purpose, the thickness of the plate-like structures must be reduced according to a polynomial shape function. As a result of the smooth adaptation of the mechanical impedance, the amplitudes of the bending waves increase while the propagation speed decreases—an optimal area for an application of passive damping measures is formed. In 2000, Krylov published a combination of shaped area with a local damping measure and named it an Acoustic Black Hole [5].

A review of the literature since then reveals that most of the studies deal with homogeneous metal structures. Many investigations were carried out experimentally on generic plate structures [6–8]. There are also few studies on industrial examples, e.g., turbine blades [9] and an engine cover [10]. In addition to measurements, several numerical investigations were carried out [11–14]. However, composites and sandwich structures were also investigated with their higher design freedom. Bowyer et al. inserted an ABH exclusively into the glass fiber cover layers, in 2012 [15]. In 2017, Dorn et al. extended this approach and integrated an ABH into the sandwich core [16]. Blech et. al. showed how this can be used in an aircraft structure [17]. Further studies on carbon and glass fiber composites can be found in [18,19]. In their papers, Zhao and Prasad [20] as well as Pelat et al. [21] provide a very good overview of the current state of ABHs and their applications.

The application of AM technologies with the highest design freedom compared to the above mentioned manufacturing techniques for design and manufacturing ABHs has successfully been demonstrated by Rothe et al. [22,23]. Even complex tube structures which act as ABHs for fluids can be produced [24]. In Figure 1 an example of an additive manufactured ABH beam with and without additional damping material is shown. The effect of the additional damping material can even be enhanced if it is designed as a constraint layer damping [25]. Such constraining layers can also be designed very efficiently by AM, e.g., by fully integrated ABHs.

Figure 1. Tip of Acoustic Black Holes (ABH) beam with (**right**) and without (**left**) additional damping material.

Chong et al. [26] give an overview of the possibilities of how ABHs can be integrated into additive manufactured structures. They additionally show experiments and simulations on additively manufactured beams, but always assuming a constant modulus of elasticity over the frequency range. In [23], a need for frequency and thickness-dependent material parameters arise. Especially when modeling ABHs, regions of different thicknessess of the base material are obtained (as shown in Figure 1 on the left). Studies in [27] assume that a varying Young's modulus and loss factor results depending on the thickness of the sample on which the homogenization is performed. A suitable and efficient procedure is required to identify these homogenized linear elastic parameters, which are difficult to determine due to anisotropy and strong dependencies on geometry and process parameters [27]. For this purpose, a methodological procedure is proposed in this article and demonstrated on additively manufactured beam structures. To the authors' knowledge, the determination and consideration of the dependencies of the material parameters on the thickness have not been considered so far and offer a new approach for more reliable vibroacoustic simulations of additively manufactured structures.

In order to investigate the influence of the printed thickness, beams of different thicknesses are manufactured. Subsequently, these are characterized vibroacoustically by laser-scanning vibrometry and compared with numerical results. For efficiency reasons, the numerical models are transformed into reduced models and used for parameter study. The aim is to determine the homogenized parameters as a function of frequency and thickness.

2. Additive Manufacturing

AM provides a vast potential for the realization of graded properties, for instance, regarding bending stiffness and, thus, for the incorporation of passive damping measures such as ABHs due to the provision of enhanced design freedom compared to other manufacturing technologies, e.g., milling or casting. This freedom in design allows the manufacturing of complex shapes or a combination of multiple materials in order to achieve the required mechanical properties. One of the most commonly used AM technologies offering processing multiple materials in one part, without an additional joining process is needed, is MEX also referred to as fused deposition modeling [28,29]. Besides prototyping, MEX is also established in manufacturing of functional parts and end-use products [30]. This AM technology uses thermoplastic polymers as feedstock material. The material is plasticized and directed in an extrusion unit in order to build up the part's geometry in lines or layers. A great variety of thermoplastic polymers and also thermoplastic elastomers and fiber-reinforced materials are available [28]. Because of its ease of processing and good mechanical properties, especially the resulting stiffness, polylactic acid (PLA) is one of the most frequently used materials for material extrusion and, thus, focused on in this contribution.

With AM, the creation of complex shapes is not limited to external geometries in order to achieve locally variable bending stiffness. The internal structure of a part can also be influenced, for instance, by using lattice structures with variable wall thickness, different raster angle orientations or integrated damping structures by using a combination of a stiff and a flexible material [28,31,32]. As the mechanical properties of additively manufactured parts arise during the manufacturing process, they are significantly determined by the geometry and the selected process parameters in comparison to conventional manufacturing processes [29]. In addition to process parameters, the anisotropy in mechanical properties of additively manufactured parts is also influenced by machine-specific factors such as the heated build platform or the leveling (distance between the nozzle and the build platform) [27,33]. On the one hand, this process- and machine-related anisotropy enhances the design freedom for a local adjustment of the mechanical properties. On the other hand, the modeling of additively manufactured structures is more challenging regarding the identification and quantification of the influencing factors that have to be considered.

In order to increase the complexity of the considered ABH systems step by step, simple beam structures made of one material without ABH and additional damping material are manufactured first (represented by the specimen at the front in Figure 2). Afterwards, they are vibroacoustically characterized and their homogenized material parameters are determined.

Figure 2. Overview of beam specimens with different system complexity: simple beam structure made of PLA (**front**), beam structure with an ABH made of PLA (**center**), and beam structure with an ABH made of PLA with an additional damping treatment made of an thermoplastic polyurethane (**back**).

In Rothe et al. [23,27] it is shown that even at this step simple homogenization models are no longer sufficient. In addition to the frequency dependence of the material parameters, a dependency on the thickness respectively the number of layers can be observed.

In Figure 2 are also shown further exemplary complexity steps for the application of ABH. In the center the consideration of an ABH shape to weaken the cross-section is presented. The next complexity step (back) would be the consideration of a second material to locally increase the damping (here in white: NinjaFlex® (thermoplastic polyurethane, TPU) from NinjaTek).

A valid dynamic simulation of such additively manufactured structures is only possible with material descriptions that consider both frequency and thickness dependence. This is the focus of the investigations in this paper, where a procedure for the identification of material parameters is presented. It is the necessary step towards the next level of complexity.

The chosen material for manufacturing the test specimens used for parameter identification are PLA from DAS FILAMENT (Emskirchen, Germany). The used process parameter set is shown in Table 1. The flow rate was set to 105 % in order to increase stiffness due to a minimized internal void fraction. The other parameters have been selected according to the recommendations of the material manufacturer. For the manufacturing, a X400 by German RepRap GmbH (Feldkirchen, Germany) with a dual extruder system and a nozzle diameter of 0.4 mm is used. All specimens were manufactured at the same ambient temperature ($23 \pm 1\,^{\circ}C$) and relative humidity (45–50 %) and the feedstock materials are dried before processing in order to ensure similar manufacturing conditions.

Table 1. Utilized process parameters for manufacturing of the test specimens.

Material	PLA
Temperature Build Platform	60 °C
Temperature Nozzle	215 °C
Layer Thickness	0.0002 m
Raster Angle	±45°
Perimeter Shells	2
Flow Rate	105%
Infill Percentage	100%
Extrusion Width	0.0004 m
Extrusion Speed	0.05 m/s

3. Experimental Investigation

The experimental investigations on beam structures provide the necessary data basis for the subsequent numerical studies. In the following, it is explained which samples are to be studied and which experimental setup is to be chosen. Finally, the results are discussed.

3.1. Setup and Specimens

Three different types of beam samples of the same length are produced under the conditions described in Section 2 with three different thicknesses but a constant thickness over the length. These are used for fitting the material parameters and for studying the influence of thickness. The samples are exemplary shown in Figure 3.

Figure 3. Overview of investigated specimen of three different thicknesses (6 mm, 3 mm, 1 mm).

Each beam type is manufactured twice to get an idea of the repetitive uncertainties in vibroacoustic behavior that may arise from manufacturing imperfections. This results in six samples with the properties summarized in Table 2.

Table 2. Known geometrical and material data of specimens.

Specimen	Length (m)	Width (m)	Thickness (m)	Mass (kg)	Effective Density (kg/m^3)
Beam_6a	0.200	0.020	0.0063	0.03081	1222.6
Beam_6b	0.200	0.020	0.0062	0.03067	1236.7
Beam_3a	0.200	0.020	0.0032	0.01613	1260.2
Beam_3b	0.200	0.020	0.0033	0.01643	1244.7
Beam_1a	0.200	0.020	0.0012	0.00588	1225.0
Beam_1b	0.200	0.020	0.0012	0.00579	1206.3

In order to minimize external influences on the vibration behavior of the beams, they are characterized contactless by means of a laser scanning vibrometer. For this purpose, the samples are mounted eccentrically in the longitudinal direction (off-set of 0.02 m from longitudinal center) on an electrodynamic shaker in order to excite as many bending modes as possible in the considered frequency range. The surface velocities are measured. Simultaneously, the excitation force is recorded with a force sensor mounted directly at the force transmission point between shaker and beam. The setup is illustrated in Figure 4.

Due to the dark and highly light-absorbing PLA, the surfaces of the beams are sprayed with a reflection spray so that measurement with the laser vibrometer is possible. This can be seen as a grey top coat on the samples in Figures 3 and 4. The examined frequency range is defined between 0 and 6000 Hz, so that several bending eigenfrequencies can also be identified for the thicker beam samples. A measuring point grid of 4×39 (=156) points is selected for all measurements.

Figure 4. Experimental setup with 6 mm beam mounted on electrodynamic shaker.

Test measurements show that PLA already has high inherent damping, compared to steel and aluminum. The broadband excitation and measurement of the entire frequency range lead to noise in the higher frequency range and overloads at lower frequencies. For this reason, the frequency range is subdivided into smaller frequency sections of 1000 Hz. Each of these frequency ranges is excited and measured separately. The resulting frequency response of the beam over the entire frequency range is subsequently combined. The individual frequency sections and the signal types used are summarized in Table 3.

Table 3. Overview of frequency boundaries and signal type for separately excited and investigated frequency ranges.

Frequency Range (Hz)	Signal Type
0–1000	Pseudo Random
1000–2000	Sweep
2000–3000	Sweep
3000–4000	Sweep
4000–6000	Sweep

3.2. *Material Parameter Identification for Static Case*

In order to determine the flexural modulus of the additively manufactured samples test specimens for the bending test are manufactured according to DIN EN ISO 178 by using the process parameters shown in Table 1. The dimensions of the specimens are set to $80 \times 10 \times 4\,\text{mm}^3$ and the span is selected to be 64 mm. In Figure 5 the test setup with the beam specimen between the supports and the compression fin is schematically illustrated.

The specimens are tested with a speed of 2 mm/min. The results of the bending test are shown in Table 4. In addition to the values determined for the flexural modulus, the standard deviation is specified. It is to be expected that the flexural modulus will change with increasing frequency. The determined static value of the flexural modulus provides an orientation value between the 3 mm and 6 mm beam at 0 Hz for the fitting of the frequency-dependent modules using numerical simulations.

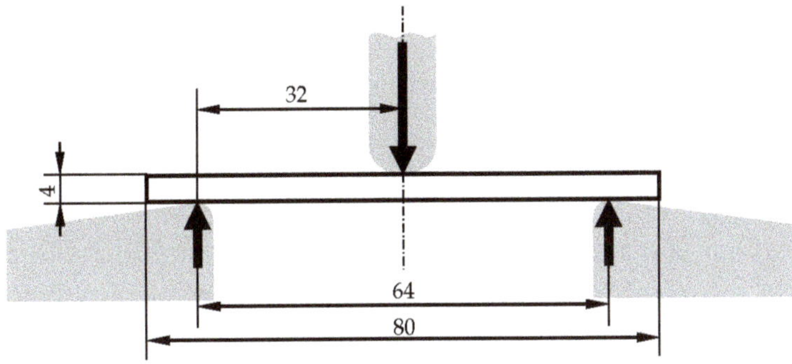

Figure 5. Schematic setup of the three-point bending test with dimensions (in mm).

Table 4. Flexural modulus (static) of PLA according to DIN EN ISO 178.

Material	**PLA**
Flexural modulus	3196.75×10^6 N/m^2
Standard deviation	68.98×10^6 N/m^2

3.3. *Results*

In this section the experimental results are presented and compared. To compare a quantity independent of the force excitation and equivalent to an energy quantity, the mean squared admittance h^2 is calculated according to Equation (1).

$$h^2(f) = 10\log_{10}\left(\sum_{i=1}^{N_p}\frac{\left(\frac{|v_i(f)|}{|F_i(f)|}\right)^2}{N_p}\right) dB \tag{1}$$

In Equation (1), v_i are the velocities of each surface point i and F_i the exciting force. These parameters are frequency-dependent, which also results in a h^2 value dependent on frequency f. N_p is the number of surface points.

In Figures 6–8, the experimentally determined frequency responses functions $h^2(f)$ of the different beam types are shown. In the diagrams, a comparison of the frequency response of two beams with similar thickness is shown, respectively. In the lower frequency range up to approximately 500 Hz, deviations between the curves can hardly be detected. In the higher frequencies, some deviations in the position of the peaks and the height are visible. Reasons are assumed in the manufacturing process and in the experimental data collection. Again, this illustrates the sensitivity of the frequency response of additively manufactured structures to process parameters and points out the necessity of a suitable procedure for material parameter identification. However, the differences of the curves seem not as significant as completely different homogenized material parameters are expected.

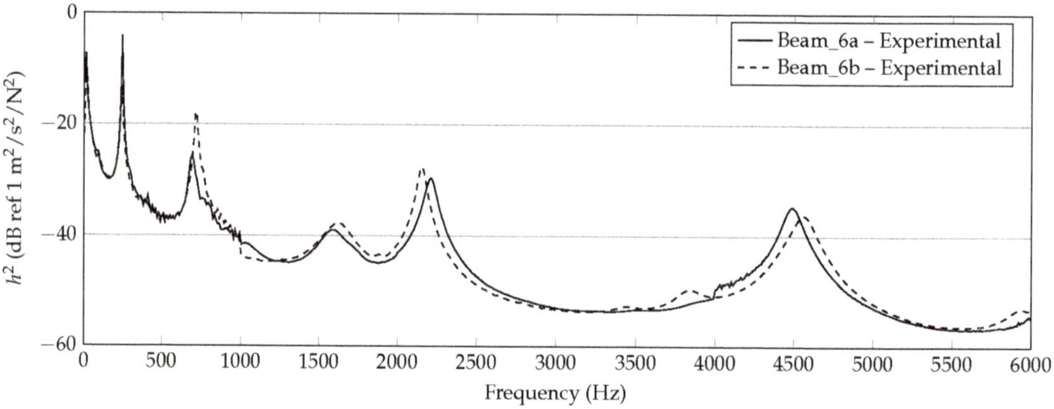

Figure 6. Experimental results of 6 mm beam measurements (Beam_6a, Beam_6b).

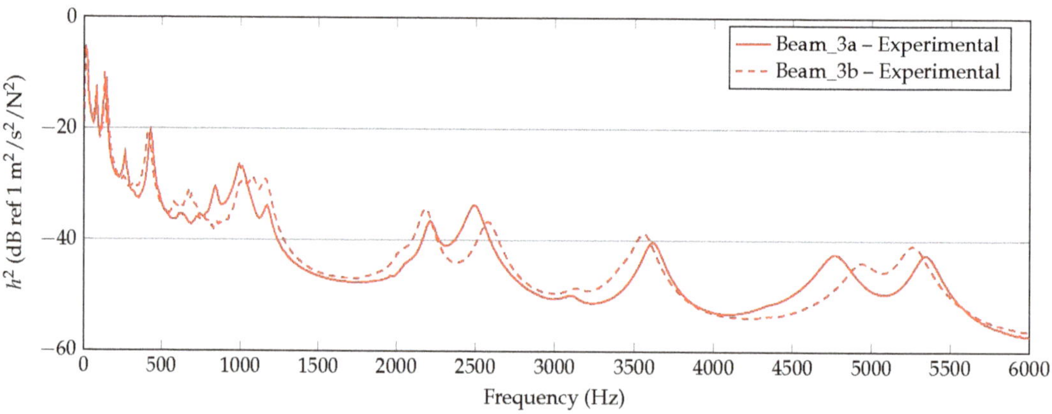

Figure 7. Experimental results of 3 mm beam measurements (Beam_3a, Beam_3b).

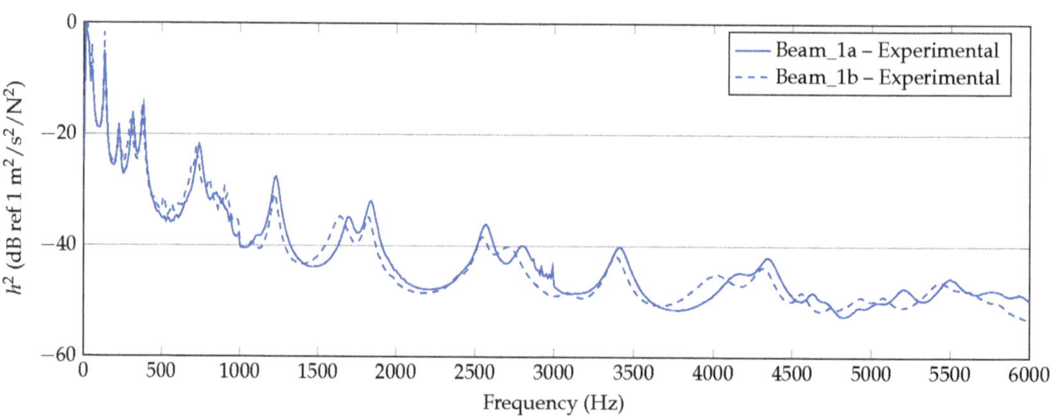

Figure 8. Experimental results of 1 mm beam measurements (Beam_1a, Beam_1b).

4. Parameter Study by Numerical Investigation

A mechanical model is built and numerically solved in Section 4.1 in order to predict the dynamic behavior of the beams. For each beam, a parameter study is conducted in Section 4.2 to yield the best fitting parameters in the frame of linear elasticity and structural damping for the different thicknesses.

4.1. Mechanical Model and Numerical Solution

The major focus of this work is laid on a potential layer-effect within additively manufactured structures with varied thickness t as required by typical ABHs. Layer-effect means that elastic material parameters (here: Young's modulus and loss factor) may change in dependency on the number of printed layers within the manufacturing process. The studies are explicitly limited to the base material of such structures which is PLA in this case. If the PLA material is thicker, different heat inputs are expected which may influence the modeling parameters to be chosen. With that background in mind, the beams are modeled by a 3D continuum with three translational degrees of freedom (dof). This way, the model is extendible to a layer-wise investigation (of the base PLA material only) in future work and a damping measure within the ABH area can be added easily. Linearity is assumed as the measurements show small deflections. Based on experiences, half the mass of the force sensor ($m_{\text{Sensor}} = 2 \times 0.012\,\text{kg}$) is considered as concentrated mass at the excitation point.

A structured mesh by 27-node hexahedrons with quadratic ansatzfunctions (Lagrangian type) is applied to the 3D continuum and solved in the frequency domain with a frequency step size of $\Delta f = 20\,\text{Hz}$ using the institute's in-house implementation elPaSo [34]. A convergence study is conducted for each beam thickness (1, 3 and 6 mm) in the higher frequency range (4000–6000 Hz). As a criterion, the maximum error of the mean squared admittance Δh^2 (Equation (1)) at all frequency steps f_i must be smaller than 0.1 dB. Under this assumption, it is expected that the material parameters are identified exactly enough in the frame of the engineering task.

By preliminary studies in [27] and the static bending test in Section 3, an estimation of the minimum flexural modulus (Young's modulus is called a flexural modulus as bending waves are dominant in the structures applied here.) for PLA is known. These values are round down to $E_0 = 3 \times 10^9\,\text{N/m}^2$ in order to receive a conservative mesh size by the convergence study. Applying E_0 to the three thickness setups $t = \{1, 3, 6\}$ mm, the mesh size is reduced systematically, until Δh^2 is smaller than 0.1 dB. In Figure 9, the maximum error is plotted in dependency on the mesh size. The chosen mesh size for each beam is further marked in the figure. Finally, this results in FE models with 12 k, 24 k and 48 k dof for the 6, 3 and 1 mm beam, respectively. The mesh is shown above Figure 9.

For the FE meshes applied for the convergence study in Figure 9, two elements (five nodes) over the thickness have been applied. In Figure 10, a comparison with three elements (seven nodes) over the thickness is plotted.

The curves do not show significant differences, hence, the mesh sizes as shown in Figure 9 are applied for the parameter fitting and two elements over the beam's thickness are considered in the model. The convergence study is a crucial basis for the parameter fitting in Section 4.2 as the authors want to exclude any significant side effects like numerical errors.

In order to speed up the parameter study itself, a reduced-order model (ROM) is derived on the basis of the full order model (FOM). For the model order reduction (MOR) process, a first-order Krylov subspace method based on moment matching is used, since the system is proportionally damped [35]. As mentioned in [35], the obtained ROM is valid for variation of the Rayleigh damping coefficients and therefore also valid for a variation of the flexural modulus, which is just a linear factor for the stiffness matrix. Multi-point moment matching is applied to construct a global basis, which yields a small error over the entire frequency domain [36]. One expansion point is set every 1000 Hz (including 0 Hz and 6000 Hz) while the matched moments are increased until the maximum error Δh^2 is smaller than 0.1 dB. A Gramm–Schmidt orthogonalization is conducted at each step to construct an orthonormal basis and perform a vector-wise deflation strategy. For the convergence tests of the ROM, a frequency step size of $\Delta f = 20\,\text{Hz}$ is chosen. The loss factor is set to a conservative value of $\eta = 0.001$. In Table 5, the resulting necessary moments are documented—the number of matched moments is kept equal for all seven development points. The parameter fitting is conducted using the ROM while the final results are again computed by the FOM.

Table 5. Number of matched moments in the reduced-order model (ROM) in order to ensure a maximum $\Delta h^2 < 0.1$ dB for the parameter identification.

Model	Moments Matched
6 mm beam	2
3 mm beam	3
1 mm beam	5

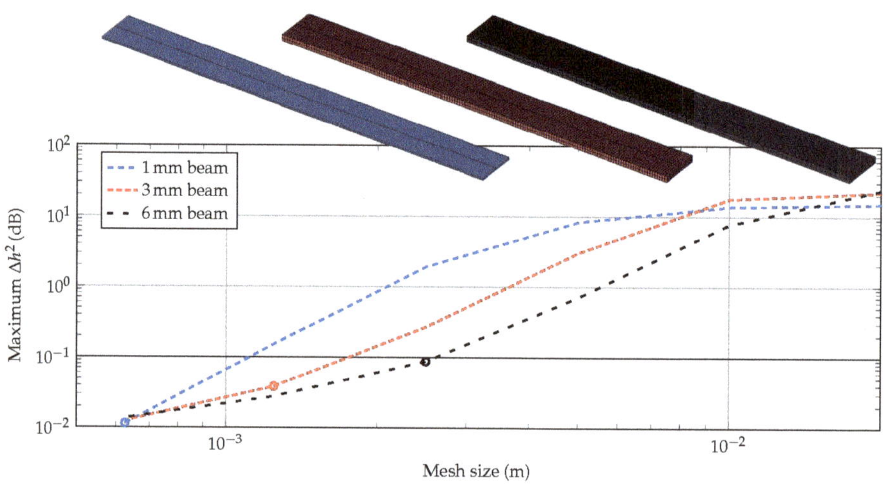

Figure 9. Maximum error in dependency on mesh sizes for each beam specimen with two elements over the beam's thickness (chosen mesh size marked with circle).

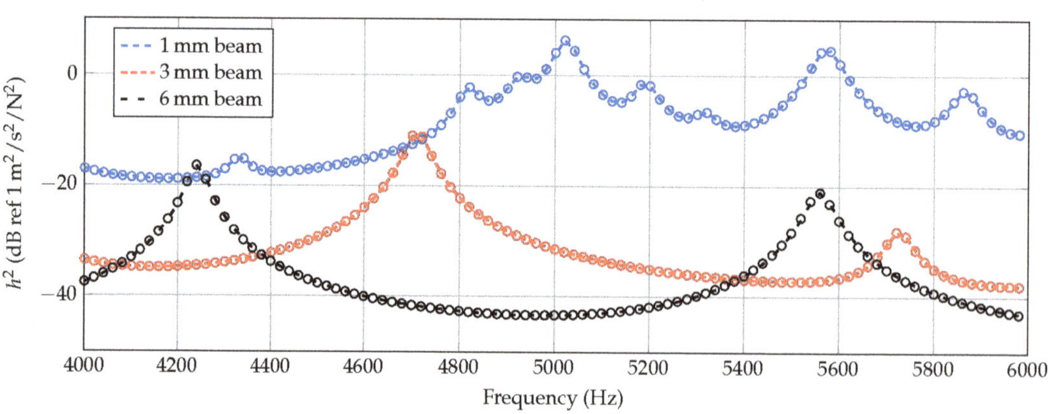

Figure 10. Comparison of h^2 for two elements (dashed line) and three elements (marks) in thickness direction for each beam.

4.2. *Parameter Identification*

As mentioned above, the parameter space is discretized and studied entirely. For E and η, the ranges are set according to Table 6.

Table 6. Parameter space including assumed constants.

Parameter		Unit	Range	Delta
Flexural modulus	E	N/m^2	1.5×10^9–6.0×10^9	0.06×10^9
Loss factor	η		0.005–0.15	0.005
Density	ρ	kg/m^3	constant (see Table 2)	
Poisson's ratio	ν		constant (0.35)	

For the identification of optimal parameters, two criteria are applied comparing the experimental curve h_α^2 and the numerical curves of h_β^2. The first criterion δ_1 is the sum of errors in dB at each frequency sampling points i of the total points N_f according to Equation (2). The optimal response is assumed to be the one leading to min(δ_1).

$$\delta_1 = \sum_{i=1}^{N_f} \left| h_{\beta,i}^2 - h_{\alpha,i}^2 \right| \tag{2}$$

As second criterion, Frequency Response Assurance Criterion (FRAC) is applied comparing two frequency responses by scalar multiplication and normalization. A FRAC value of 1 identifies curves with identical course. The absolute level is ignored by FRAC. In Equation (3), the criterion is defined according to [37] and adopted for the application here. The energy quantity h^2 is directly taken for FRAC with $H_{\alpha/\beta} = 10^{h^2/10}$.

$$\text{FRAC} = \frac{\left(H_\alpha(f)^T \cdot H_\beta(f)\right)^2}{\left(H_\alpha(f)^T \cdot H_\alpha(f)\right)\left(H_\beta(f)^T \cdot H_\beta(f)\right)}. \tag{3}$$

4.2.1. Frequency-Independent Parameters

As the first step, constant parameters E and η are applied to the entire frequency range. As criterion, δ_1 is applied. In Figure 11, the results are shown for all six specimens separately. For each beam, a best-fitting combination of constant E and constant η can be identified which are marked, respectively.

In Table 7, the resulting parameter combinations are listed. A large variation of the resulting constant parameter combinations can be clearly seen which may be a result by the thickness difference or process parameters in general. However, slight tendencies in dependence on the thickness can be observed. With decreasing thickness, the homogenized flexural modulus E and the homogenized loss factor η seem to be decreased as well.

Table 7. Frequency-independent identified parameter combinations with minimum δ_1.

Specimen	Thickness (m)	Flexural Modulus (N/m^2)	Loss Factor
Beam_6a	0.0063	4.20×10^9	0.090
Beam_6b	0.0062	4.44×10^9	0.095
Beam_3a	0.0032	4.38×10^9	0.080
Beam_3b	0.0033	3.96×10^9	0.060
Beam_1a	0.0012	3.42×10^9	0.055
Beam_1b	0.0012	3.24×10^9	0.065

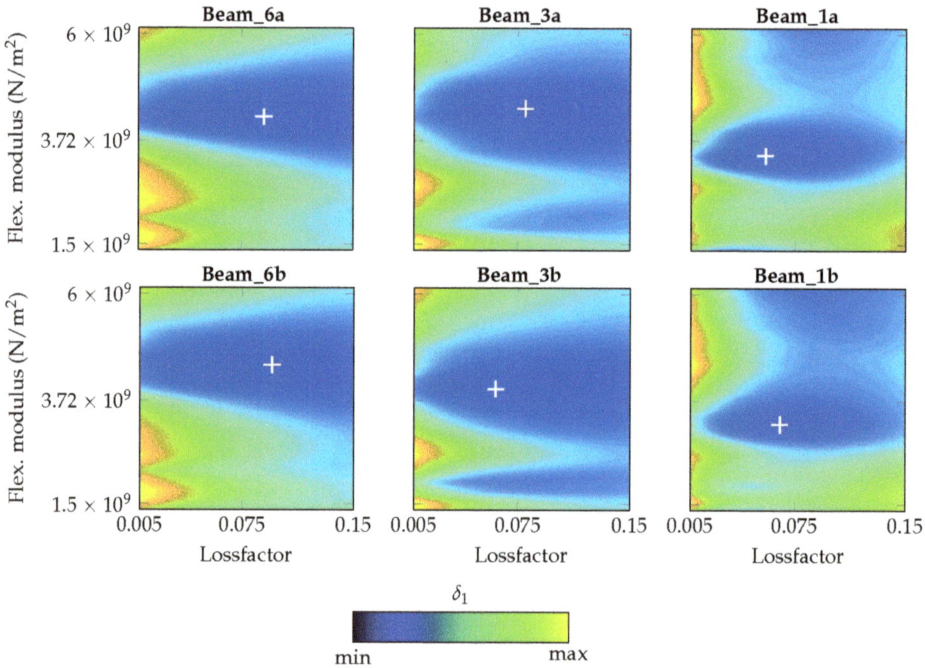

Figure 11. δ_1 contour plot for each specimen with identified optima under variation of E and η (frequency-independent, constant values).

4.2.2. Frequency-Dependent Parameters

The frequency-independent material parameters for each specimen (Table 7) are set as basis for the frequency-dependent material parameter identification for E. This means for the following parameter study, a constant η is continuously applied for each beam. Similar to the previous studies, E is varied, but now, FRAC is applied as the criterion. Every 500 Hz, a sectionwise FRAC is determined using a range of $\pm$500 Hz. For each sample point (every 500 Hz), the maximum FRAC value is assumed to indicate the best-fitting flexural modulus. Discontinuities in $E(f)$ are suppressed in the identification process by allowing only a maximum difference of 0.6 $\times 10^9$ N/m^2 from one sample to the next. The motivation is to avoid non-physical jumps. Finally, the contour plots in Figure 12 are created which show the FRAC distribution over frequency and flexural modulus. The identified values are marked by the $+$. On the first view, the procedure works quite well with the exception of Beam_1a and Beam_3a. For these specimens, the identified values are not laying on a recognizable curve as FRAC is indicating several best-fitting flexural moduli for one frequency sample.

However, in Figures 13–15, the identified data points are plotted for each beam thickness, respectively. By use of the data points, a linear curve fit is applied in order to receive a practicable mean curve. For Beam_6a and Beam_6b, the two resulting curves are quite similar which indicates a robust manufacturing and identification process. For Beam_3a/b and Beam_1a/b, the two curve are different. According to Figure 12, some outliers and the explanation above may be reasons for this deviation.

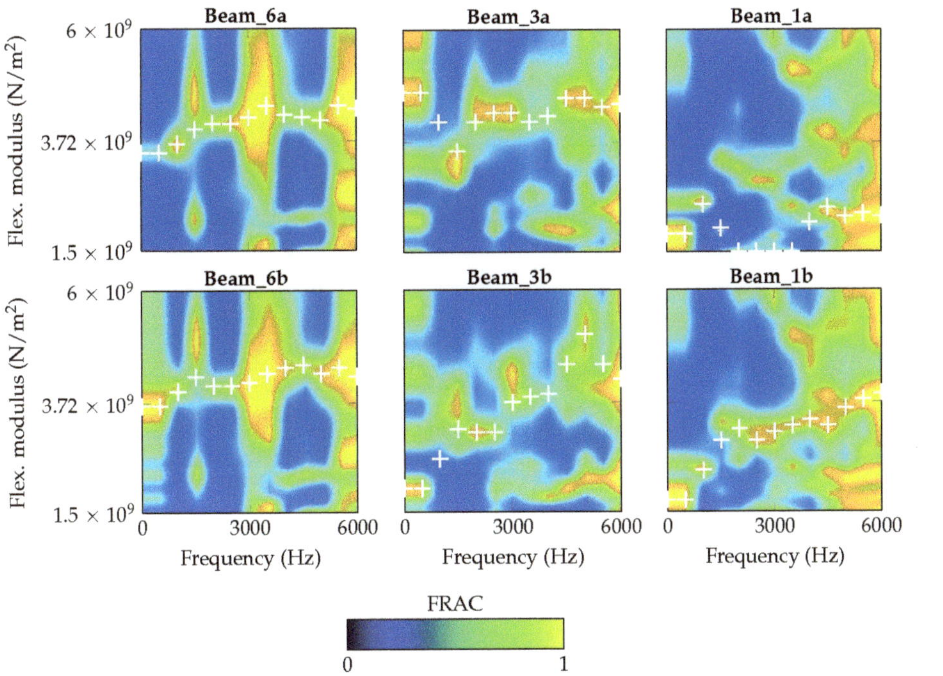

Figure 12. FRAC contour plot for each specimen with identified optima under variation of E (frequency-dependent, constant value for η).

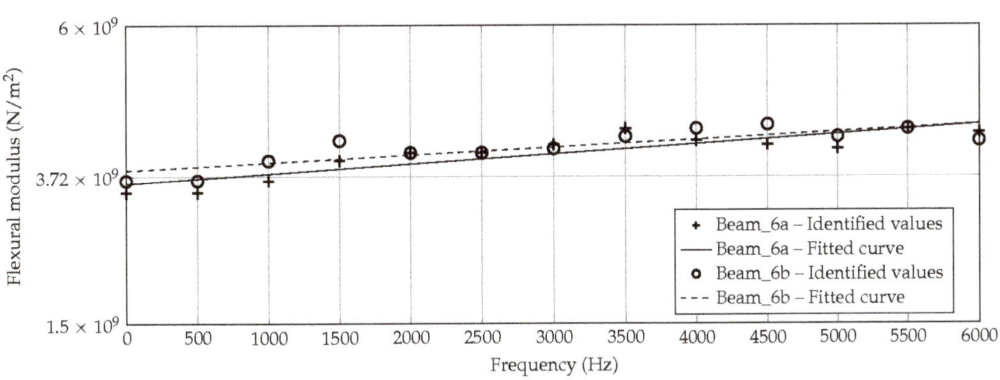

Figure 13. Identified flexural moduli by Figure 12 for Beam_6a/b with fitted linear curve.

The identification process could be improved by a more robust application of the criterion, for instance by a consideration of several dof as intended by [37]. This way, deflection shapes may be matched better and flexural moduli matching different peaks in the h^2 curve are excluded inherently. Keeping the problems by the identification process in mind, nevertheless an increasing flexural modulus can be observed with increasing frequency for all three thicknesses and all six specimens. Furthermore, the authors exclude a significantly high error due to the measurement setup, as Figures 7 and 8 show a quite well agreement of the two specimen's h^2 curves, respectively.

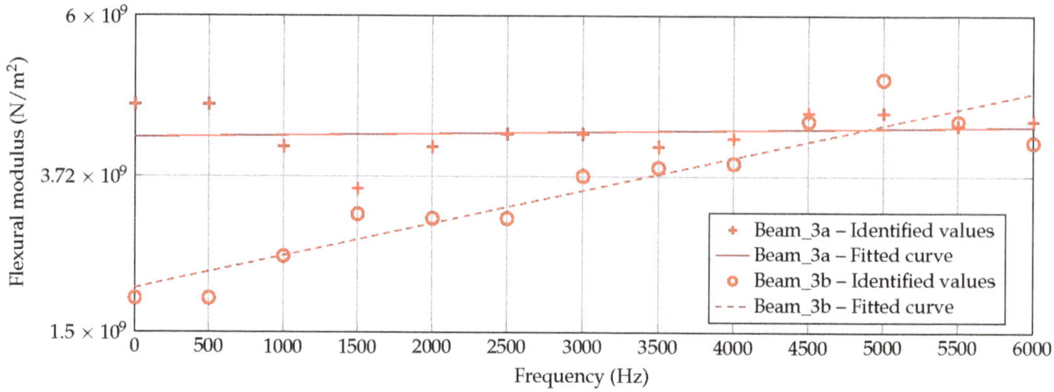

Figure 14. Identified flexural moduli by Figure 12 for Beam_3a/b with fitted linear curve.

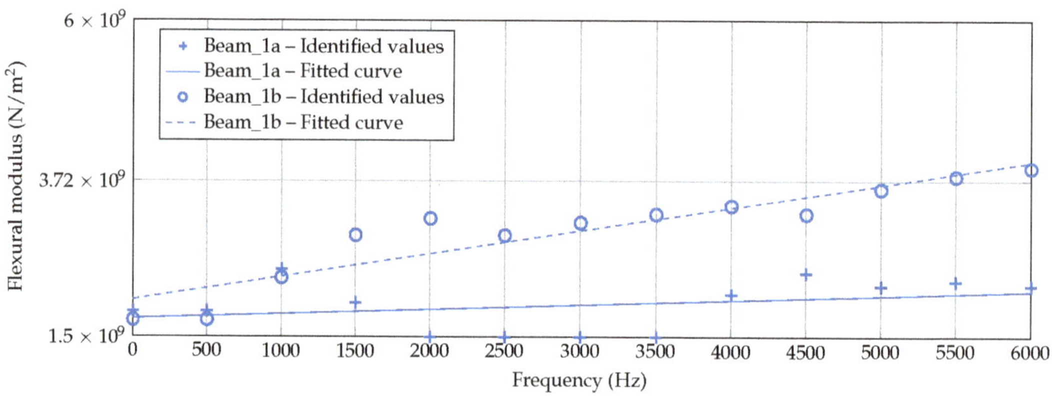

Figure 15. Identified flexural moduli by Figure 12 for Beam_1a/b with fitted linear curve.

Finally, all sampling points by Figure 12 are used for a fitting per thickness. In Figure 16, the overall fitted linear curves (a) and the distribution of E in dependency on frequency and thickness (b) are shown. For the contour plot (b), a piecewise cubic interpolation including extrapolation is applied for the thickness-dimension. Differently, a linear fitting as shown in Figure 16a is applied for the frequency dimension. As motivated in Section 1, a dependency of the flexural modulus on the additively manufactured thickness can be clearly observed. Generally speaking, the flexural modulus decreases with the decreased thickness of the structure. For 0 Hz, the range is between $1.9 \times 10^9\,\mathrm{N/m^2}$ and $3.7 \times 10^9\,\mathrm{N/m^2}$ which is about twice the value due to a 6 times thicker beam. The sensitivity reduces comparing the 3 mm beam and the 6 mm beam which indicates a non-linear behavior of the flexural modulus in dependency on the thickness. For ABH structures, besides the usual stiffness reduction by a lowered thickness, an additional effect can be expected by the manufacturing process itself. Considering this effect in mechanical models, a more precise and reliable prediction may be the result.

In the second step, the loss factors η are identified by the use of the frequency-dependent flexural moduli. For each significant peak in the experimental curve, the frequency range around ($\pm 10\%$) is investigated with a $\Delta f = 1\,\mathrm{Hz}$. By optical checks and manual adjustment, the loss factors are adapted in a way that the peak levels fit. An automatized routine using δ_1 as the criterion is hardly realizable as a slight shift in frequency lead to high differences though the amplitude of close peaks might be similar. This means the results shown in Figure 17 may give only tendencies.

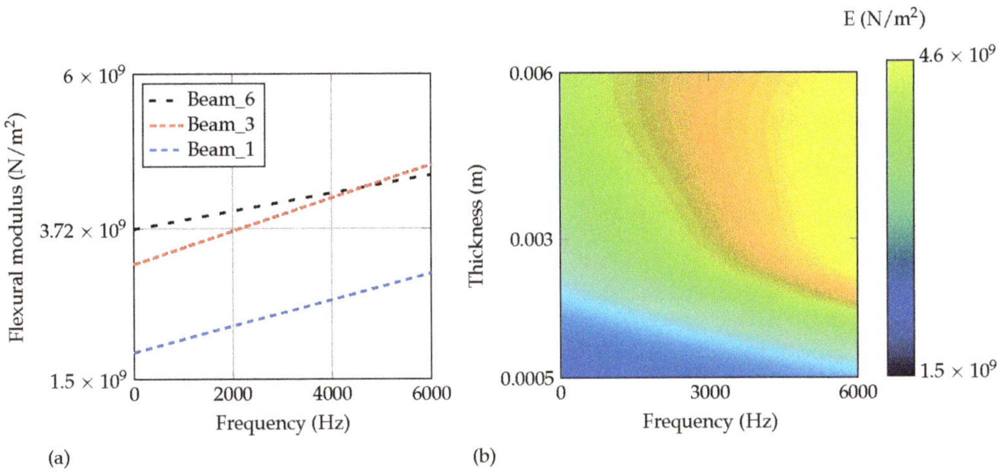

Figure 16. Overall fitted (**a**) linear curves for the three different manufactured thicknesses and (**b**) flexural modulus in dependency on frequency and thickness.

From 2000 Hz, similar loss factors between 0.02 and 0.06 are yielded by all six beams, independent of their thickness. The tendency of all curves shows a slight decrease towards higher frequencies. Below 2000 Hz, a spreading of the values are visible in dependency on thickness. This spreading cannot be identified to be systematic. For example, for both of the 3 mm beams, the homogenized η is increasing up to 0.1 towards lower frequencies.

In opposite, η shows a maximum for both 6 mm beams at around 1500 Hz. This behavior has not been expected and may be assigned to unknown effects based on the micro-structure. Due to the manufacturing process, the layers might act similar to a constraint layer damping. However, at this stage this cannot be further emphasized as only two 6 mm beams are available. In addition, the experimental setup might significantly influence some of the peaks.

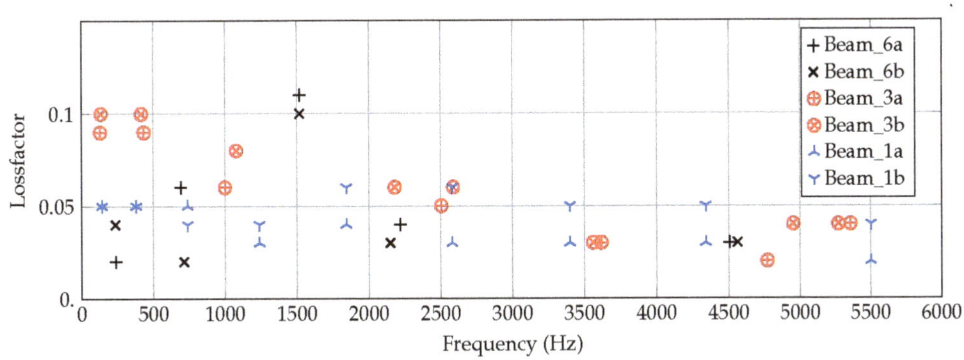

Figure 17. Identified loss factors η for all six specimens in dependency on frequency f.

4.3. Final Results

The identified frequency-dependent material parameters by Section 4.2 are applied to each beam specimen and solved with $\Delta f = 20$ Hz. For each specimen, the corresponding fitting curve by Figures 13–15 are considered. In Figure 18, h^2 is plotted in comparison with the experiment for each specimen, respectively. The results show a quite good agreement for Beam_6a/b, but with decreasing thickness (Beam_3 and Beam_1) the agreement gets worse. Thinner beams are expected to show higher sensitivities to the manufacturing process and the material parameter identification. Nevertheless, the tendency of a decreased flexural modulus with decreased thickness seems to be appropriate for the problem.

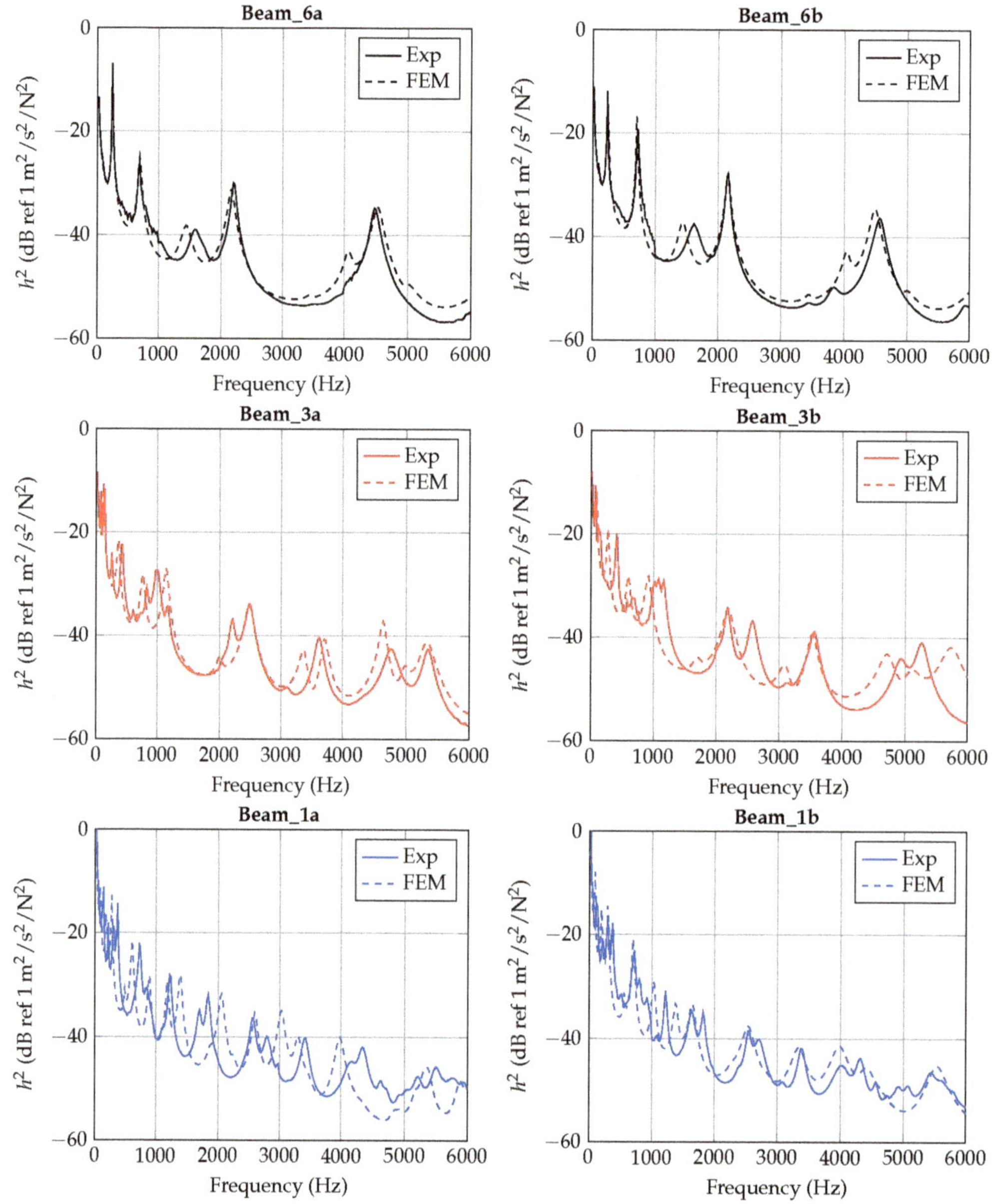

Figure 18. Experimental and numerical response of each specimen with frequency-dependent material parameters E and η.

5. Conclusions

The aim of the paper is to provide a procedure to identify material parameters of additively manufactured structures for robust and reliable dynamic simulations. AM provides great design freedom but at the same time makes valid mechanical modeling more challenging. As an illustrative example, the procedure is shown in the context of 3D-printed ABHs by using material extrusion. As ABH structures require a continuously adapted thickness profile, the dependency of homogenized material parameters on the thickness of additively manufactured beam structures is studied.

A dependency of Young's modulus and the loss factor on frequency and thickness can be observed based on the parameter fitting of the 3D model. The homogenized Young's modulus is decreased with a decreased thickness of the printed structure. Quantitatively, a doubling of the value can be identified due to a change from 1 mm to 3 mm thickness. A change from 3 mm to 6 mm induces a slight change which is no longer systematically. It is assumed that the flexural modulus converges with the thickness as the heat input becomes more homogeneous. The main findings can be summarized as follows:

- dependence of homogenized material parameter (Young's modulus, loss factor) on frequency and thickness
- Young's modulus decreases significantly with decreasing thickness

In order to improve the material parameter identification, further studies should focus on an improved criterion comparing the responses and the deflection shapes and an investigation of the inherent uncertainties by measurements and the manufacturing process. For this purpose, a larger number of samples should be analyzed. This way, mechanical models considering uncertain parameters may be applied for a robust design process of additively manufactured structures.

The method presented here is also universally applicable to other additive manufactured materials. Due to the combination with the model order reduction, even more complex fittings can be handled. In this way, it is possible to use the manifold possibilities of AM to optimize the performance of acoustic measures.

Author Contributions: S.R. identified the research idea, advised the specimen manufacturing, conducted all experiments and wrote the manuscript. C.B. conducted all computations, contributed to the research idea and the writing of the part for parameter identification. H.W. conducted all manufacturing of test specimens, contributed to the research idea and the writing of the part for additive manufacturing. T.V. and S.C.L. provided input and reflection on the research idea and improvements of the manuscript. All authors have read and agreed to the published version of the manuscript.

Funding: This research received no external funding.

Data Availability Statement: Data sharing is not applicable to this article.

Acknowledgments: We acknowledge the support by the Faculty of Mechanical Engineering of the Technische Universität Braunschweig with the 12plus6 funding initiative and the Open Access Publication Funds of the Technische Universität Braunschweig.

Conflicts of Interest: The authors declare no conflict of interest.

Abbreviations

The following abbreviations are used in this manuscript:

3D	Three-dimensional
ABH	Acoustic Black Hole
AM	Additive manufacturing
δ_1	First error criterion
dof	degrees of freedom
E	Young's modulus

elPaSo	Elementary parallel solver (in-house code)
η	Loss factor
f	Frequency
F	Exciting Force
FOM	Full order model
FRAC	Frequency Response Assurance Criterion
h^2	Mean squared admittance
H	Frequency response function
MEX	Material extrusion
MOR	Model order reduction
MSA	Mean squared admittance
N_f	Number of frequency points
N_p	Number of surface points
ν	Poisson's ratio
PLA	Polyactic acid
ϱ	Density
ROM	Reduced order model
t	Thickness of beam specimens
TPU	Thermoplastic polyurethane
v	Surface velocity

References

1. Setaki, F.; Tenpierik, M.; Turrin, M.; van Timmeren, A. Acoustic absorbers by additive manufacturing. *Build. Environ.* **2014**, *72*, 188–200. [CrossRef]
2. Ring, T.P.; Langer, S.C. Design, Experimental and Numerical Characterization of 3D-Printed Porous Absorbers. *Materials* **2019**, *12*, 3397. [CrossRef] [PubMed]
3. Bowyer, E.P. Acoustic black hole manufacturing for practical applications and the effect of geometrical and material imperfection. In *INTER-NOISE and NOISE-CON Congress and Conference Proceedings*; Institute of Noise Control Engineering: Hamburg, Germany, 2016.
4. Mironov, M. Propagation of a flexural wave in a plate whose thickness decreases smoothly to zero in a finite interval. *Sov. Phys. Acoust. USSR* **1988**, *34*, 318–319.
5. Krylov, V.V.; Shuvalov, A.L. Propagation of localised flexural vibrations along plate edges described by a power law. *Proc. Inst. Acoust.* **2000**, *22*, 263–270.
6. Krylov, V.V.; Winward, R.E.T.B. Experimental investigation of the acoustic black hole effect for flexural waves in tapered plates. *J. Sound Vib.* **2007**, *300*, 43–49. [CrossRef]
7. Georgiev, V.B.; Cuenca, J.; Gautier, F.; Simon, L.; Krylov, V.V. Damping of structural vibrations in beams and elliptical plates using the acoustic black hole effect. *J. Sound Vib.* **2011**, *330*, 2497–2508. [CrossRef]
8. Bowyer, E.P.; Krylov, V.V. Sound radiation of rectangular plates containing tapered indentations of power-law profile. In *Proceedings of Meetings on Acoustics 164ASA*; Acoustical Society of America: Melville, NY, USA, 2012; Volume 18, p. 030002.
9. Bowyer, E.P.; Krylov, V.V. Damping of flexural vibrations in turbofan blades using the acoustic black hole effect. *Appl. Acoust.* **2014**, *76*, 359–365. [CrossRef]
10. Bowyer, E.P.; Krylov, V.V. A review of experimental investigations into the acoustic black hole effect and its applications for reduction of flexural vibrations and structure-borne sound. In Proceedings of the INTER-NOISE 2015, San Franscisco, CA, USA, 9–12 August 2015.
11. Conlon, S.C.; Fahnline, J.B.; Semperlotti, F. Numerical analysis of the vibroacoustic properties of plates with embedded grids of acoustic black holes. *J. Acoust. Soc. Am.* **2015**, *137*, 447–457. [CrossRef]
12. Unruh, O.; Blech, C.; Monner, H.P. Numerical and experimental study of sound power reduction performance of acoustic black holes in rectangular plates. *SAE Int. J. Passeng. Cars Mech. Syst.* **2015**, *8*, 956–963. [CrossRef]
13. Rothe, S.; Ghaffari Mejlej, V.; Langer, S.C.; Vietor, T. Optimal adaptation of acoustic black holes by evolutionary optimization algorithms. *PAMM* **2016**, *16*, 625–626. [CrossRef]
14. Rothe, S.; Dorn, M.; Langer, S.C. Numerical study on shape functions for optimal exploitation of the acoustic black hole effect. *PAMM* **2020**, *20*.
15. Bowyer, E.P.; Nash, P.; Krylov, V.V. Damping of flexural vibrations in glass fibre composite plates and honeycomb sandwich panels containing indentations of power-law profile. *PAMM* **2012**, *18*, 030004.
16. Dorn, M.; Blech, C.; Langer, S.C. Numerical studies on the optimal exploitation of the acoustic black hole effect in curved panels. In Proceedings of the INTER-NOISE and NOISE-CON Congress and Conference, Hong Kong, China, 27–30 August 2017.
17. Blech, C.; Langer, S.C. Aircraft cabin noise reduction by means of acoustic black holes. In Proceedings of the INTER-NOISE and NOISE-CON Congress and Conference, Hong Kong, China, 27–30 August 2017.

18. Bowyer, E.P.; Krylov, V.V. Experimental investigation of damping flexural vibrations in glass fibre composite plates containing one-and two-dimensional acoustic black holes. *Compos. Struct.* **2014**, *107*, 406–415. [CrossRef]
19. Bowyer, E.P.; Krylov, V.V. Slots of power-law profile as acoustic black holes for flexural waves in metallic and composite plates. *Structures* **2016**, *6*, 48–58. [CrossRef]
20. Zhao, C.; Prasad, M.G. Acoustic Black Holes in Structural Design for Vibration and Noise Control. In *Acoustics*; Multidisciplinary Digital Publishing Institute: Basel, Switzerland, 2019; Volume 1, pp. 220–251.
21. Pelat, A.; Gautier, F.; Conlon, S.C.; Semperlotti, F. The acoustic black hole: A review of theory and applications. *J. Sound Vib.* **2020**, *476*, 115316. [CrossRef]
22. Rothe, S.; Watschke, H.; Langer, S.C. Study on the producibility of additively manufactured acoustic black holes. In Proceedings of the International Congress on Sound and Vibration, London, UK, 23–27 July 2017.
23. Rothe, S.; Blech, C.; Watschke, H.; Vietor, T.; Langer, S.C. Numerical modelling of additively manufactured acoustic black holes. In Proceedings of the Inter-Noise 2018 47th International Congress and Exposition on Noise Control Engineering, Chicago, IL, USA, 26–29 August 2018.
24. El-Ouahabi, A.A.; Krylov, V.V.; O'Boy, D.J. Investigation of the acoustic black hole termination for sound waves propagating in cylindrical waveguides. In Proceedings of the 44th International Congress and Exposition on Noise Control Engineering, San Franscisco, CA, USA, 9–12 August 2015.
25. Deng, J.; Zheng, L.; Zeng, P.; Zuo, Y.; Guasch, O. Passive constrained viscoelastic layers to improve the efficiency of truncated acoustic black holes in beams. *Mech. Syst. Signal Process.* **2019**, *118*, 461–476. [CrossRef]
26. Chong, B.M.P.; Tan, L.B.; Lim, K.M.; Lee, H.P. A Review on Acoustic Black-Holes (ABH) and the Experimental and Numerical Study of ABH-Featured 3D Printed Beams. *Int. J. Appl. Mech.* **2017**, *9*, 1750078. [CrossRef]
27. Rothe, S.; Blech, C.; Watschke, H.; Vietor, T.; Langer, S.C. Layer-effect by additive manufacturing of acoustic black holes. In Proceedings of the INTER-NOISE and NOISE-CON Congress and Conference, Madrid, Spain, 16–19 June 2019.
28. Gibson, I.; Rosen, D.; Stucker, B. *Additive Manufacturing Technologies: 3D Printing, Rapid Prototyping, and Direct Digital Manufacturing*; Springer: New York, NY, USA, 2015.
29. Gebhardt, A.; Hötter, J.S. *Additive Manufacturing: 3D Printing for Prototyping and Manufacturing*; Carl Hanser Verlag: Munich, Germany, 2016.
30. Wohlers, T.T. *Wohlers Report 2018: 3D printing and additive manufacturing state of the industry, Annual Worldwide Progress Report*; Technical Report; Wohlers Associates: Fort Collins, CO, USA, 2018.
31. Mohamed, O.A.; Masood, S.H.; Bhowmik, J.L. Optimization of fused deposition modeling process parameters: A review of current research and future prospects. *Adv. Manuf.* **2015**, *3*, 42–53. [CrossRef]
32. Kumke, M.; Watschke, H.; Hartogh, P.; Bavendiek, A.K.; Vietor, T. Methods and tools for identifying and leveraging additive manufacturing design potentials. *Int. J. Interact. Des. Manuf. IJIDeM* **2018**, *12*, 481–493. [CrossRef]
33. Wang, J.Y.; Xu, D.D.; Sun, W.; Du, S.M.; Guo, J.J.; Xu, G.J. Effects of nozzle-bed distance on the surface quality and mechanical properties of fused filament fabrication parts. In *IOP Conference Series: Materials Science and Engineering*; IOP Publishing: London, UK, 2019; Volume 479, p. 12094.
34. Schauer, M.; Langer, S.C.; Roman, J.E.; Quintana-Orti, E.S. Large scale simulation of wave propagation in soils interacting with structures using FEM and SBFEM. *J. Comput. Acoust.* **2011**, *19*, 75–93. [CrossRef]
35. Eid, R.; Salimbahrami, B.; Lohmann, B.; Rudnyi, E.B.; Korvink, J.G. *Parametric Order Reduction of Proportionally Damped Second Order Systems*; Technical Report; Lehrstuhl für Regelungstechnik: München, Germany, 2006.
36. Hetmaniuk, U.; Tezaur, R.; Farhat, C. Review and assessment of interpolatory model order reduction methods for frequency response structural dynamics and acoustics problems. *Int. J. Numer. Methods Eng.* **2012**, *90*, 1636–1662. [CrossRef]
37. Heylen, W.; Lammens, S. FRAC: A Consistent way of Comparing Frequency Response Functions. In Proceedings of the International Conference on Identification in Engineering, Swansea, UK, 1 March 1996.

materials

MDPI

Article

A Practical Finite Element Modeling Strategy to Capture Cracking and Crushing Behavior of Reinforced Concrete Structures

Alexandre Mathern [1,2,*,†] and Jincheng Yang [3,*,†]

1 Division of Structural Engineering, Concrete Structures, Department of Architecture and Civil Engineering, Chalmers University of Technology, SE-412 96 Gothenburg, Sweden
2 Research and Innovation, NCC, Lilla Bomen 3c, SE-411 04 Gothenburg, Sweden
3 Division of Structural Engineering, Steel and Timber Structures, Department of Architecture and Civil Engineering, Chalmers University of Technology, SE-412 96 Gothenburg, Sweden
* Correspondence: alexandre.mathern@chalmers.se (A.M.); jincheng.yang@chalmers.se (J.Y.)
† These authors contributed equally to this work.

Abstract: Nonlinear finite element (FE) analysis of reinforced concrete (RC) structures is characterized by numerous modeling options and input parameters. To accurately model the nonlinear RC behavior involving concrete cracking in tension and crushing in compression, practitioners make different choices regarding the critical modeling issues, e.g., defining the concrete constitutive relations, assigning the bond between the concrete and the steel reinforcement, and solving problems related to convergence difficulties and mesh sensitivities. Thus, it is imperative to review the common modeling choices critically and develop a robust modeling strategy with consistency, reliability, and comparability. This paper proposes a modeling strategy and practical recommendations for the nonlinear FE analysis of RC structures based on parametric studies of critical modeling choices. The proposed modeling strategy aims at providing reliable predictions of flexural responses of RC members with a focus on concrete cracking behavior and crushing failure, which serve as the foundation for more complex modeling cases, e.g., RC beams bonded with fiber reinforced polymer (FRP) laminates. Additionally, herein, the implementation procedure for the proposed modeling strategy is comprehensively described with a focus on the critical modeling issues for RC structures. The proposed strategy is demonstrated through FE analyses of RC beams tested in four-point bending—one RC beam as reference and one beam externally bonded with a carbon-FRP (CFRP) laminate in its soffit. The simulated results agree well with experimental measurements regarding load-deformation relationship, cracking, flexural failure due to concrete crushing, and CFRP debonding initiated by intermediate cracks. The modeling strategy and recommendations presented herein are applicable to the nonlinear FE analysis of RC structures in general.

Keywords: reinforced concrete; finite element analysis; crack band; strain localization; post-peak softening; viscoplastic regularization; convergence; mesh sensitivity; bond–slip; flexural behavior

check for updates

Citation: Mathern, A.; Yang, J. A Practical Finite Element Modeling Strategy to Capture Cracking and Crushing Behavior of Reinforced Concrete Structures. *Materials* **2021**, *14*, 506. https://doi.org/10.3390/ma14030506

Received: 28 December 2020
Accepted: 15 January 2021
Published: 21 January 2021

1. Introduction

Finite element (FE) analysis is effective for investigating the nonlinear behavior of reinforced concrete (RC) structures and performing parametric studies at lower costs than experimental tests. The nonlinear FE analysis of RC members has been extensively reported, and good agreement is often achieved between numerical and experimental results [1–3]. However, the comparison and application of the existing FE models are difficult owing to the differences in the adopted modeling strategies, which involve a considerable number of options, e.g., regarding the concrete constitutive models, critical parameters, bond between the concrete and the steel reinforcement, and numerical analysis procedures. Nonlinear FE analyses for blind predictions of the ultimate capacity and cracking of simple RC structural members have been associated with large uncertainty [4,5].

The major challenges in the nonlinear FE analysis of RC structures include the following: (1) defining the concrete tensile and compressive behaviors with the proper consideration of the strain localization in fracture zones [6], (2) efficiently assigning proper bond–slip behavior between the concrete and the steel reinforcement [2], (3) solving convergence difficulties commonly observed in the modeling of concrete with high nonlinearity [7,8], and (4) the misinterpretation of the processing logic of the FE software in the derivation and definition of input data. These challenges, which are described in detail below, render nonlinear FE analyses of RC complex and time-consuming; thus, such nonlinear analyses are rarely performed by practicing engineers.

Concrete cracking in tension: The cracking of tensile concrete is usually modeled by either discrete or smeared crack approach in practical FE analyses. In the discrete crack approach, physical cracks are modeled as displacement discontinuities in a concrete continuum. Although it allows the precise prediction of localized deformation at cracks, the discrete crack approach requires pre-defining tensile fracture zones when the finite element is generated. However, the position of the cracks is not known beforehand for most structural analyses. To overcome this limitation, automated re-meshing techniques are required to adapt the configuration of finite elements in accordance with the propagating cracks [9,10]. Although recent research has been devoted to developing discrete crack models allowing arbitrary crack initiation and propagation (e.g., the extended FE method [11]), the sophisticated modeling methods are not suitable for practical application by engineering practitioners. The ease of application motivates the wide use of the smeared crack approach in practical FE analyses. In the smeared crack approach, a crack width is transformed into an equivalent cracking strain smearing over a certain length (referred to as the "smeared length" in this paper). The numerical results correspond to reality only if the widths of the simulated fracture process zones (i.e., the cracking regions) are equal to the assumed smeared length [6,12,13]. To assure the reliability of the numerical results, the smeared length should be properly determined and incorporated into the definition of the constitutive law of concrete in tension.

Concrete crushing in compression: The constitutive model for concrete in uniaxial compression is usually provided as a stress–strain relationship in design codes, e.g., Model Codes [14,15] and ACI 318 [16]. However, such as a compressive strain, i.e., the "mean strain" obtained by smearing the measured deformation over the length of the standard test specimen, cannot describe the local strain-softening behavior in the critical fracture damaged zone of concrete. Studies since the 1980s [17–19] have investigated the effects of the strain localization on deriving the concrete compressive constitutive models. The challenge faced when considering the strain localization in FE analyses is that the actual size of simulated fracture zones is not known in advance, but it must be determined and used to modify the constitutive model of concrete as input data for the FE analyses. Zandi Hanjari et al. [20] modified the post-peak branch of the stress–strain relationship proposed by Thorenfeldt et al. [21], assuming that concrete crushing occurred in one row of concrete elements in the FE analyses of RC members. To properly define the concrete compressive behavior and reliably predict the capacity of RC members governed by concrete crushing, it is necessary to (1) clarify the principle of modifying the concrete compressive constitutive model with consideration of the strain localization and (2) develop a practical approach for determining the size of the fracture zones.

Bond–slip behavior between concrete and steel reinforcement: In the nonlinear FE analyses of RC structures, the definition of the bond–slip behavior between the concrete and the steel reinforcement is critical for the accurate prediction of the structural responses, crack patterns, and crack widths [2,3]. However, there is a lack of guidelines and different methods are used in the literature to assign the steel-concrete interaction, requiring different input data and workarounds to overcome implementation difficulties, which are often not described in detail [22,23]. Therefore, it is important for the research community to evaluate the existing approaches [2,24,25] for assigning the bond–slip behavior and to develop new approaches involving simple application procedures.

Convergence difficulties: The softening behavior and stiffness degradation of cracking or crushing concrete cause severe convergence difficulties in the static analysis of concrete [8, 26]. Instead of solving the problem in static analyses, researchers implemented dynamic analysis procedures adopting implicit [7] or explicit [1] integration methods. However, such a dynamic approach requires additional effort to carefully select, e.g., the time integration algorithm, loading scheme, loading time, damping ratio, and time increment size, to achieve a good balance between minimizing the inertial forces (for a better approximation of the static problem) and reducing the computational time (by using a shorter time to model the static event in an accelerated manner). Therefore, it remains important to develop a simple solution strategy to perform the static analysis procedure with a high convergence rate.

Misinterpretation: Misinterpretation refers to the users' misunderstanding of the processing logic of FE software packages. Misinterpretation may cause the incorrect definition of input data, which may lead to errors in numerical results or aborted analyses [27]. For instance, in nonlinear FE analyses of concrete structures, the definition of the post-peak softening behavior of concrete constitutive laws or the stiffness degradation of damaged concrete is not straightforward. Furthermore, such analyses require to assign a lot of input data and to make a great number of modeling choices, which are rarely reported in a very detailed way in the literature. This is explained by the fact that such details do not constitute the focus of the investigation and are usually software specific. Nevertheless, if the approach for obtaining critical input data is not reported, it can undermine the reliability and reproducibility of the FE analyses.

In light of these challenges, the objective of this study was to develop a robust and reliable modeling strategy to capture the tensile cracking and compressive crushing behavior of RC structures associated with low computational costs and ease of implementation, based on the well-established constitutive relations from fib Model Codes [14,15]. The strategy was implemented to simulate the flexural behavior of an RC beam as reference and another identical RC beam strengthened with an externally bonded carbon-reinforced polymer (CFRP) laminate; both beams were tested in four-point bending until failure. Modeling of crack opening after the reinforcement yielding stage and ultimate concrete crushing were carefully studied on the reference beam to ensure a reliable basis for the modeling of the strengthened beam. The nonlinear FE analyses presented herein were performed using the concrete damaged plasticity (CDP) model implemented in the commercial software ABAQUS [26], as it is widely used in both academia and industry to analyze RC structures [28–32]. The focus of this paper was to provide reliable, practical, and computationally cost-efficient implementation guidelines for nonlinear FE analyses of concrete structures, which can be used as a basis for more complex cases and support the application of nonlinear analyses to real-world engineering problems, e.g., for load-bearing assessment, strengthening assessment, structural health monitoring, and damage identification of building and civil engineering structures. For instance, the use of externally bonded FRP laminates for strengthening and rehabilitation of concrete or masonry structural members [33–35] has emerged as an effective technique and found strong interest in both research and practice, which supported the consideration of such a case in this work. The experimental setup is shown in Section 2. In Section 3, the modeling procedures and recommendations for overcoming the aforementioned challenges are presented in detail. In Section 4, the proposed modeling strategy is demonstrated, and modeling choices are validated by parametric studies considering the reference RC beam. In Section 5, the numerical results of the reference beam are compared with the experimental measurements regarding load-deformation relationship, cracking, flexural failure due to concrete crushing, and CFRP debonding initiated by intermediate cracks.

2. Experimental Test

The RC members modeled in the present FE study included two slender RC beams subjected to four-point bending in the laboratory, see Figure 1a. The RC beams were cast at a workshop using C35/45 concrete. One beam served as reference; the other one, with identical dimensions, was strengthened with an externally bonded CFRP laminate on the tensile side of the beam. The cross-sectional dimensions of the beams are shown in Figure 1b, including main steel rebars, shear reinforcement, and externally bonded CFRP plate (for the strengthened beam only). A two-component epoxy adhesive (StoPox SK41, StoCretec GmbH, Kriftel, Germany) was applied to bond the CFRP plate; whose layer design thickness was 1 mm. Mechanical properties of the above-mentioned materials used for the FE analyses are listed in Table 1: the elastic modulus E_c, compressive strength f_c, and tensile strength f_{ct} of concrete C35/45 at the age of 287 days were estimated according to Eurocode 2 [36]; the Poisson's ratio of concrete ν_c was defined according to Model Code 2010 [15]; the elastic modulus E_s, yield strength f_{sy}, ultimate strength f_{su}, and ultimate strain ε_{su} of steel reinforcement were measured by laboratory tests on bars with a diameter of 16 mm (Φ16) according to ASTM A615 [37]; the elastic modulus E_f and ultimate tensile strain ε_{fu} of the CFRP plate were measured according to standard tensile tests as reported in [38]; the elastic modulus E_a, tensile strength f_a, and Poisson's ratio ν_a of the epoxy adhesive were reported in [39].

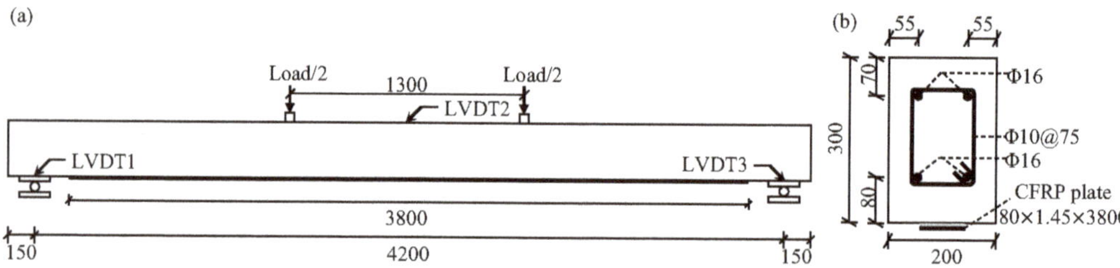

Figure 1. (**a**) Reinforced concrete (RC) beams subjected to four-point bending tests until failure (unit: mm); (**b**) cross-sectional dimension of the RC beams (with a carbon-reinforced polymer (CFRP) plate for the strengthened beam only).

Table 1. Material properties of concrete, steel rebars, CFRP plate, and cured epoxy adhesive.

Concrete C35/45		**Steel Rebars B500C**		**CFRP Plate**		**Adhesive**	
E_c	36.9 GPa	E_s	201 GPa	E_f	214 GPa	E_a	7.1 GPa
f_c	51.1 MPa	f_{sy}	510 MPa	ε_{fu}	12.7‰	f_a	34 MPa
f_{ct}	3.6 MPa	f_{su}	617 MPa	–	–	ν_a	0.3
ν_c	0.2	ε_{su}	12.0%	–	–	–	–

In the four-point bending tests, the beams were simply supported on two movable steel supports, giving an effective span of 4.2 m. The steel support at each end consisted of two identical steel plates (170 × 30 × 200 mm^3) and one steel cylinder roller (diameter of 50 mm and length of 200 mm). External loading was applied via two synchronized hydraulic jacks using displacement control. The foot of each hydraulic jack rested on a steel plate (50 × 50 × 200 mm^3) to distribute the load to the RC beam. The beams were loaded to failure. In the reference beam, flexural failure after yielding of the reinforcement was governed by concrete crushing in the compressive side of the beam; the failure of the strengthened beam was due to premature debonding of the CFRP plate initiated by intermediate flexural cracks.

During the test, strain gauges and linear variable differential transducers (LVDTs) were used to monitor the beam specimens. Two strain gauges were installed at the midspan cross-section on the tensile steel reinforcement, and three LVDTs were used to obtain the net deflection at the midspan, as shown in Figure 1. Cracks in the RC beams were monitored during the test; the crack widths were measured at the height of the tensile steel reinforcement using a digital handheld microscope (AM4115ZT, Dino-Lite Europe,

Almere, The Netherlands) at load levels of 15, 30, 45, 55 (reference beam only), and 70 kN (strengthened beam only).

3. FE Modeling Strategy

The proposed modeling strategy is discussed in detail, in this section, with special focus on proper modeling of RC considering strain-softening in fracture zones and bond–slip between steel reinforcement and concrete to ensure the reliable prediction of cracking and crushing. The strategy is adapted to the modeling of the RC beams introduced in Section 2. However, common critical issues in the nonlinear FE modeling of RC structures are addressed in a general manner, which makes these recommendations applicable to other types of RC beam and frame structures. The nonlinear FE analyses were conducted using the commercial FE package ABAQUS/CAE, version 6.14 [26].

Considering that the beam geometry and the test configuration were symmetric about the midspan, only one half of the RC beam was modeled in the current FE analyses to reduce the computational cost. The vertical load acting on the beam was defined as a boundary condition in the FE model, inducing a vertical displacement on the top of the steel plate between the load and the beam. The interaction between the steel plate and the beam was set as surface-to-surface contact, which defined the interfacial constraints in the normal direction (i.e., "hard" contact) and friction in the tangential direction. The coefficient of tangential friction was assumed to be 0.57 according to a previous recommendation [40]. The same contact settings were used at the interfaces between the beam and steel support. On the bottom side of the movable steel support, boundary conditions were defined at the middle point to constrain the degree of freedom in the vertical direction but allow translation in the horizontal direction. The concrete beam, steel plate under the loading point, and movable steel support were modeled with 2D shell, discretized into structured meshes, and assigned with element type CPS4 (4-node plane stress quadrilateral elements with four integration points). Steel reinforcement and CFRP plate, modeled as one-dimensional (1D) wire, were assigned with truss (T2D2) and beam (B21) elements, respectively. Material properties of the concrete, steel reinforcement, and CFRP were defined according to the values in Table 1. Details about the modeling of concrete, the interaction at the concrete and steel reinforcement interface and the concrete and CFRP plate interface, and the numerical solution strategy are described below according to the proposed modeling strategy.

3.1. *Modeling of Concrete*

The concrete material is defined in the CDP model implemented in ABAQUS [26], including the definition of the concrete plasticity, the tensile behavior, the compressive behavior, and the damage evolution of the stiffness.

3.1.1. Concrete Plasticity

The concrete plasticity parameters to be defined in the CDP model include (1) dilation angle φ and eccentricity factor ϵ related to the flow potential given by the Drucker–Prager hyperbolic function; (2) factors σ_{b0}/σ_{c0} and K_c related to the yield surface based on the function presented by Lubliner et al. [41] with the modifications proposed by Lee and Fenves [42] to account for the evolution of strength in tension and compression; and (3) the viscosity parameter μ to introduce viscoplastic regularization. The values of these parameters defined in the reference FE model are presented in Table 2. Default values of ϵ, σ_{b0}/σ_{c0}, and K_c are assigned according to the design manual of ABAQUS [26]; the values of φ and μ are defined based on the validation discussed in Sections 4.1 and 4.2.

Table 2. Plasticity parameters defined in the concrete damaged plasticity (CDP) model.

Categories	Plastic Flow Potential		Yield Surface		Viscosity Parameter
Parameters	φ	ϵ	σ_{b0}/σ_{c0}	K_c	μ
Values	35°	0.1	1.16	2/3	1×10^{-6}

3.1.2. Concrete Tensile Behavior

The tensile behavior of concrete is characterized by a linear elastic stress–strain relationship before the concrete reaches the tensile strength f_{ct} (Figure 2a) and a bilinear stress σ_t-crack width w relationship for the post-peak softening behavior according to Model Code 2010 [15]. The Model Code relationship, as shown in Figure 2b, is determined by the tensile strength f_{ct} and the fracture energy G_F. The fracture energy G_F describes the amount of energy required to propagate a tensile crack of unit area; for normal-strength concrete, G_F (in N/m or J/m^2) can be estimated using Equation (1) according to Mode Code 2010 [15]:

$$G_F = 73 f_c^{\,0.18}, \tag{1}$$

where f_c represents the mean compressive strength of concrete in MPa.

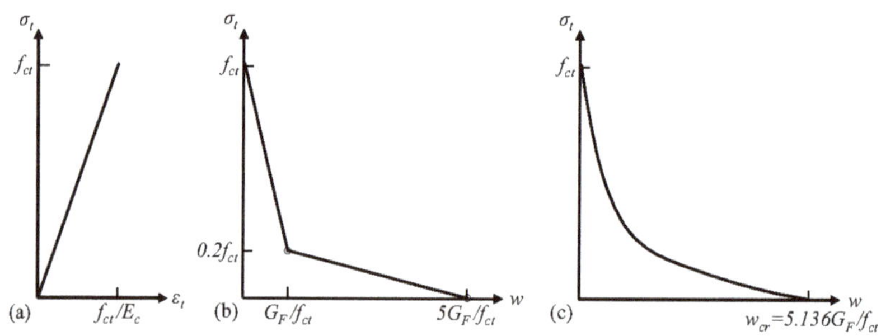

Figure 2. Constitutive models for concrete in tension: (**a**) linear elastic stress–strain relationship for uncracked concrete; (**b**) bilinear tensile stress–crack width relationship according to Model Code 2010 [15]; (**c**) exponential tensile stress–crack width relationship proposed by Hordijk [43].

To validate the adopted concrete tensile behavior, the effects of different modeling choices on the numerical results were investigated and are discussed in Section 4.3. This included (1) a comparison between the bilinear Model Code relation and another commonly used post-peak softening model, i.e., the exponential descending σ_t-w relation (Figure 2c) proposed by Hordijk [43], and (2) parametric studies of the assumed tensile strength f_{ct} and fracture energy G_F. The Hordijk σ_t-w relation is expressed by Equation (2):

$$\frac{\sigma_t}{f_{ct}} = \left[1 + \left(c_1 \frac{w}{w_{cr}}\right)^3\right] e^{-c_2 \frac{w}{w_{cr}}} - \frac{w}{w_{cr}}\left(1 + c_1^3\right) e^{-c_2}, \tag{2}$$

where, $c_1 = 3.0$, $c_2 = 6.93$, $w_{cr} = 5.136\frac{G_F}{f_{ct}}$.

Smeared crack method: To simulate the cracks in concrete, the CDP model adopted in the present FE study employs the smeared crack method, where the cracking concrete is treated as a continuum and a physical crack opening w_{cr} is characterized as an equivalent cracking strain ε_{cr} smearing over a certain length (the smeared length l_s). Thus, the post-peak tensile behavior defined in the CDP model complies with stress–cracking strain relationship, which is converted from the selected stress–crack width model given $\varepsilon_{cr} = w_{cr}/l_s$. The numerical results correspond to reality only if the width of the simulated fracture process zone l_F is equal to the assumed smeared length l_s. As indicated by Equation (3), the inelastic deformation $\delta(\sigma)$ of cracking concrete at a certain stress level σ is not dependent on the assumed l_s but determined by the selected stress–crack width relation in the present FE analyses.

$$\delta(\sigma) = l_F \varepsilon_{cr}(\sigma) = l_F \frac{w_{cr}(\sigma)}{l_s} = \{l_F = l_s\} = w_{cr}(\sigma). \tag{3}$$

Crack band approach: To define the smeared length l_s in accordance with the size of the simulated fracture zone l_F, the crack band approach—a simple technique for practical engineering computations—is adopted. In this approach, it is assumed that the strain-softening of cracking concrete is localized into a clear band of elements running across the concrete mesh and thus forming a "crack band." Thus, the size of the simulated fracture zone l_F becomes the width of the crack band h_b, and h_b can be estimated and assigned to the smeared length l_s to adjust the strain-softening behavior of the concrete in the post-peak regime [6]. The crack band approach, which is widely applicable and utilized in many FE packages, is based on pioneering studies performed in the 1980s [12,13,44,45]. Theoretically, the width of crack bands h_b is a function of not only the element topology, e.g., the type, shape, size, and integration scheme, but also the crack band orientation [6]. However, the width of crack bands h_b implemented in common FE packages, e.g., ABAQUS [26], is simply estimated as the square root of the element area (for two-dimensional elements) or the cubic root of the element volume (for three-dimensional elements). This simplified estimation may induce substantial error and mesh sensitivity for elements with large aspect ratios. It is recommended to use elements having aspect ratios close to 1 (e.g., square or cubic elements) to reduce the mesh sensitivity [26]. Even for square or cubic elements, there may be errors if the crack band is not aligned with the mesh line. For instance, for a two-dimensional mesh of square elements with side length a, it is reasonable to estimate the band width $h_b = \sqrt{A} = a$ automatically in ABAQUS only if the crack band is parallel to the element sides. If the crack band runs along the element diagonal, the appropriate width of the band is $h_b = \sqrt{2}a = \sqrt{2A}$ instead of $h_b = \sqrt{A}$. Accordingly, when using the crack band approach in FE modeling, it is recommended to define the post-peak tensile behavior of concrete by the input data of the stress–cracking strain relationship, which allows users to evaluate and determine the crack band width h_b. For a detailed discussion regarding the estimation of the crack band width with consideration of the element topology and crack band orientation, readers are referred to [12,46,47].

In the present FE analyses, the crack bands developed in the concrete mesh mainly ran parallel to the mesh lines, owing to the predominant bending effect on the beam. Thus, the crack band width h_b was determined as the width of square-shaped concrete elements for deriving the stress–cracking strain input data.

3.1.3. Concrete Compressive Behavior

For the FE analysis of RC beams in bending, the concrete compressive behavior is widely defined according to a stress–strain relationship obtained from uniaxial compression tests of standard concrete cylinders. However, the strain in the standard compressive test is the "mean" strain obtained by smearing the measured displacement over the total length of the specimen. Considering that the compressive failure of concrete is initiated by a local shear band formed in one of the fracture zones and the post-peak deformation mainly arises from such a local zone, the "mean" strain naturally underestimates the strain in the critical fracture zone. If the post-peak deformation is expressed by the mean strain, the strain-softening curve tends to depend on the geometry of the specimen [17,18]—the descent of the post-peak branch is faster for a longer specimen. For instance, the compressive stress–strain relationship provided in Model Codes [14,48] is reasonably accurate for a concrete specimen length of approximately 200 mm.

In the present FE study, the relationship based on Model Code 1990 [14] and Model Code 2010 [15] is selected as the reference constitutive model for concrete in compression. Model Code 1990 provides the part of the descending branch with strains exceeding the concrete ultimate/limit strain $\varepsilon_{c,lim}$. To highlight the differences among existing constitutive models for concrete compressive behavior and the impacts of these differences on the predicted ultimate crushing failure of the concrete beam, two other commonly used models are also studied for comparison (see Section 4.5): the Thorenfeldt relationship and the Saenz relationship. As shown in Table 3, the Thorenfeldt relationship is based on previous studies by Tomaszewicz [49] and Thorenfeldt et al. [21], with the modifications proposed by Collins and Porasz in CEB Bulletin 193 [50]. The Saenz relationship includes modifications [51] based on a previous discussion of the compressive stress–strain equation [52].

Table 3. Three commonly used constitutive models for concrete in compression.

Model Code	$0 < \varepsilon_c < \varepsilon_{c,\lim}$ $\sigma_c = f_c\left[k\frac{\varepsilon_c}{\varepsilon_{c1}} - \left(\frac{\varepsilon_c}{\varepsilon_{c1}}\right)^2\right] / \left[1 + (k-2)\frac{\varepsilon_c}{\varepsilon_{c1}}\right]$ $\varepsilon_c > \varepsilon_{c,\lim}$ $\sigma_c = f_c\left[\left(\frac{1}{e}\xi - \frac{2}{e^2}\right)\left(\frac{\varepsilon_c}{\varepsilon_{c1}}\right)^2 + \left(\frac{4}{e} - \xi\right)\frac{\varepsilon_c}{\varepsilon_{c1}}\right]^{-1}$	where $k = E_{ci}/E_{c1}$ $e = \varepsilon_{c,\lim}/\varepsilon_{c1}$ $\xi = 4\frac{[e^2(k-2) + 2e - k]}{[e(k-2) + 1]^2}$ $\varepsilon_{c,\lim} = \varepsilon_{c,1}\left[\frac{1}{2}\left(\frac{k}{2} + 1\right) + \sqrt{\frac{1}{4}\left(\frac{k}{2} + 1\right)^2 - \frac{1}{2}}\right]$
Thorenfeldt	$\sigma_c = \frac{E_c\varepsilon_c}{1 + \left(\frac{E_c\varepsilon_{c1}}{f_c} - 2\right)\left(\frac{\varepsilon_c}{\varepsilon_{c1}}\right) + \left(\frac{\varepsilon_c}{\varepsilon_{c1}}\right)^2}$	-
Saenz	$\sigma_c = f_c\frac{\varepsilon_c}{\varepsilon_{c1}}\left(\frac{n}{n - 1 + \left(\frac{\varepsilon_c}{\varepsilon_{c1}}\right)^{nk}}\right)$	where $n = 0.80 + \frac{f_c}{17}$ $k = \begin{cases} 1, & 0 < \varepsilon_c < \varepsilon_{c1} \\ 0.67 + \frac{f_c}{62}, & \varepsilon_c > \varepsilon_{c1} \end{cases}$

Note: $E_{ci} = E_c$ represents the initial tangent modulus, $E_{c1} = f_c/\varepsilon_{c1}$ represents the secant modulus from the origin to the peak compressive stress, $\varepsilon_{c1} = 1.60(f_c/10\ \text{MPa})^{0.25}/1000$ represents the strain corresponding to the peak compressive stress [48,53], and $\varepsilon_{c,\lim}$ represents the ultimate compressive strain of the concrete.

To obtain reliable predictions of the ultimate capacity and the crushing failure of the beam in flexural loading, the post-peak strain localization should be considered in the definition of the constitutive law. The original constitutive model based on the mean strain—mainly the post-peak softening branch—must be modified to better describe the localized strain softening in the critical fracture zone with a reasonable size. The assumed size should be verified according to the size of the simulated crushing zone.

Modified compressive behavior considering strain localization: The procedure for modifying the post-peak branch is presented in Figure 3. If the stress–mean strain relationship and the length L of the tested specimen are known, the increased post-peak deformation $\delta|_{c1}^{cs}$ can be obtained from the stress–strain relationship using Equation (4):

$$\delta|_{c1}^{cs} = L(\varepsilon_{cs} - \varepsilon_{c1}), \tag{4}$$

where, ε_{c1} and ε_{cs} represent the strains (mean strains smeared over the whole specimen) corresponding to the concrete compressive strength f_c and the stress level σ_{cs} in the softening branch, respectively.

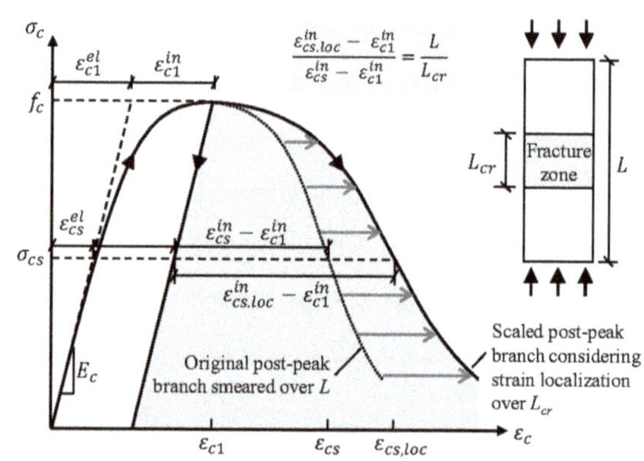

Figure 3. Modification of the post-peak softening branch of the constitutive model for concrete in compression originally expressed by the mean strain smearing over the whole length of specimen L to consider the strain localization in the critical fracture zone with the length of L_{cr}.

Considering the strain localization, instead of using Equation (4), the increased deformation after the peak stress $\delta|_{c1}^{cs}$ is reached can be calculated using Equation (5), taking into account the strain-softening in the fracture zone with the length of L_{cr} and the elastic unloading due to the reduction in the compressive stress from the peak stress f_c to σ_{cs}:

$$\delta|_{c1}^{cs} = (L - L_{cr})(\varepsilon_{cs}^{el} - \varepsilon_{c1}^{el}) + L_{cr}[(\varepsilon_{cs}^{el} + \varepsilon_{cs.loc}^{in}) - (\varepsilon_{c1}^{el} + \varepsilon_{c1}^{in})] = L(\varepsilon_{cs}^{el} - \varepsilon_{c1}^{el}) + L_{cr}(\varepsilon_{cs.loc}^{in} - \varepsilon_{c1}^{in}), \quad (5)$$

where, $\varepsilon_{cs}^{el} - \varepsilon_{c1}^{el}$ represents the change in elastic strain due to the stress reduction from f_c to σ_{cs} in the post-peak regime, $\varepsilon_{cs.loc}^{in}$ represents the inelastic strain localized in the critical fracture process zone, and ε_{c1}^{in} represents the inelastic strain at peak stress.

The transitivity between Equations (4) and (5) implies that,

$$L(\varepsilon_{cs} - \varepsilon_{c1}) = L\left(\varepsilon_{cs}^{el} - \varepsilon_{c1}^{el}\right) + L_{cr}\left(\varepsilon_{cs.loc}^{in} - \varepsilon_{c1}^{in}\right), \quad (6)$$

which can be rewritten as,

$$\varepsilon_{cs.loc}^{in} = \frac{L}{L_{cr}}\left(\varepsilon_{cs}^{in} - \varepsilon_{c1}^{in}\right) + \varepsilon_{c1}^{in}. \quad (7)$$

The scaling rule for determining the localized inelastic strain $\varepsilon_{cs.loc}^{in}$ in the critical fracture zone is expressed by Equation (7), as shown in Figure 3.

Identifying the size of crushing zone: The modification of the post-peak strain-softening behavior also requires a proper assumption of the size of the critical fracture zone L_{cr}, which should be validated according to the size of the simulated crushing zone in the FE analysis. Herein, an iterative procedure is proposed for identifying the value of L_{cr} via a reasonable approach. As indicated by the flowchart of Figure 4, the assumed value $L_{cr.input}$ used to modify the post-peak branch of the compressive behavior as input data should be verified according to the observed length $L_{cr.output}$ of the simulated crushing zone in the numerical result. $L_{cr.output}$ can be visualized by contour plots highlighting the regions with compressive strains larger than $\varepsilon_{c,lim}$.

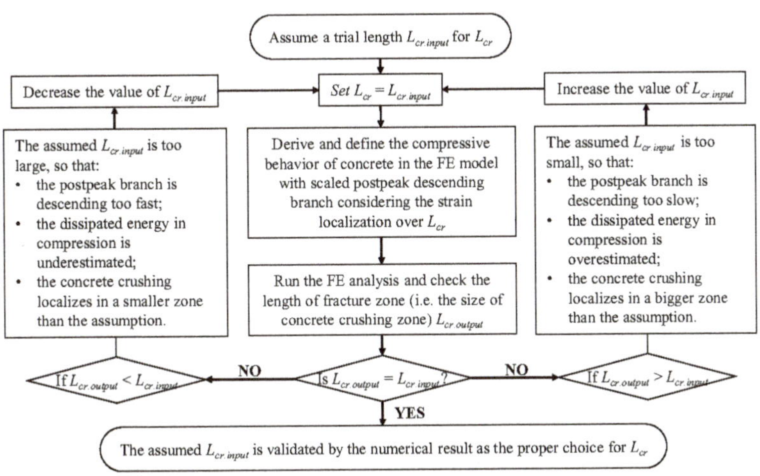

Figure 4. Proposed iterative procedure for identifying the reasonable length L_{cr} of the critical fracture zone.

As the reference constitutive model adopted in the present FE analysis, the Model Code relation is reasonably accurate for specimens with a length of approximately 200 mm [14], providing a fair benchmark to modify the post-peak branch with the specified length $L = 200$ mm according to the scaling rule in Figure 3. The reasonable length of the critical fracture zone is identified as $L_{cr} = 100$ mm via the proposed iterative procedure shown in Figure 4. Details regarding the identification of L_{cr} are presented in Section 4.5.

3.1.4. Concrete Damage Evolution

In the adopted CDP model, the concrete damage evolution is characterized by the degradation of the material stiffness (i.e., the elastic modulus of concrete) in the post-peak regime of the constitutive law. Figure 5 presents a generic stress–strain relationship (Equation (8)) for concrete in uniaxial tension or compression:

$$\varepsilon(\sigma) = \varepsilon_{\rm el}(\sigma) + \varepsilon_{\rm in}(\sigma) = \frac{\sigma}{E_{\rm c0}} + \varepsilon_{\rm in}(\sigma), \tag{8}$$

where $\varepsilon_{\rm el}(\sigma)$, $\varepsilon_{\rm in}(\sigma)$, and $\varepsilon(\sigma)$ represents the elastic, inelastic, and total strains, respectively, at a given stress σ; the inelastic strain is $\varepsilon_{\rm in}(\sigma) = 0$ when $\sigma < \sigma_{\rm y}$ (the inelastic strain $\varepsilon_{\rm in}$ is commonly referred to as the cracking strain $\varepsilon_{\rm cr}$ in the description of concrete in tension). $E_{\rm c0}$ represents the initial elastic modulus of undamaged concrete.

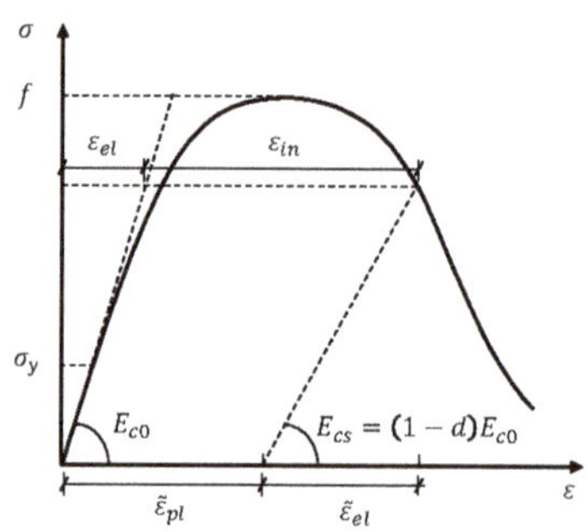

Figure 5. Damage evolution and degradation of the material stiffness beyond the peak stress of concrete.

Considering the damage evolution beyond the peak stress, the initial elastic strain $\varepsilon_{\rm el}(\sigma)$ in Figure 5 changes to $\widetilde{\varepsilon}_{\rm el}(\sigma)$ owing to the degradation of the elastic modulus from $E_{\rm c0}$ to $E_{\rm cs}$ according to Equation (9).

$$E_{\rm cs} = E_{\rm c0}(1-d), \tag{9}$$

where, d is the damage factor to be defined for characterizing the damage evolution.

In the present FE analyses, the damage model proposed by Lubliner et al. [41] is adopted, which assumes that in the post-peak regime, the degraded material stiffness is proportional to the residual cohesion of the material. Considering that the material cohesion can be correlated to the stress state, this leads to:

$$\frac{E_{\rm cs}}{E_{\rm c0}} = \frac{c}{c_{\rm max}} = \frac{\sigma}{f}, \tag{10}$$

where c and $c_{\rm max}$ represent the material cohesion in the yield criteria corresponding to the stress level σ and peak stress f, respectively. For concrete, f represents the concrete tensile strength $f_{\rm ct}$ or compressive strength $f_{\rm c}$.

Substituting Equation (10) into Equation (9) yields the damage variable, as follows:

$$d = 1 - \frac{\sigma}{f}. \tag{11}$$

Importantly, the defined damage variable must satisfy the condition that the equivalent plastic strain $\widetilde{\varepsilon}_{\rm pl}$ should not decrease as the damage variable increases.

3.2. Modeling of Steel Reinforcement

The steel reinforcement, including the longitudinal reinforcing bars and the transverse stirrups, is modeled as a one-dimensional wire, to which the element type *truss* is assigned. The material properties of the steel reinforcement are defined according to the results of standard tensile tests performed in the laboratory. Figure 6 shows the tensile stress–strain relationship defined for the steel reinforcement in the FE analyses, which is characterized by the elastic modulus $E_s = 201$ GPa, the yielding stress $f_{sy} = 510$ MPa, the ultimate tensile strength $f_{su} = 617$ MPa, the ultimate strain $\varepsilon_{su} = 12.0\%$, and the rupture strain $\varepsilon_{smax} = 15.6\%$.

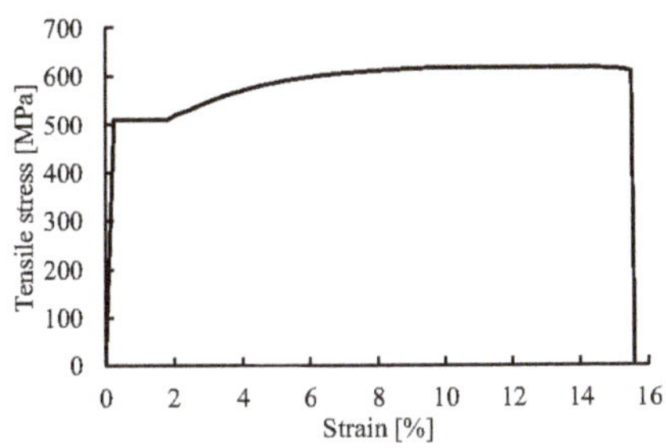

Figure 6. Tensile behavior defined for the steel reinforcement.

3.3. Interaction between Concrete and Steel Reinforcement

The proper modeling of the bond–slip behavior between the concrete and the embedded steel reinforcement is critical for obtaining clear discrete crack bands in the simulated beam model and reasonable flexural responses after the cracking point. In the present FE analyses, the bond between the longitudinal reinforcement and the concrete was assumed to be in good condition, and the bond–slip relationship according to Model Code 2010 [15] is adopted. The Model Code bond–slip relationship is shown in Figure 7, where $\tau_{b.max} = 17.9$ MPa, $\tau_{b.f} = 0.4\tau_{b.max}$, $s_1 = 1.0$ mm, $s_2 = 2.0$ mm, and $s_3 = 5.0$ mm. The interaction between the stirrups and the concrete was modeled as embedded, assuming a perfect bond with no relative slip (a simplification with negligible effects).

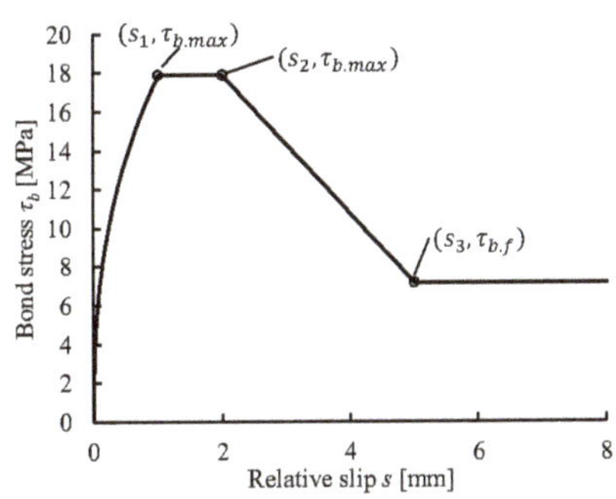

Figure 7. Bond–slip relation between the (longitudinal) steel reinforcement and the concrete in good bond condition according to Model Code 2010 [15].

To implement the bond–slip behavior in the FE model, the use of connectors to build the node-to-node bond is proposed as a reference method. Another commonly applied method using cohesive elements [2,54] is implemented for comparison. These two methods are referred to as the connector method and the cohesive method, and corresponding schemes are presented in Figure 8. An intermediary part is created as a copy of the steel reinforcement wire but with a significantly lower material stiffness (e.g., 0.1% of

E_s). The intermediary part is embedded (i.e., nodes are fully constrained) in the concrete continuum at the position of the steel reinforcement. The real steel reinforcement wire is connected to the intermediary part instead of the concrete continuum, using either connectors or cohesive elements. The chosen bond–slip relationship is finally assigned to the connectors or cohesive elements. The main benefit of introducing such an additional intermediary part is that the interaction properties assigned to the connectors or cohesive elements are not affected by the mesh refinement of the concrete continuum.

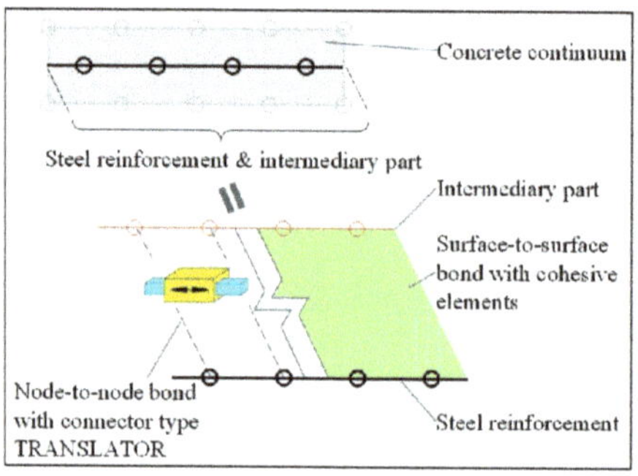

Figure 8. Illustration of the bond between the steel reinforcement and the concrete built using the connector method or cohesive method.

3.3.1. Node-to-Node Connector Method

In the connector method, the bond–slip interaction is created using connectors (type: translator) between the nodes of the steel reinforcement and the nodes of the concrete. This method is adopted in the FE analyses as the reference modeling choice to assign the bond–slip behavior. To facilitate the assignment of node-to-node connectors, a Python script is developed to automatically implement multiple wire features between the neighboring nodes of the reinforcement and concrete. These wires can be efficiently defined using a suitable type of connector called translator, which allows uniaxial translation only in the direction of the steel reinforcement between the connected nodes. Considering that the bond–slip behavior is realized by discrete node-to-node connections, the bond force V_b, rather than the bond stress τ_b, should be derived to define the bond at a given relative slip s between the steel reinforcement and the concrete, as indicated by Equation (12). When the connector method is used, for obtaining an accurate simulation of the nodal bond forces and crack widths, the connector spacing (and the length of the steel reinforcement elements) should not exceed the size of the concrete elements.

$$V_b = \tau_b C_s s_{conn}, \tag{12}$$

where, C_s represents the total circumference of the steel reinforcement, and s_{conn} represents the distance between two neighboring node-to-node connectors.

3.3.2. Surface-to-Surface Cohesive Method

The cohesive method in ABAQUS can model the interfacial bond behavior in either cohesive-contact or cohesive-element approach. The cohesive-contact approach defines the cohesive behavior as part of a contact model with zero interface thickness. For instance, this approach is used in [24,25] to efficiently define multiple interfacial responses in Mode I and Mode II. As an alternative, the cohesive-element approach utilizes cohesive elements to model the bond interface with a finite thickness; thus the interfacial response is characterized by the constitutive behavior assigned to the cohesive elements [2]. The cohesive-element approach allows to easily track damage evolution in the interface and visualize the bond failure by removing the damaged cohesive elements. This is critical to capture the debonding process of the CFRP plate in the current study (see Section 3.4). Therefore, the cohesive-element approach is also used here for modeling the steel-concrete interaction in order to compare it with the proposed node-to-node connector method. The layer of cohesive elements has a negligible thickness (1 μm in the current FE models) between the steel reinforcement wire and the concrete continuum. The constitutive response of cohesive elements was defined to represent the bond stress–relative slip relationship in Figure 7; damage evolution was introduced to characterize the nonlinear response of softening and degradation of elasticity.

3.4. Interaction between CFRP Plate and Concrete

In the modeling of the CFRP-strengthened beam, the bond between the CFRP plate and the concrete was modeled in the cohesive-element approach, as described in Section 3.3.2, in order to capture the debonding process induced by intermediate cracks. The 1 mm-thick adhesive layer was modeled with cohesive elements. The constitutive response of the cohesive elements was defined to represent the bond–slip model proposed by Lu et al. [55], see Figure 9.

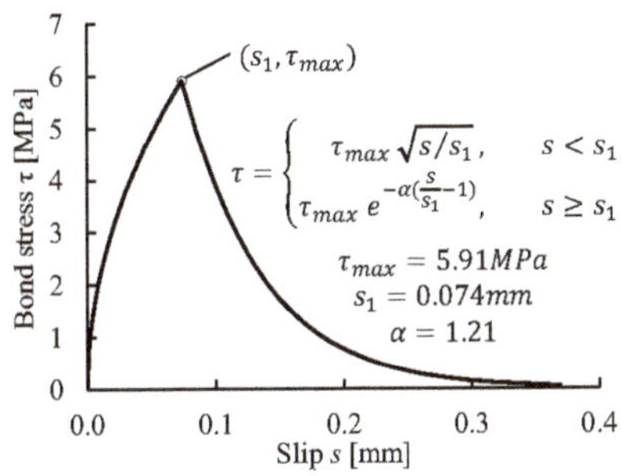

Figure 9. Bond–slip model assigned between the CFRP plate and concrete.

3.5. Numerical Solution Strategy

The static analysis procedure is used for the reference FE model to solve the nonlinear response of the RC beam subjected to monotonic loading. To overcome severe convergence difficulties in the static analysis [8,26], the technique of viscoplastic regularization is implemented and recommended (see Section 4.2). As an alternative to the static method, numerical analyses can be performed in a dynamic procedure adopting an implicit or explicit solution method. In Section 4.7, the dynamic analysis approach based on the implicit Hilber–Hughes–Taylor–α solution method proposed by Chen et al. (2015) [7] is implemented in comparison with the static analysis solution for validation.

4. Validation of Modeling Choices and Parametric Studies

According to the proposed modeling strategy, the modeling choices adopted in the reference FE model were validated and analyzed via multiple groups of parametric studies. Additionally, through the parametric studies, the effects of available modeling alternatives were investigated, and the impacts of essential parameters were quantified.

4.1. Dilation Angles

The dilation angle—one of the parameters defining the concrete plasticity—should be positive and in the range of 0°–56.3°. Malm (2009) [8] performed a parametric study of the dilation angle in FE analyses of an RC beam subjected to a four-point bending test. The results indicated that a dilation angle between 30° and 40° yielded converging load–deflection curves having a good agreement with the experimental flexural behavior. Jankowiak and Lodygowski (2005) [56] performed flexural tests on RC concrete beams in a laboratory to identify the reasonable value of the dilation angle required in the CDP model. A dilation angle of $\varphi = 38°$ (and $\epsilon = 1.0$) resulted in the best fit between the simulated flow potential surface and the experimental results of concrete beams ($f_{ct} = 2.8$ MPa and $f_c = 50$ MPa). Therefore, in the proposed reference FE model, the dilation angle φ was set as 35°. This value was also used in previous studies of RC beams [2,8,57].

Additionally, a parametric study was performed in the present study to investigate the effects of the dilation angle on the numerical results. Figure 10 shows the bending responses for different dilation angles. The FE results indicated that $\varphi = 35°$ was a suitable choice providing fast convergence and a reasonable response comparable to the experimental result. $35° \leq \varphi \leq 45°$ appeared to be a reasonable range, as the predicted load–deflection curves converged to a similar flexural response and closely matched the experimental response. However, a value lower ($\varphi = 25°$) or higher ($\varphi = 55°$) than this range likely caused underestimation of the concrete resistance to crushing failure, leading to failure (maximum load) at a smaller deflection. According to this parametric study and the results of previous studies [2,8,57], the dilation angle of $\varphi = 35°$ was validated and recommended for the modeling of similar slender concrete beams subjected to bending.

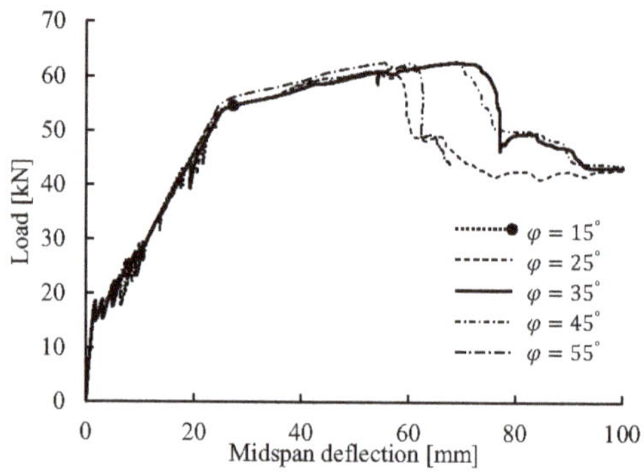

Figure 10. Load–midspan deflection relations for different dilation angles ranging from 15° to 55°.

4.2. Viscosity Parameter

As mentioned in Section 3.1.1, the viscosity parameter can be prescribed for viscoplastic regularization in the FE analysis to overcome the severe convergence difficulties in nonlinear concrete problems using the static analysis procedure [26,58]. The viscoplasticity introduced by μ permits stresses to be outside of the yield surface, improving the convergence rate for the damaged concrete in the strain-softening regime. The default setting is $\mu = 0$, indicating that no viscoplastic regularization is introduced.

To appropriately introduce the viscoplastic regularization without compromising the results, it is important to define an appropriate value of the viscosity parameter μ, which is theoretically smaller than the characteristic increment in the step of the nonlinear solution [26]. According to the checking of the automatically divided size of increments in the loading step, a parametric study of the viscosity parameter was performed (in the range of 10^{-3}–10^{-9} with a tenfold decrease) to identify the reasonable value for the reference FE model. The simulated load–deflection curves and crack patterns are shown in Figure 11, indicating the following: (1) a μ value that is too large (e.g., $\mu = 10^{-3}$ or 10^{-4}) tends to reduce the accuracy of the numerical results, leading to overestimation of the bending response of the beam and the inability to obtain clear crack patterns with localized crack bands; (2) a μ value that is too small (e.g., $\mu = 10^{-8}$ or 10^{-9}), although not affecting the simulated flexural behavior, can reduce the convergence rate and even abort the analysis owing to convergence difficulties; and (3) the suitable μ range appears to be 10^{-5}–10^{-7}. Within such a range, the FE analyses not only effectively overcome the convergence problems arising from concrete cracking and crushing but also yield converging results with negligible differences regarding the load–deflection curves and the crack patterns.

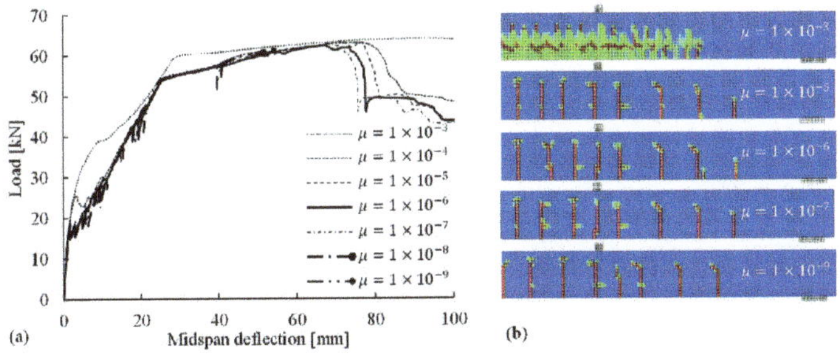

Figure 11. Numerical results: (**a**) load–midspan deflection relations; (**b**) crack patterns in the RC beam given the viscosity parameter μ ranging from 10^{-3} to 10^{-9}.

According to the parametric study, the value of the viscosity parameter was selected as 10^{-6} in the reference FE model. The use of the viscoplastic regularization technique to solve the convergence problems in the static analysis procedure is highly recommended. However, the viscosity parameter for a specific model should be carefully selected according to parametric studies.

4.3. Concrete Tensile Behavior

The concrete tensile behavior is mainly characterized by the concrete tensile strength f_{ct}, the fracture energy G_F, and the post-peak behavior (i.e., the shape of strain-softening curve after f_{ct} is reached). Considering the uncertainty of the parameters (i.e., f_{ct} and G_F) and the differences among the available post-peak constitutive models (see Figure 12), the modeling choices adopted for the reference FE model were validated through parametric studies of their impacts on the flexural responses (Figure 13) and cracking patterns.

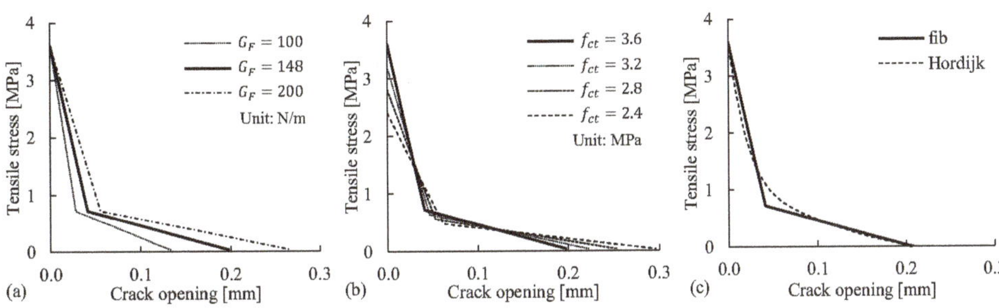

Figure 12. Concrete tensile behaviors considering (**a**) the uncertainty of the fracture energy G_F, (**b**) pre-cracking at a lower tensile strength f_{ct}, and (**c**) the differences among post-peak softening models, e.g., between Model Code 2010 [15] and Hordijk [43].

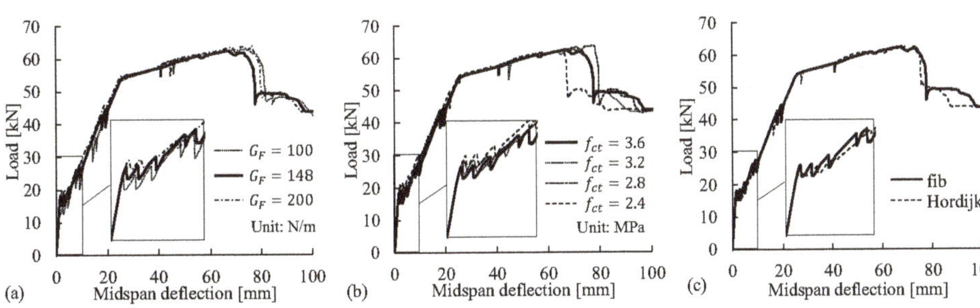

Figure 13. Numerical results for the load–midspan deflection relations obtained in the parametric studies on the modeling choices for (**a**) the fracture energy G_F, (**b**) the tensile strength f_{ct}, and (**c**) the post-peak softening behavior.

Although the fracture energy G_F in the reference model was defined as 148 N/m according to Equation (1), the parametric study covered the wide range of 100–200 N/m for the concrete material with a strength of $f_c = 51.1$ MPa, according to fib Bulletin 42 [48]. The concrete tensile strength $f_{ct} = 3.6$ MPa was defined in the reference model, and lower tensile strengths were assumed in the parametric study to investigate their effects, considering that cracks were observed before the testing date. In the parametric study of the post-peak softening behaviors, the commonly used stress–crack opening model proposed by Hordijk [43] was implemented and compared with the bilinear softening model adopted in the reference FE model according to Model Code 2010 [15]. The numerical results for the load–deflection curves (Figure 13) and crack patterns were reviewed. The results indicated that in general, the modeling choices for the three variables had negligible effects on the simulated load–deflection curves and cracking patterns, with the following exceptions: (1) there was a small difference in the first load drop after the cracking point, and (2) the number of cracks decreased from 8 to 7 when f_{ct} was reduced to 2.4 MPa.

4.4. Mesh Sensitivity Analysis

As discussed in Section 3.1.2, the crack band approach was implemented to address the sensitivity of the numerical results to the concrete element sizes when the smeared crack method was applied to model cracks in the concrete continuum. Therefore, it was necessary to perform a mesh sensitivity analysis to verify the size-independence of the simulated results. As shown in Figure 14, the concrete continuum of each model was discretized into square elements with sizes (i.e., side lengths) ranging from 10 to 40 mm. Clearly, the simulated load–deflection curves were not sensitive to the mesh discretization. With regard to the accuracy and computational efficiency, the element size of 20 mm (20 × 20 in Figure 14) appeared to be suitable for the reference FE model, as (1) the element size of 10 mm required a longer computational time and (2) mesh sizes of >30 mm (e.g., 40 × 40 in Figure 14) resulted in a smaller number of cracks and led to the underestimation of the resistance to crushing failure in the concrete compressive zone, as the height of the compressive zone after the yielding of the steel reinforcement was less than 40 mm.

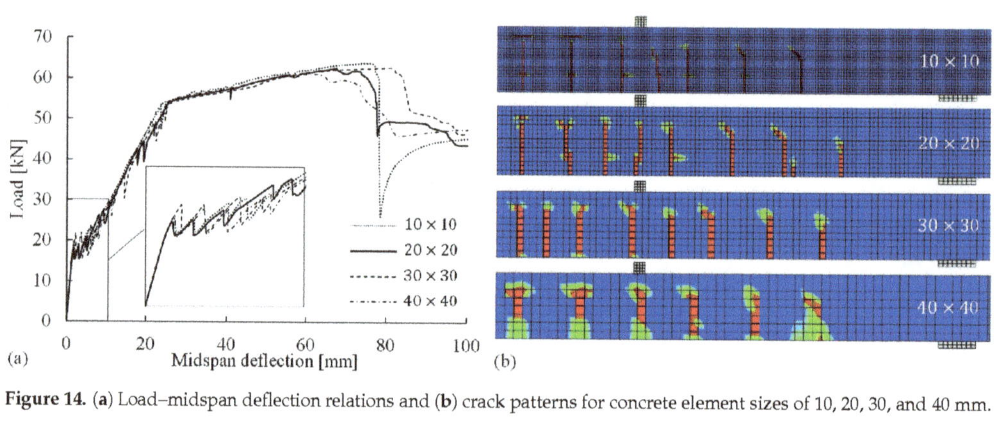

Figure 14. (**a**) Load–midspan deflection relations and (**b**) crack patterns for concrete element sizes of 10, 20, 30, and 40 mm.

4.5. Strain-Softening Behavior of Concrete in Compression

In the modeling of the concrete compressive behavior, it is important to (1) consider the strain localization in the critical fracture zone and (2) accordingly modify the post-peak strain-softening behavior of the concrete in compression with the verified size of the crushing zone L_{cr}.

As mentioned in Section 3.1.3, the stress–strain models characterized by the mean strain tend to be size-dependent in the post-peak regime and cannot account for the strain localization in the fracture zone. For instance, the commonly used stress–strain relationships in Table 3 have different post-peak softening branches, as shown in Figure 15a. The differences in the softening behaviors significantly affect the simulated flexural responses, particularly in the ultimate state, as shown in Figure 15b. Thus, properly modifying the softening behavior is necessary for accurately simulating the flexural failure of beams governed by concrete crushing.

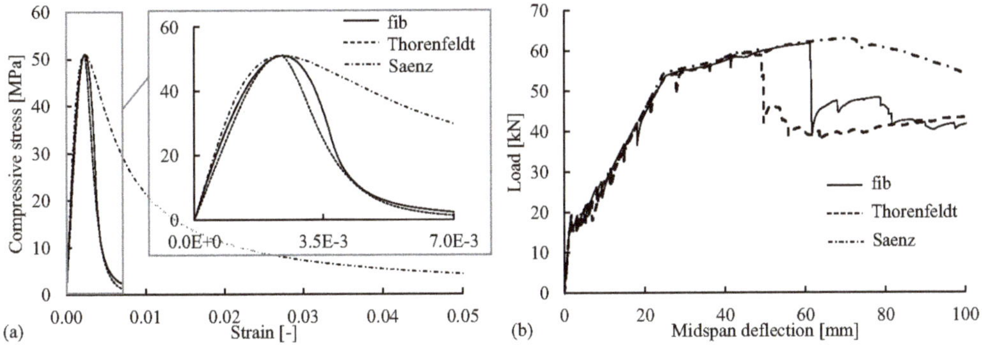

Figure 15. (**a**) Stress–strain models—Model Code [14,15], Thorenfeldt [21], and Saenz [51]—for the concrete compressive behavior with different post-peak softening branches; (**b**) load–deflection curves for FE analyses based on the stress–strain models without modification.

The iterative procedure proposed in Figure 4 was followed to determine the actual length of the critical crushing zone L_{cr} for the reference FE model. Figure 16a presents the Model Code relation, including the original post-peak branch ($L_{cr.input}$ = 200 mm) and modified post-peak branches given an $L_{cr.input}$ of 40–100 mm. Given the iterative assumption of $L_{cr.input}$ to modify the compressive behavior, the simulated flexural responses are shown in Figure 16b. The contour plot in Figure 16d shows the field output of strains within the constant-moment region of the beam model, where the concrete crushing zones are represented by the black areas with compressive-strain values > $\varepsilon_{c,lim}$. The lengths of the simulated crushing zones $L_{cr.output}$ (in the compressive direction) were evaluated and compared with the assumed length $L_{cr.input}$ in each analysis, as shown in Figure 16c. When the size of the critical crushing zone was assumed to be 100 mm ($L_{cr.input}$ = 100 mm), the assumption was verified by the simulated crushing zone ($L_{cr.output}$ = 100 mm). Therefore, in the reference FE model, the actual length of the critical fracture zone (i.e., compressive crushing zone) was selected as L_{cr} = 100 mm, and this value was used to modify the post-peak strain-softening behavior of the concrete in compression.

4.6. Bond between Steel Reinforcement and Concrete

The modeling choices to be validated for the bond between the steel reinforcement and the surrounding concrete include the method of assigning the bond and the bond–slip relationship.

The proposed connector method for assigning the steel-concrete bond was implemented in the reference model and compared with the cohesive-element method for validation. The numerical results for the flexural responses and crack patterns are shown in Figure 17a, indicating that the proposed connector method can provide comparable results to the cohesive method. No obvious difference was observed in the computational time between these two bond methods.

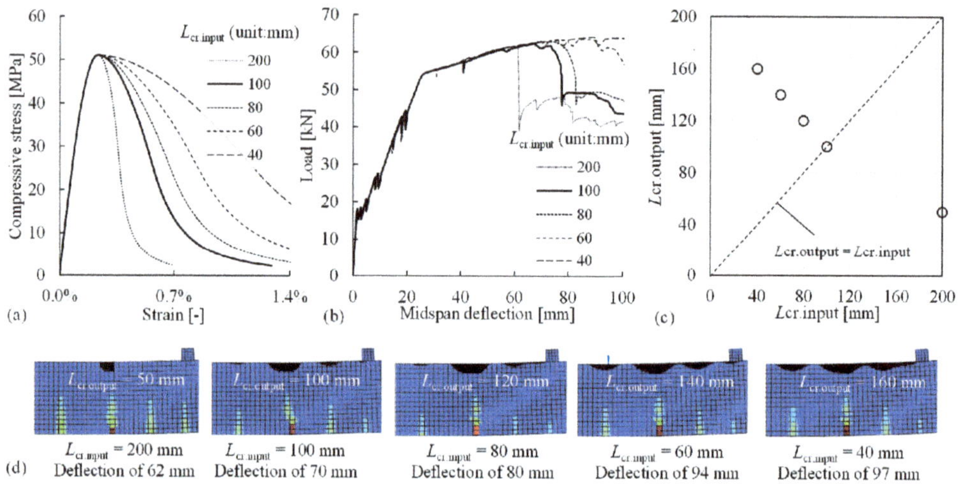

Figure 16. (**a**) Concrete compressive behavior based on the Model Code relation with the modified post-peak strain-softening branches given different $L_{cr.input}$ values ranging from 40–200 mm; (**b**) simulated load–deflection relations corresponding to each value of $L_{cr.input}$; (**c**) comparison of the assumed $L_{cr.input}$ and the observed size of the critical crushing zone $L_{cr.output}$ for each FE analysis (represented by a circle); (**d**) contour plot of the beam models within the constant-moment region for visualizing the crushing zones and evaluating $L_{cr.output}$ given different values of $L_{cr.input}$.

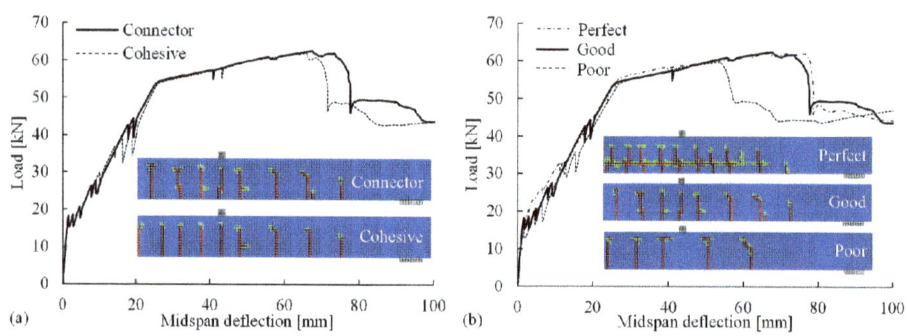

Figure 17. Comparisons of (**a**) methods for assigning the steel-concrete bond and (**b**) bond–slip behaviors based on the numerical results for the load–deflection relations and crack patterns.

The effects of the bond–slip behaviors (perfect, good, and poor bonds) on the flexural behavior and crack pattern in the FE analyses were investigated. In a perfect bond, all the degrees of freedom of the reinforcement nodes were constrained to the nearby concrete nodes, and there was no relative slip. The good and poor bond behaviors refer to the two bond–slip relationships provided in Model Code 2010 [15] corresponding to "good" and "other" bond condition, respectively. As shown in Figure 17b, a comparison of the flexural responses revealed that a weaker bond (poor bond) caused the earlier occurrence of flexural failure due to concrete crushing, whereas the perfect bond led to a stronger response after the cracking point. The significant differences in the crack patterns indicate that selecting a suitable bond stress–slip relation is critical for the accurate simulation of cracks, e.g., a poor bond relation led to the underestimation of the number of cracks. Thus, the Model Code relationship for good condition (good bond) is a better option than the one for poor bond in order to deliver flexural responses and crack patterns comparable to the experimental observations.

4.7. Static and Dynamic Analysis Procedures

The static analysis procedure (aided by the viscoplastic regularization technique) was defined in the reference FE model. It was compared with an FE model using dynamic implicit analysis based on the approach proposed by Chen et al. [7] for validation. In the dynamic analysis, critical choices and parameters included (1) ramp loading scheme, (2) load time $t_0 = 180T_1$, where $T_1 = 0.033$ s is the period of the fundamental vibration mode of the beam found from an eigenvalue analysis of the FE model period, (3) viscous damping ratio $\xi = 0.05$, and (4) time increment size as $T_1/100$. The corresponding numerical results are presented in Figure 18, which shows well-matched flexural responses until the maximum load of 62 kN with a deflection of 67 mm. The two methods predicted the same number of cracks and similar crack distributions, although there were minor differences in the positions of the cracks formed in the shear region. Thus, the static analysis procedure adopted in the reference FE model was validated, as the predicted flexural response and crack pattern were similar to those obtained via the dynamic approach.

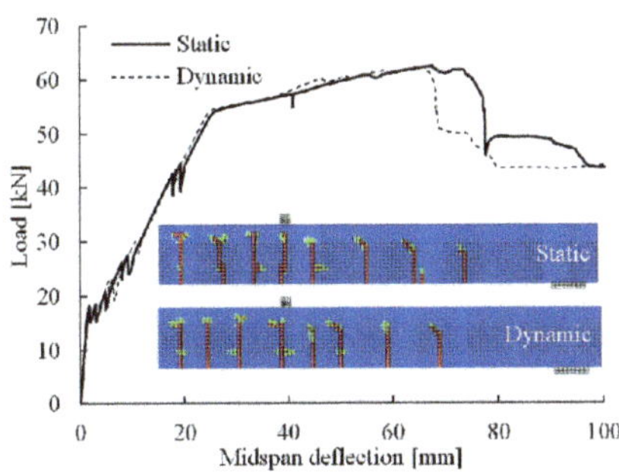

Figure 18. Comparison of the solution strategies between the static general analysis and the dynamic implicit analysis for the load–deflection relations and crack patterns.

5. Comparison of FE Predictions with Experimental Results

The FE models of the reference RC beam and the CFRP-strengthened beam followed the modeling strategy introduced in Section 3 and adopted the reference modeling choices for RC validated in Section 4. The numerical results of the reference beam were compared with the experimental measurements to examine the reliability of the predicted flexural response, crack pattern, crack widths, and ultimate failure due to concrete crushing. The accurate simulation of the reference beam laid the foundation for simulating the critical failure of CFRP debonding initiated by intermediate flexural cracks. The FE analysis of the CFRP-strengthened beam was chosen to show the applicability of the proposed modeling strategy for RC members with additional complexity.

5.1. Reference RC Beam

The flexural responses are expressed as load–deflection behaviors in Figure 19. In general, the load–deflection curve obtained from FE analysis matched the experimental measurements; however, there were small differences in the cracking point and the ultimate failure governed by concrete crushing. The weaker response measured in the cracking stage of the experimental test was attributed to minor cracks present in the RC beam before the test because of the concrete shrinkage and unexpected loading during the transportation from the workshop to the laboratory. In the ultimate state, the FE analysis predicted that the flexural failure would be initiated at a maximum load of 62.5 kN (midspan deflection of 68 mm), whereas the loading process in the experimental test stopped (at 65.0 kN) when concrete crushing was observed at the deflection of 79 mm. Although the FE analysis slightly underestimated the ultimate capacity of the flexural failure, the degradation of the flexural response was simulated well without convergence problems.

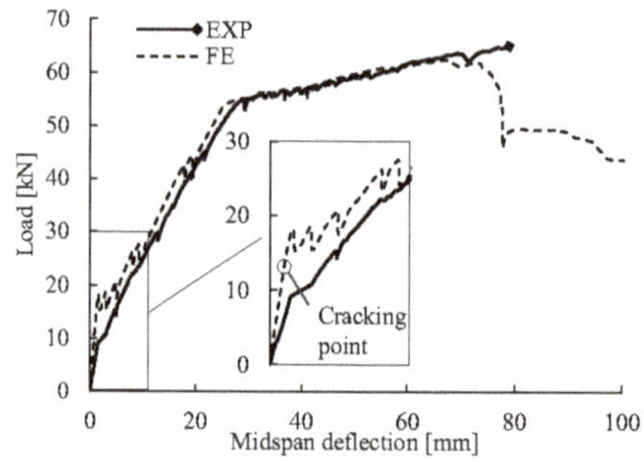

Figure 19. Load–deflection curves of the reference RC beam from experimental measurements (EXP) and FE analysis (FE).

Crack patterns and crack widths: The crack pattern documented at the load of 55 kN (the last measurement round) is shown in Figure 20c. Although the FE analysis predicted one less crack in half of the constant-moment region, the simulated crack pattern, in general, was comparable to the distribution of cracks in the experimental test.

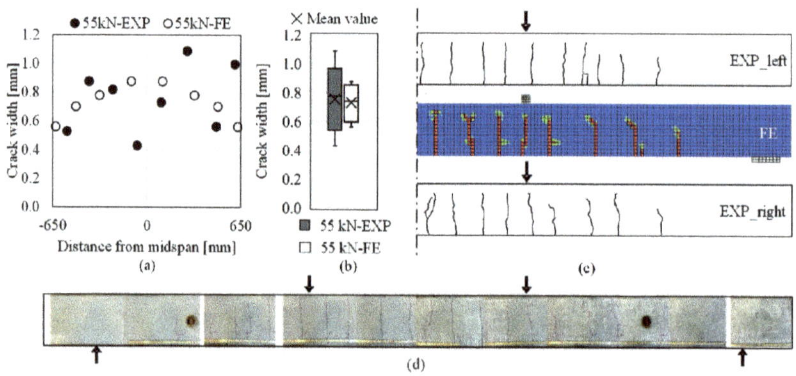

Figure 20. Comparison of cracks formed in the reference RC beam at the load of 55 kN regarding (**a**) crack pattern; (**b**) crack widths in the constant moment region (EXP: experimental results, FE: numerical results); (**c**) crack widths compared in box-and-whisker plots and (**d**) assembled photos depicting the crack pattern after the flexural test.

The experimental and numerical results of crack widths within the constant moment region are presented in Figure 20a and compared in box-and-whisker plots in Figure 20b depicting the quartiles, the variability, the median and mean values of the datasets. The boxplot of the crack widths at the load of 55 kN indicated that the mean and median values were similar between the experimental and numerical results, although there was larger variability in the experimentally measured crack widths. After reaching 55 kN, no new cracks occurred until the end of loading period; Figure 20d depicts the crack pattern on the beam after the flexural test.

5.2. CFRP-Strengthened RC Beam

Reliable prediction of the fast-growing crack openings after the yield point is crucial, as it is the foundation of advanced modeling with a focus on the consequential issues triggered by the opening of cracks. This was demonstrated by the modeling of the CFRP-strengthened beam; numerical results of the strengthened beam were compared to the experimental measurements with respect to its flexural response, concrete cracking, and CFRP debonding.

Flexural behavior and CFRP debonding: The flexural responses of the CFRP-strengthened beam are depicted with load–deflection curves in Figure 21a. To be highlighted, the FE analysis predicted the debonding of CFRP plate induced by intermediate cracks (IC debonding) at a maximum load of 104 kN (deflection 44 mm), which matched well with the experimental observation of IC debonding at 105 kN (deflection 45 mm). The development of IC debonding captured in the numerical simulation is shown in Figure 21b, including the initiation at the intermediate flexural crack and the evolution visualized by the removal of fully damaged adhesive (modeled as cohesive elements).

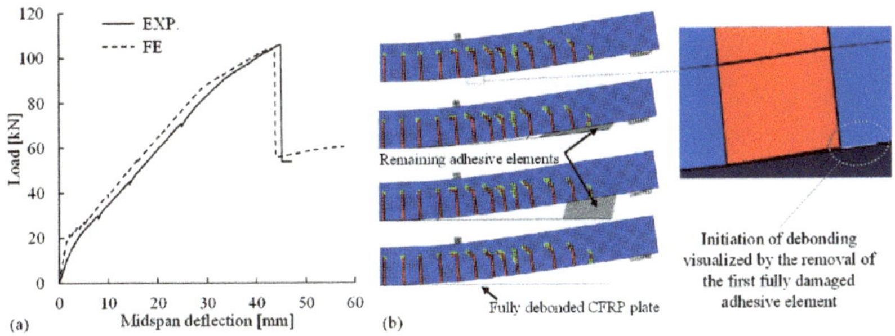

Figure 21. (**a**) Comparison of the flexural responses between experimental measurements (EXP) and numerical results (FE) and (**b**) intermediate cracks (IC) debonding of CFRP plate simulated in the FE analysis.

Crack patterns and cracks widths: Reliable prediction of the IC debonding is highly dependent on the accurate simulation of cracks in the RC beam based on proper modeling options. Crack pattern and crack widths in the FE analysis were checked with the experimental measurements at different load levels. For example, Figure 22a–c demonstrates the cracks experimentally measured at the load of 70 kN (the last measurement round) in comparison to the numerical results at the same load. After 70 kN, no new cracks occurred until the end of loading period; Figure 22d depicts the crack pattern on the beam after the flexural test.

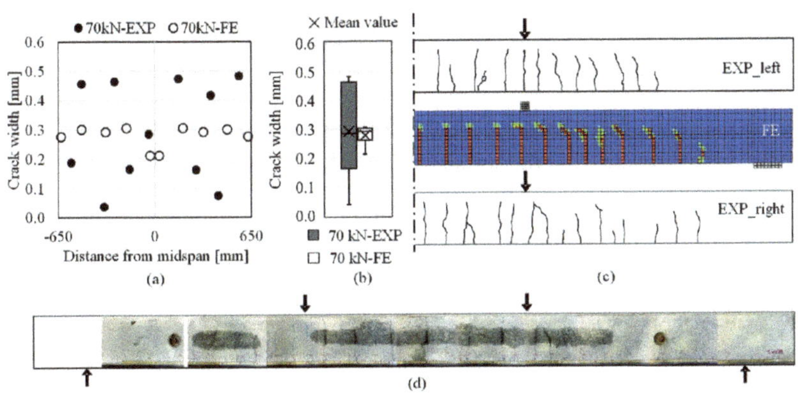

Figure 22. Comparison of cracks formed in the CFRP-strengthened RC beam at the load of 70 kN regarding (**a**) crack pattern; (**b**) crack widths in the constant moment region (EXP: experimental results, FE: numerical results); (**c**) crack widths compared in box-and-whisker plots and (**d**) assembled photos depicting the crack pattern after the flexural test.

As shown in Figure 22c, localized crack bands were formed and distributed in a clear pattern, which was comparable to the experimental observation. Crack openings in the constant moment region are shown in Figure 22a–b; the crack widths less than 0.1 mm are corresponding to the measurement of two minor cracks on each side of the beam. Although the crack widths measured in the experiment showed larger variability than the simulated ones, both the median and the mean value of the experimental measurements were well captured in the numerical simulation. The reliable simulation of crack openings laid the foundation for capturing the IC debonding of CFRP in Figure 21.

6. Summary and Conclusions

In this paper, a robust modeling strategy for reliable nonlinear FE analyses of RC structures was proposed. The modeling choices adopted for the reference FE model were validated through parametric studies and comparisons with other commonly used options; the effects of these choices on the numerical results were investigated. The contributions of this study are summarized as follows:

- Consistent approaches for deriving and defining the stress–strain relationships for concrete in tension and compression with consideration of the damage evolution and strain localization in the fracture zones of cracking and crushing concrete, respectively, were comprehensively described;
- Principles and recommendations for appropriately determining crack band width in structurally meshed concrete continuum to adjust the tensile stress–strain relationships were discussed and proved effective in avoiding the mesh sensitivity problem;
- An iterative implementation procedure was proposed to modify the concrete compressive stress–strain relationship in the post-peak regime according to the verified size of the crushing zone for considering the strain-localization effect;
- A simple and robust method was proposed for assigning the steel-concrete reinforcement interaction in ABAQUS using node-to-node connectors, and its accuracy was confirmed via a comparison with the commonly used method based on cohesive elements;
- Viscoplastic regularization using a properly defined viscosity parameter exhibited the capability to overcome the convergence difficulties encountered in the simulation of cracking or crushing concrete and thus significantly reduced the computational time;
- The proposed modeling strategy, as exemplified through the FE analyses of a reference RC beam, provided a reliable simulation of nonlinear responses including the development of cracks and resistance to concrete crushing. The ability to capture these effects on this simple case laid the foundation for the accurate modeling of CFRP debonding induced by intermediate flexural cracks in the strengthened beam, which provided an example of modeling RC structures with additional complexity.

Most of the recommendations presented in this paper for the nonlinear concrete model are general and applicable to FE analyses of other quasi-brittle material models with different FE software. The proposed modeling strategies can be directly used to model RC beam and frame structures with different dimensions, loading configuration, and boundary conditions. The current study also paves the way to FE analyses with higher degrees of complexity.

Author Contributions: Conceptualization, A.M. and J.Y.; methodology, A.M. and J.Y.; software, A.M. and J.Y.; validation, A.M. and J.Y.; formal analysis, A.M. and J.Y.; investigation, A.M. and J.Y.; resources, A.M. and J.Y.; data curation, A.M. and J.Y.; writing—original draft preparation, A.M. and J.Y.; writing—review and editing, A.M. and J.Y.; visualization, A.M. and J.Y.; supervision, A.M. and J.Y.; project administration, A.M. and J.Y. All authors have read and agreed to the published version of the manuscript.

Funding: This work was financially supported by the European Union's 7th Framework program [grant number 31109806.0009]; the Swedish Transport Administration [grant number BBT-2017-037 and BBT-2018-011]; Sweden's Innovation Agency [grant number 2017-03312]; the Swedish Wind Power Technology Centre (SWPTC); and NCC.

Institutional Review Board Statement: Not applicable.

Informed Consent Statement: Not applicable.

Data Availability Statement: The data that support the findings of this study are contained within the article.

Conflicts of Interest: The authors declare no conflict of interest.

References

1. Malm, R.; Holmgren, J. Cracking in deep beams owing to shear loading. Part 2: Non-linear analysis. *Mag. Concr. Res.* **2008**, *6*, 381–388. [CrossRef]
2. Chen, G.M.; Teng, J.G.; Chen, J.F. Finite-element modeling of intermediate crack debonding in FRP-plated RC beams. *J. Compos. Constr.* **2011**, *15*, 339–353. [CrossRef]
3. Grassl, P.; Johansson, M.; Leppänen, J. On the numerical modelling of bond for the failure analysis of reinforced concrete. *Eng. Fract. Mech.* **2018**, *189*, 13–26. [CrossRef]
4. Hendriks, M.A.N.; de Boer, A.; Belletti, B. *Validation of the Guidelines for Nonlinear Finite Element Analysis of Concrete Structures—Part: Reinforced Beams*; Report RTD:1016-3A:2017; Rijkswaterstaat Centre for Infrastructure, Rijkswaterstaat Centre for Infrastructure: Brussels, Belgium, 2017.
5. Schlune, H.; Plos, M.; Gylltoft, K. Safety formats for non-linear analysis of concrete structures. *Mag. Concr. Res.* **2012**, *64*, 563–574. [CrossRef]
6. Jirásek, M.; Bažant, Z.P. *Inelastic Analysis of Structures*; Wiley: Hoboken, NJ, USA, 2002; ISBN 0471987166.
7. Chen, G.M.; Teng, J.G.; Chen, J.F.; Xiao, Q.G. Finite element modeling of debonding failures in FRP-strengthened RC beams: A dynamic approach. *Comput. Struct.* **2015**, *158*, 167–183. [CrossRef]
8. Malm, R. *Predicting Shear Type Crack Initiation and Growth in Concrete with Non-Linear Finite Element Method*; Royal Institute of Technology: Stockholm, Sweden, 2009.
9. Camacho, G.T.; Ortiz, M. Computational modelling of impact damage in brittle materials. *Int. J. Solids Struct.* **1996**, *33*, 2899–2938. [CrossRef]
10. May, S.; de Borst, R.; Vignollet, J. Powell-Sabin B-splines for smeared and discrete approaches to fracture in quasi-brittle materials. *Comput. Methods Appl. Mech. Eng.* **2016**, *307*, 193–214. [CrossRef]
11. Wells, G.N.; Sluys, L.J. A new method for modelling cohesive cracks using finite elements. *Int. J. Numer. Methods Eng.* **2001**, *50*, 2667–2682. [CrossRef]
12. Bažant, Z.P.; Oh, B.H. Crack band theory for fracture of concrete. *Mater. Struct.* **1983**, *16*, 155–177. [CrossRef]
13. Rots, J.G.; Nauta, P.; Kuster, G.M.A.; Blaauwendraad, J. Smeared crack approach and fracture localization in concrete. *HERON* **1985**, *30*, 48.
14. CEB-FIP. *CEB-FIP Model Code 1990: Design Code*; Thomas Telford: Lausanne, Switzerland, 1993.
15. Fib. *Fib Model Code for Concrete Structures 2010*; International Federation for Structural Concrete: Lausanne, Switzerland, 2013; ISBN 978-3-433-03061-5.
16. American Concrete Institute. *ACI 318-14 Building Code Requirements for Structural Concrete and Commentary (Metric)*; American Concrete Institute: Farmington Hills, MI, USA, 2014.
17. Van Mier, J.G.M. Strain-Softening of Concrete under Multiaxial Loading Conditions. Ph.D. Thesis, Eindhoven University of Technology, Eindhoven, The Netherlands, 20 November 1984.
18. Bažant, Z.P. Identification of strain-softening constitutive relation from uniaxial tests by series coupling model for localization. *Cem. Concr. Res.* **1989**, *19*, 973–977. [CrossRef]
19. Jansen, D.C.; Shah, S.P. Effect of length on compressive strain softening of concrete. *J. Eng. Mech.* **1997**, *123*, 25–35. [CrossRef]
20. Zandi Hanjari, K.; Kettil, P.; Lundgren, K. Modelling the structural behaviour of frost-damaged reinforced concrete structures. *Struct. Infrastruct. Eng.* **2013**, *9*, 416–431. [CrossRef]
21. Thorenfeldt, E.; Tomaszewicz, A.; Jensen, J.J. Mechanical properties of high-strength concrete and applications in design. In Proceedings of the Symposium on Utilization of High-Strength Concrete, Stavanger, Norway, 15–18 June 1987; pp. 149–159.
22. Shu, J.; Fall, D.; Plos, M.; Zandi, K.; Lundgren, K. Development of modelling strategies for two-way RC slabs. *Eng. Struct.* **2015**, *101*, 439–449. [CrossRef]
23. Dassault Systèmes. *Abaqus 6.14—Abaqus Analysis User's Guide*; Dassault Systèmes: Vélizy-Villacoublay, France, 2014.
24. Carloni, C.; D'Antino, T.; Sneed, L.H.; Pellegrino, C. Three-dimensional numerical modeling of single-lap direct shear tests of frcm-concrete joints using a cohesive damaged contact approach. *J. Compos. Constr.* **2018**, *22*, 04017048. [CrossRef]
25. Ombres, L.; Verre, S. Experimental and numerical investigation on the steel reinforced grout (SRG) composite-to-concrete bond. *J. Compos. Sci.* **2020**, *4*, 182. [CrossRef]
26. Dassault Systèmes. *Abaqus 6.14 Theory Manual*; Dassault Systèmes: Vélizy-Villacoublay, France, 2014.
27. Malm, R. *Guideline for FE Analyses of Concrete Dams*; Energiforsk: Stockholm, Sweden, 2016; ISBN 9789176732700.
28. Hanif, M.U.; Ibrahim, Z.; Jameel, M.; Ghaedi, K.; Aslam, M. A new approach to estimate damage in concrete beams using non-linearity. *Constr. Build. Mater.* **2016**, *124*, 1081–1089. [CrossRef]
29. Alfarah, B.; López-Almansa, F.; Oller, S. New methodology for calculating damage variables evolution in plastic damage model for RC structures. *Eng. Struct.* **2017**, *132*, 70–86. [CrossRef]

30. Kmiecik, P.; Kaminski, M. Modelling of reinforced concrete structures and composite structures with concrete strength degradation taken into consideration. *Arch. Civ. Mech. Eng.* **2011**, *11*, 623–636. [CrossRef]
31. Labizadeh, M.; Hamidi, R. Effect of stress path, size and shape on the optimum parameters of a brittle-ductile concrete model. *Eng. Struct. Technol.* **2017**, *9*, 195–206. [CrossRef]
32. Obaidat, Y.T.; Heyden, S.; Dahlblom, O. The effect of CFRP and CFRP/concrete interface models when modelling retrofitted RC beams with FEM. *Compos. Struct.* **2010**, *92*, 1391–1398. [CrossRef]
33. Triantafillou, T.C. Shear strengthening of reinforced concrete beams using epoxy-bonded FRP composites. *ACI Struct. J.* **1998**, *95*, 107–115. [CrossRef]
34. Hollaway, L.C. A review of the present and future utilisation of FRP composites in the civil infrastructure with reference to their important in-service properties. *Constr. Build. Mater.* **2010**, *24*, 2419–2445. [CrossRef]
35. Cascardi, A.; Dell'Anna, R.; Micelli, F.; Lionetto, F.; Aiello, M.A.; Maffezzoli, A. Reversible techniques for FRP-confinement of masonry columns. *Constr. Build. Mater.* **2019**, *225*, 415–428. [CrossRef]
36. CEN. EN 1992-1-1:2004. *Eurocode 2: Design of Concrete Structures—Part 1-1: General Rules and Rules for Buildings*; European Committee for Standardization: Brussels, Belgium, 2004; p. 191.
37. ASTM. *A615/A615M Standard Specification for Deformed and Plain Carbon-Steel Bars for Concrete Reinforcement*; ASTM International: West Conshohocken, PA, USA, 2016.
38. Yang, J.; Haghani, R.; Al-Emrani, M. Innovative prestressing method for externally bonded CFRP laminates without mechanical anchorage. *Eng. Struct.* **2019**, *197*. [CrossRef]
39. Heshmati, M.; Haghani, R.; Al-Emrani, M. Durability of bonded FRP-to-steel joints: Effects of moisture, de-icing salt solution, temperature and FRP type. *Compos. Part B Eng.* **2017**, *119*, 153–167. [CrossRef]
40. Rabbat, B.G.; Russell, H.G. Friction coefficient of steel on concrete or grout. *J. Struct. Eng.* **1985**, *111*, 505–515. [CrossRef]
41. Lubliner, J.; Oliver, J.; Oller, S.; Oñate, E. A plastic-damage model for concrete. *Int. J. Solids Struct.* **1989**, *25*, 299–326. [CrossRef]
42. Lee, J.; Fenves, G.L. Plastic-damage model for cyclic loading of concrete structures. *J. Eng. Mech.* **1998**, *124*, 892–900. [CrossRef]
43. Hordijk, D.A. Local Approach to Fatigue of Concrete. Ph.D. Thesis, Delft University of Technology, Delft, The Netherlands, 29 October 1991.
44. Pietruszczak, S.; Mróz, Z. Finite element analysis of deformation of strain-softening materials. *Int. J. Numer. Methods Eng.* **1981**, *17*, 327–334. [CrossRef]
45. De Borst, R. Non-Linear Analysis of Frictional Materials. Ph.D. Thesis, Delft University of Technology, Delft, The Netherlands, 22 April 1986.
46. Rots, J.G. Computational Modeling of Concrete Fracture. Ph.D. Thesis, Delft University of Technology, Delft, The Netherlands, 26 September 1988.
47. Oliver, J. A consistent characteristic length for smeared cracking models. *Int. J. Numer. Methods Eng.* **1989**, *28*, 461–474. [CrossRef]
48. Fib. *Bulletin 42: Constitutive Modelling of High Strength/High Performance Concrete*; International Federation for Structural Concrete (fib): Lausanne, Switzerland, 2008.
49. Tomaszewicz, A. *Betongens Arbeidsdiagram (Stress-Strain Relationship for Concrete), Report No. STF65 A84065*; Sintef: Trondheim, Norway, 1984.
50. Collins, M.P.; Porasz, A. Shear design for high strength concrete. In *CEB Bulletin D'information no. 193*; Comité Euro-International du Béton: Lausanne, Switzerland, 1989; pp. 77–83.
51. Saenz, L.P. Discussion of equation for the stress-strain curve of concrete-by Desayi, P. and Krishan, S. *J. Am. Concr. Inst.* **1964**, *61*, 1229–1235.
52. Desayi, P.; Krishnan, S. Equation for the stress-strain curve of concrete. *J. Am. Concr. Inst.* **1964**, *61*, 345–350.
53. Popovics, S. A numerical approach to the complete stress-strain curve of concrete. *Cem. Concr. Res.* **1973**, *3*, 583–599. [CrossRef]
54. Yao, L.Z.; Wu, G. Nonlinear 2D finite-element modeling of RC beams strengthened with prestressed NSM CFRP reinforcement. *J. Compos. Constr.* **2016**, *20*, 1–18. [CrossRef]
55. Lu, X.Z.; Teng, J.G.; Ye, L.P.; Jiang, J.J. Bond-slip models for FRP sheets/plates bonded to concrete. *Eng. Struct.* **2005**, *27*, 920–937. [CrossRef]
56. Jankowiak, T.; Lodygowski, T. Identification of parameters of concrete damage plasticity constitutive model. *Found. Civ. Environ. Eng.* **2005**, *6*, 53–69.
57. Esmaeeli, E. Development of Hybrid Composite Plate (HCP) for the Strengthening and Repair of RC Structures. Ph.D. Thesis, University of Minho, Braga, Portugal, 11 December 2015.
58. Lee, J.; Fenves, G.L. A plastic-damage concrete model for earthquake analysis of dams. *Earthq. Eng. Struct. Dyn.* **1998**, *27*, 937–956. [CrossRef]

MDPI
St. Alban-Anlage 66
4052 Basel
Switzerland
Tel. +41 61 683 77 34
Fax +41 61 302 89 18
www.mdpi.com

Materials Editorial Office
E-mail: materials@mdpi.com
www.mdpi.com/journal/materials

www.ingramcontent.com/pod-product-compliance
Lightning Source LLC
Chambersburg PA
CBHW061841210926
51850CB00006B/940